INTRODUCTION TO HARMONIC ANALYSIS ON REDUCTIVE P-ADIC GROUPS

Based on lectures by Harish-Chandra at
The Institute for Advanced Study, 1971-73

by Allan J. Silberger

Princeton University Press
and
University of Tokyo Press

Princeton, New Jersey

1979

Published in Japan Exclusively
by University of Tokyo Press
in other parts of the world by
Princeton University Press

Printed in the United States of America
by Princeton University Press, Princeton, New Jersey

Library of Congress Cataloging in Publication Data will
be found on the last printed page of this book

FOREWORD

These notes represent the writer's attempt to organize and comprehend the mathematics communicated to him by Harish-Chandra, both in public lectures and private conversations, during the years 1971-1973. They offer the reader an ab initio introduction to the theory of harmonic analysis on reductive p-adic groups. Besides laying the foundations for a theory of induced representations by presenting Jacquet's theory, the Bruhat theory, the theory of the constant term, and the Maass-Selberg relations, these notes develop the theory of the Schwartz space on a p-adic group and the theory of the Eisenstein integral in complete detail. They also give the construction of the algebras of wave packets as orthogonal components of Schwartz space and prove Plancherel's formula for induced series, Harish-Chandra's commuting algebra theorem, and the sufficiency of the tempered spectrum for rank one groups. Most notable among omissions from these notes is Harish-Chandra's completeness theorem (i.e., that for arbitrary rank, the tempered spectrum suffices) announced in [7f] and his theory of the characters of admissible representations.

The reader will find a summary of a part of the contents of this work given in Harish-Chandra's Williams College lectures ([7e]).

ACKNOWLEDGMENTS

The writer would like to express his great appreciation to Roger Howe for his many perceptive comments regarding the manuscript. He also wishes to thank Mark Krusemeyer, Paul Sally, and Nolan Wallach for pointing out mistakes in the first draft of these notes. Evelyn Laurent typed and several times corrected the manuscript. The writer owes her a great debt for her patience and efficiency. The preparation of these notes was supported by funds from the National Science Foundation.

September 1, 1978

TABLE OF CONTENTS

CONTENTS (cont'd) Page

CONTENTS (cont'd) Page

Chapter 0. On the Structure of Reductive p-adic Groups.

The theory to be developed in the five later chapters of these notes depends upon the structure theory for reductive groups with points in a p-adic field. Fortunately, this theory has been worked out in detail (cf., [2a], [2c] for the essentially algebraic aspects and [4c] for the directly related topological part). In this chapter we very briefly review only those facts from the structure theory which we shall need later. The reader may consult the references both for proofs and more details.

If G is any group and H a subgroup, we write $Z_G(H)[N_G(H)]$ for the centralizer [normalizer] of H in G. Given $x, g \in G$, we write $x^g = gxg^{-1}$ and $H^g = gHg^{-1}$.

We write $[X]$ for the cardinality of a finite set X.

For any ring R we write $R^\times$ for its group of units. If n is a positive integer and R is a commutative ring, we write $GL_n(R)$ for the group of all $n \times n$ matrices $(c_{ij})_{1 \leq i,\, j \leq n}$ with entries $c_{ij} \in R$ and determinant in $R^\times$. We write $\det(x)$ or $\det(c_{ij})$ to denote the determinant of a matrix x or (c_{ij}).

We write $\mathbb{Z}$ for the ring of ordinary (rational) integers, $\mathbb{Q}$, $\mathbb{R}$, and $\mathbb{C}$ for the fields, respectively, of rational, real, and complex numbers.

§0.1. Some Definitions and Facts.

Let Ω be a field and $\bar{\Omega}$ an algebraic closure of Ω. For any positive integer m the space $\bar{\Omega}^m$ of all m-vectors with components in $\bar{\Omega}$ carries the Zariski and Ω topologies: A subset S of $\bar{\Omega}^m$ is called Zariski closed [Ω-closed] if S is the zero set of some finite set of polynomials in $\bar{\Omega}[x_1, \ldots, x_m]$[$\Omega[x_1, \ldots, x_m]$]. The complement of a Zariski closed [Ω-closed] subset of $\bar{\Omega}^m$ is termed Zariski open [Ω-open]. Obviously, the Zariski topology on $\bar{\Omega}^m$ is finer than the Ω-topology. A subset S_1 of a Zariski closed [Ω-closed] set $S_2 \subset \bar{\Omega}^m$ is called Zariski dense [Ω-dense] in S_2 if every Zariski closed [Ω-closed] subset of $\bar{\Omega}^m$ which contains S_1 also contains S_2.

Under an obvious identification, we may, for any positive integer n, regard $GL_n(\bar{\Omega})$ as the Ω-open subset $\det(x_{ij}) \neq 0$ of $\bar{\Omega}^{n^2}$ or, more conveniently for what follows, the Ω-closed subset $y \det(x_{ij}) = 1$ of $\bar{\Omega}^{n^2+1}$. A linear algebraic group or l.a.g. is a subgroup $\underset{\sim}{G} \subset GL_n(\bar{\Omega})$ which is Zariski closed as a subset of $\bar{\Omega}^{n^2+1}$. If $\underset{\sim}{G}$ is an l.a.g., the group law $(x, y) \mapsto x^{-1}y$ for $\underset{\sim}{G}$ is given by a polynomial mapping $\bar{\Omega}^{n^2} \times \bar{\Omega}^{n^2} \to \bar{\Omega}^{n^2}$. An l.a.g. $\underset{\sim}{G}$ is said to be defined over Ω or called an Ω-group if $\underset{\sim}{G}$ is Ω-closed in $\bar{\Omega}^{n^2+1}$ and if the group law is given by polynomials with coefficients in Ω.

We will denote linear algebraic groups by Roman letters with wavy underlines. If $\underset{\sim}{G}$ is an l.a.g. and R is a subring of $\bar{\Omega}$, we write $\underset{\sim}{G}(R)$ for $\underset{\sim}{G} \cap GL_n(R)$. If $\underset{\sim}{G}$ is an Ω-group, we sometimes write $G = \underset{\sim}{G}(\Omega)$.

A morphism of l.a.g.'s is a group homomorphism which is at the same time a polynomial mapping. An Ω-morphism is a morphism of Ω-groups in which the polynomials defining the mapping have coefficients in Ω.

An l.a.g. is called *connected* if it contains no l.a.g. as a subgroup of finite index. An l.a.g. $\mathbf{G}$ contains a maximal connected subgroup, denoted $\mathbf{G}^0$. If $\mathbf{G}$ is an Ω-group, then so is $\mathbf{G}^0$.

Given an l.a.g. $\mathbf{G}$, we write $\mathbf{X}(\mathbf{G})$ for the group of all *rational characters* of $\mathbf{G}$, i.e., the group of all morphisms of $\mathbf{G}$ to $\mathbf{GL}_1 = GL_1(\bar{\Omega}) = \bar{\Omega}^{\times}$. If $\mathbf{G}$ is connected, $\mathbf{X}(\mathbf{G})$ is a free abelian group. If $\mathbf{G}$ is an Ω-group, we write $X(G)$ for the subgroup of $\mathbf{X}(\mathbf{G})$ which consists of all Ω-morphisms. Under these conditions, if $\chi \in X(G)$, we frequently abuse language to regard χ as a homomorphism $\chi : G \to \Omega^{\times}$.

The *radical* $\mathbf{R}_{\mathbf{G}}$ [*unipotent radical* $\mathbf{N}_{\mathbf{G}}$] of an l.a.g. $\mathbf{G}$ is the maximal connected normal solvable [unipotent] subgroup of $\mathbf{G}$. If $\mathbf{N}_{\mathbf{G}} = (1)$, we call $\mathbf{G}$ a *reductive group*. If $\mathbf{R}_{\mathbf{G}} = (1)$, we call $\mathbf{G}$ a *semisimple group*. The derived group $\mathcal{D}\mathbf{G}$ of a reductive group is semisimple; in fact, the radical of a reductive group is central and $\mathcal{D}\mathbf{G} \cap$ (center of $\mathbf{G}$) is finite.

A commutative, connected, and reductive l.a.g. is called a *torus*. The *dimension* of a torus is the rank of $\mathbf{X}(\mathbf{T})$. A torus which is an Ω-group is called an Ω-*torus*. The radical of a reductive Ω-group is the maximal Ω-torus lying in its center. If $\mathbf{T}$ is an Ω-torus and T is isomorphic to the diagonal subgroup $D_n(\Omega) \subset GL_n(\Omega)$ for some n, then we call T an Ω-*split torus*; this is the case if and only if $\mathbf{X}(\mathbf{T}) = X(T)$. If $\mathbf{T}$ is an Ω-torus, then there is a finite separable extension Ω' of Ω with respect to which $\mathbf{T}$ is an Ω'-split torus.

A reductive Ω-group is called *anisotropic* (*over* Ω) if it contains no Ω-split torus.

§0.2. Cartan Subgroups and Split Tori in Reductive Groups.

For the remainder of the chapter assume that $\underset{\sim}{G}$ is a connected and reductive Ω-group.

A *Cartan subgroup* of $\underset{\sim}{G}$ is a maximal torus of $\underset{\sim}{G}$. All Cartan subgroups of $\underset{\sim}{G}$ are conjugate, hence have the same dimension. There exist Cartan Ω-subgroups of $\underset{\sim}{G}$. The group $G = \underset{\sim}{G}(\Omega)$ operates by conjugation on the set of Cartan Ω-subgroups of $\underset{\sim}{G}$. If char $\Omega = 0$, the number of orbits is finite--otherwise, it may be infinite.

The group G operates transitively by conjugation on the set of maximal Ω-split tori of $\underset{\sim}{G}$. Let $\underset{\sim}{Z}$ denote the maximal Ω-split torus lying in the center of $\underset{\sim}{G}$. The dimension of a maximal Ω-split torus of the group $\underset{\sim}{G}/\underset{\sim}{Z}$ is called the *reduced, semisimple,* or *split rank* of $\underset{\sim}{G}$.

For any Ω-group $\underset{\sim}{X}$ a maximal Ω-split torus lying in the radical of $\underset{\sim}{X}$ is called a *split component* of $\underset{\sim}{X}$.

If $\underset{\sim}{T}$ is an Ω-torus of $\underset{\sim}{G}$, then $Z_{\underset{\sim}{G}}(\underset{\sim}{T})$ is a connected and reductive Ω-subgroup of $\underset{\sim}{G}$. Thus, $\underset{\sim}{T}$ is a split component of $Z_{\underset{\sim}{G}}(\underset{\sim}{T})$ if and only if $\underset{\sim}{T}$ is the maximal Ω-split torus lying in the center of $\underset{\sim}{T}$.

If $\underset{\sim}{\Gamma}$ is a Cartan Ω-subgroup of $\underset{\sim}{G}$, then $Z_{\underset{\sim}{G}}(\underset{\sim}{\Gamma})/\underset{\sim}{\Gamma}$ is obviously anisotropic. In particular, if $\underset{\sim}{\Gamma}/\underset{\sim}{Z}$ is anisotropic, then $Z_{\underset{\sim}{G}}(\underset{\sim}{\Gamma}) = \underset{\sim}{\Gamma}$.

§0.3. Parabolic Subgroups of Reductive Groups.

A Borel subgroup B of G is a maximal connected solvable subgroup of G; equivalently, G/B is a projective variety and B is minimal with this property. All Borel subgroups of G are conjugate. A parabolic subgroup of G is a subgroup which contains a Borel subgroup. Every parabolic subgroup of G is a connected l.a.g.

Let P be a parabolic Ω-subgroup (or p-subgroup) of G. The unipotent radical $N = N_P$ of P is also an Ω-subgroup of P. A connected and reductive Ω-subgroup M of P is called a Levi subgroup or Levi factor of P if $P = M \cdot N$, an Ω-semidirect product of Ω-groups--this means, in particular, that P is Ω-isomorphic to $M \times N$ as Ω-varieties. A subgroup M of G is a Levi subgroup of P if and only if $M = Z_G(A)$ for some split component A of P. The group N acts transitively and freely on the set of split components of P, consequently, also on the set of Levi subgroups of P. The choice of a split component A or Levi subgroup $M = Z_G(A)$ determines a Levi decomposition $P = MN$ for P.

Let (P, A) be a parabolic pair or p-pair of G, i.e., a pair consisting of a parabolic Ω-subgroup P of G and a split component A of P. The codimension of Z, the split component of G, in A is called the parabolic rank or p-rank of P or (P, A).

An Ω-split torus A of G is called a special torus of G if A is a split component of some p-subgroup of G. We note that A is a special torus if and only if A is the split component of $M_A = Z_G(A)$. A special subtorus of a maximal Ω-split torus A_0 of G is called an A_0-standard torus of G.

Given two p-pairs (P, A) and (P', A') of G, we write $(P, A) \succ (P', A')$ if $P \supset P'$ and $A' \supset A$. A p-pair (P_0, A_0) of G is called a minimal p-pair of G if P_0 is a minimal parabolic Ω-subgroup of G; in this case, A_0 is necessarily a maximal Ω-split torus of G. The pair (G, Z) is a p-pair; however, by convention we refer to a maximal proper p-pair, i.e., a p-pair of p-rank one, as a maximal p-pair.

Fix a minimal p-pair (P_0, A_0) of G. A p-pair (P, A) is called standard (with respect to (P_0, A_0)) if $(P, A) \succ (P_0, A_0)$, semistandard (with respect to A_0) if $A \subset A_0$. The set of semistandard p-pairs is finite and will be described in detail in §0.5. For any p-pair (P, A) there is one and only one standard p-pair (P_1, A_1) such that P is conjugate to P_1; in fact, there exists $x \in G$ such that $(P_1^x, A_1^x) = (P, A)$. In particular, any two minimal p-pairs are conjugate in this strong sense.

Let A be a special torus of G. Write $\mathcal{P}(A)$ for the set of parabolic Ω-subgroups of G with A as split component. The finitely many elements of $\mathcal{P}(A)$ are called associated (or A-associated) p-subgroups. §0.5 will also characterize $\mathcal{P}(A)$, when A is standard.

Let P_1 and P_2 be parabolic Ω-subgroups of G. We say that P_1 and P_2 are opposite parabolic subgroups of G if $P_1 \cap P_2$ is a Levi factor of both. For any $P \in \mathcal{P}(A)$ there is exactly one opposite parabolic subgroup $\overline{P} \in \mathcal{P}(A)$. In this case, we also say that (P, A) and $(\overline{P}, A)$ are opposite p-pairs.

Let (P, A) $(P = MN)$ be a standard p-pair of G. There is a one-one correspondence between [semi-standard] (standard) p-pairs (P', A') $(P' = M'N')$ of G such that $(P, A) \succ (P', A')$ and [semi-standard] (standard) p-pairs of M. This correspondence is defined as follows. Given (P', A') as above, set $({}^*\underset{\sim}{P}, \underset{\sim}{A}') = (\underset{\sim}{P}' \cap \underset{\sim}{M}, \underset{\sim}{A}')$. Then $({}^*\underset{\sim}{P}, \underset{\sim}{A}')$ is a p-pair of $\underset{\sim}{M}$ with the Levi decomposition ${}^*\underset{\sim}{P} = {}^*\underset{\sim}{M}\,{}^*\underset{\sim}{N} = \underset{\sim}{M}' \cdot \underset{\sim}{M} \cap \underset{\sim}{N}'$. For this correspondence, $({}^*\underset{\sim}{P}_0, \underset{\sim}{A}_0)$ $({}^*\underset{\sim}{P}_0 = \underset{\sim}{M}_0 \cdot {}^*\underset{\sim}{N}_0 = \underset{\sim}{M}_0 \cdot \underset{\sim}{N}_0 \cap \underset{\sim}{M})$ is the "standard" minimal p-pair of $\underset{\sim}{M}$.

§0.4. On the Rational Points of p-adic Reductive Groups.

A p-adic field is a topological field which, as a topological space, is a nondiscrete totally disconnected space in the sense of §1.1. Concretely, any p-adic field is a completion with respect to a discrete valuation of either a function field in one variable over a finite field of constants or of a number field. From here on, let Ω denote a p-adic field. We normalize the absolute value function on Ω such that, if O is any nonempty compact open subset of Ω, μ is a Haar measure on Ω, and $a \in \Omega^\times$, then $\mu(aO) = |a|\mu(O)$. In this case, if a is a prime element of Ω (i.e., if $|a| < 1$ and $|a|$ generates the group of values), then $|a| = q^{-1}$, where q is the module of Ω.

The space of m-vectors Ω^m has the product p-adic topology. It is easy to see that the intersection with Ω^m of any Ω-closed subset of $\overline{\Omega}^m$ is closed in the p-adic topology. In particular, the group of Ω-points of any Ω-group is a t.d. group (§1.2).

If $\underset{\sim}{X}$ is a _______ Ω-subgroup of $\underset{\sim}{G}$, then we call X a _______ subgroup of G. The blank can contain any of the terms defined in the preceding two sections. If $\underset{\sim}{T}$ is an Ω-split torus of $\underset{\sim}{G}$, we call T a split torus of G. If $(\underset{\sim}{P}, \underset{\sim}{A})$ is a p-pair of $\underset{\sim}{G}$ with $\underset{\sim}{P} = \underset{\sim}{M}\underset{\sim}{N}$ its associated Levi decomposition we call

(P, A) a p-pair of G and $P = MN$ its Levi decomposition.

All the groups $\underset{\sim}{X}$ considered in the previous two sections have the property that X is Zariski dense in $\underset{\sim}{X}$.

If $\underset{\sim}{G}$ is anisotropic, then G is compact.

Given $\chi \in X(G)$ and $x \in G$, we set $\langle \chi, H_G(x) \rangle = \log_q |\chi(x)|$. This defines a continuous homomorphism $H_G : G \to \mathrm{Hom}(X(G), \mathbb{Z})$. Set ${}^{\circ}G = \bigcap_{\chi \in X(G)} \ker|\chi| = \ker H_G$. Then ${}^{\circ}G$ is an open normal subgroup of G which contains every compact subgroup of G. Indeed, the factor group $G/{}^{\circ}G$ is a free abelian group.

Lemma 0.4.1. Let $\underset{\sim}{G}$ be a connected and reductive Ω-group and let $\underset{\sim}{Z}$ be the maximal Ω-split torus in the center of $\underset{\sim}{G}$. There is a natural injection $r^* : X(G) \hookrightarrow X(Z)$. The factor group $X(Z)/X(G)$ is finite.

Proof. The natural map r^* is restriction. We have only to show that r^* maps $X(G)$ injectively to a subgroup of the same rank as $X(Z)$. Note first that the semisimple subgroup $\mathfrak{D}\underset{\sim}{G}$ is a connected normal Ω-subgroup of $\underset{\sim}{G}$. Set $\underset{\sim}{Z}' = \underset{\sim}{Z}/\underset{\sim}{Z} \cap \mathfrak{D}\underset{\sim}{G}$ and $\underset{\sim}{G}' = \underset{\sim}{G}/\mathfrak{D}\underset{\sim}{G}$. Then $\dim \underset{\sim}{Z}' = \dim \underset{\sim}{Z}$ and $\underset{\sim}{Z}'$ is a maximal Ω-split torus in $\underset{\sim}{G}'$. Since $X(G) = X(G') = X(Z')$, which is a subgroup of finite index in $X(Z)$, the lemma is true.

Corollary 0.4.2. The subgroup ${}^{\circ}G \cdot Z$ is of finite index in G.

Proof. Observe that $G/{}^{\circ}G \supset {}^{\circ}GZ/{}^{\circ}G$ and both groups are isomorphic to lattices of the same rank.

Remark. Let $G_1 = G \cap \mathfrak{D}\underset{\sim}{G}$. If $\mathrm{char}\,\Omega = 0$, then $[G : G_1 \cdot Z] < \infty$; however, if char $\Omega > 0$, this is not always true. Note that ${}^{\circ}G$ is not necessarily the group of Ω-rational points of an l.a.g. (e.g., ${}^{\circ}\Omega^{\times} = \{x \in \Omega^{\times} \mid |x| = 1\}$). However, G/G_1Z is compact and abelian.

§0.5. Lie Algebras, Roots, and Weyl Groups.

Let A be a special torus of G and let $M = Z_G(A)$. Then M is a Levi subgroup of P for all $P \in \mathcal{P}(A)$. We define the Weyl group of A (relative to G) as $W(G/A) = W(A) = N_{\underset{\sim}{G}}(\underset{\sim}{A})/Z_{\underset{\sim}{G}}(\underset{\sim}{A}) = N_G(A)/Z_G(A)$. We note that $Z_{\underset{\sim}{G}}(\underset{\sim}{A}) = N_{\underset{\sim}{G}}(\underset{\sim}{A})^{\circ} = \underset{\sim}{M}$, which implies that $W(A)$ is a finite group. More generally, if A_1 and A_2 are special tori, we write $W(A_2|A_1)$ for the set of homomorphisms $s : A_1 \to A_2$ which are induced by inner automorphisms of G.

There is a natural action of $W(A)$ on A and, dually, on $X(A)$. Given $a \in \underset{\sim}{A}$ and $s \in W(A)$, set $s \cdot a = a^s = a^y = yay^{-1}$, where $y = y(s) \in N_{\underset{\sim}{G}}(\underset{\sim}{A})$ represents s; for $\chi \in X(A)$ define χ^s such that $\chi^s(s \cdot a) = \chi(a)$ $(a \in \underset{\sim}{A})$.

Define the real Lie algebra of A as $\mathfrak{a} = \mathrm{Hom}(X(A), \mathbb{Z}) \otimes_{\mathbb{Z}} \mathbb{R}$ and its dual $\mathfrak{a}^* = X(A) \otimes_{\mathbb{Z}} \mathbb{R}$, also the complexifications $\mathfrak{a}_{\mathbb{C}} = \mathfrak{a} \otimes \mathbb{C}$ and $\mathfrak{a}^*_{\mathbb{C}} = \mathfrak{a}^* \otimes \mathbb{C}$. Notice that, canonically, $\mathfrak{a} = \mathrm{Hom}(X(M), \mathbb{Z}) \otimes \mathbb{R}$ and $\mathfrak{a}^* = X(M) \otimes \mathbb{R}$. The mapping H_M of §0.4 imbeds $M/{}^{\circ}M$ as a lattice in $\mathfrak{a}$; each element $\chi \in X(A)$ corresponds to a unique element of $\mathfrak{a}^*$, called the associated weight. We usually denote rational characters and the canonically associated weights by the same Greek letter, depending upon the context to indicate the intended meaning. The pairing $\langle\ ,\ \rangle$ of §0.4 extends to a pairing of $\mathfrak{a}^* \times \mathfrak{a}$ to $\mathbb{R}$, of $\mathfrak{a}^*_{\mathbb{C}} \times \mathfrak{a}_{\mathbb{C}}$ to $\mathbb{C}$. The group $W(A)$ operates on both $\mathfrak{a}$ and $\mathfrak{a}^*$.

Next we let the action of $\underset{\sim}{G}$ on $\underset{\sim}{G}$ by inner automorphisms induce the adjoint representation $\mathrm{Ad} : \underset{\sim}{G} \to \mathrm{Aut}(\underset{\sim}{\mathfrak{g}})$, where $\underset{\sim}{\mathfrak{g}}$ is the Lie algebra of $\underset{\sim}{G}$ and $\mathrm{Aut}(\underset{\sim}{\mathfrak{g}})$ is its automorphism group. The group $\mathrm{Aut}(\underset{\sim}{\mathfrak{g}})$ is an Ω-group and Ad is an Ω-morphism. We write $\mathrm{Ad}_{\underset{\sim}{A}}$ for the restriction of Ad to $\underset{\sim}{A}$. Since $\underset{\sim}{A}$ is Ω-split, $\mathrm{Ad}_{\underset{\sim}{A}}$ is diagonalizable (over Ω). We call a nontrivial rational character of $\underset{\sim}{A}$ which occurs in $\mathrm{Ad}_{\underset{\sim}{A}}$ a <u>root character.</u> The weights of the root characters are called the <u>roots</u> or A-<u>roots</u> with respect to G. An A-root is called <u>reduced</u> if $t\alpha$ an A-root with $t \in \mathbb{Q}$ implies $t \in \mathbb{Z}$. We write $\Sigma(\underset{\sim}{G}, \underset{\sim}{A})$ or $\Sigma(G, A)$ [$\Sigma_r(\underset{\sim}{G}, \underset{\sim}{A})$ or $\Sigma_r(G, A)$] for the set of A-roots [reduced A-roots] with respect to G. We have the direct sum decomposition $\underset{\sim}{\mathfrak{g}} = \underset{\sim}{\mathfrak{m}} \oplus \underset{\sim}{\mathfrak{g}}_\alpha$ $(\alpha \in \Sigma(G, A))$, where $\underset{\sim}{\mathfrak{m}}$ is the Lie algebra of $\underset{\sim}{M}$ and $\underset{\sim}{\mathfrak{g}}_\alpha$ the eigenspace in $\underset{\sim}{\mathfrak{g}}$ associated to the root character $\alpha \in X(A)$.

To each pair $\pm\alpha \in \Sigma_r(G, A)$ there corresponds an orthogonal hyperplane $H_\alpha = \{a \in \mathfrak{a} \mid \langle\alpha, a\rangle = 0\}$. The connected components of the space $\mathfrak{a} - \cup H_\alpha = \mathfrak{a}'$ $(\alpha \in \Sigma(G, A))$ are called <u>chambers.</u> Choosing a chamber $\mathfrak{c} \subset \mathfrak{a}$, we obtain a set $\Sigma_{\mathfrak{c}} = \{\alpha \in \Sigma(G, A) \mid \langle\alpha, \mathfrak{c}\rangle > 0\}$. There is also a unique set $\Sigma^0_{\mathfrak{c}} \subset \Sigma_{\mathfrak{c}}$ of <u>simple roots</u> such that the elements of $\Sigma^0_{\mathfrak{c}}$ are linearly independent and every element of $\Sigma_{\mathfrak{c}}$ is a positive integer combination of elements of $\Sigma^0_{\mathfrak{c}}$. Let $\underset{\sim}{\mathfrak{n}} = \oplus\, \underset{\sim}{\mathfrak{g}}_\alpha$ $(\alpha \in \Sigma_{\mathfrak{c}})$. Then $\underset{\sim}{\mathfrak{n}}$ is the Lie algebra of the unipotent radical $\underset{\sim}{N}$ of a parabolic subgroup $P = MN$ of G. We also write $\Sigma(P, A) = \Sigma_{\mathfrak{c}}$, $\Sigma^0(P, A) = \Sigma^0_{\mathfrak{c}}$ and $\Sigma_r(P, A) = \Sigma_r(G, A) \cap \Sigma_{\mathfrak{c}}$. Observe that $-\Sigma_{\mathfrak{c}} = \Sigma_{-\mathfrak{c}}$ and that $\Sigma(G, A) = \Sigma_{\mathfrak{c}} \cup \Sigma_{-\mathfrak{c}}$. The chamber $-\mathfrak{c}$ corresponds to the opposite parabolic subgroup $\overline{P} = M\overline{N} \in \mathcal{P}(A)$. We have the following one-one correspondences:

$$\mathcal{P}(A) \longleftrightarrow \{\mathfrak{c} \mid \mathfrak{c} \subset \mathfrak{a}\} \longleftrightarrow \{\Sigma_{\mathfrak{c}} \mid \Sigma_{\mathfrak{c}} \subset \Sigma(G, A)\} \longleftrightarrow \{\Sigma^0_{\mathfrak{c}}\}.$$

Let $(P, A)(P = MN)$ be a p-pair of G. To any subset $F \subset \Sigma^0(P, A)$ there corresponds a special torus $\underset{\sim}{A}' = (\bigcap_{\alpha \in F} \ker \alpha)^0$ (the connected component containing the identity of the intersection of the kernels of all the root characters corresponding to roots $\alpha \in F$) and a parabolic subgroup $P' = Z_G(A')N$ such that $(P', A') = (P, A)_F \succ (P, A)$. The correspondence $F \longleftrightarrow (P, A)_F$ is one-one between subsets of $\Sigma^0(P, A)$ and p-pairs (P', A') of G such that $(P', A') \succ (P, A)$. For any F we may regard $\Sigma^0(P, A) - F$, by restriction, as $\Sigma^0(P, A)_F$.

Now let A_0 be a maximal split torus of G and set $M_0 = Z_G(A_0)$. Recalling that A_0 is unique up to conjugacy, we note that $W(G/A_0) = W_0$ is unique up to isomorphism. We call W_0 the <u>(relative) Weyl group of</u> G. The set of A_0-roots $\Sigma(G, A_0) \subset \mathfrak{a}_0^*$ is a not necessarily reduced root system with Weyl group W_0. Corresponding to each pair $\pm\alpha$ of reduced A_0-roots there is a reflection $s_\alpha \in W_0$ with respect to the hyperplane H_α. For any $P_0 \in \mathcal{P}(A_0)$ the set of reflections $\{s_\alpha | \alpha \in \Sigma^0(P_0, A_0\}$ is a set of generators for W_0. The Weyl group W_0 acts simply transitively on the set of chambers $\mathfrak{C} \subset \mathfrak{a}_0$, i.e., on the set of minimal parabolic subgroups $\mathcal{P}(A_0)$.

The reader may easily verify that, if A is an A_0-standard torus, then $W(A)$ is, in a natural way, a subquotient of W_0. However, the group $W(A)$ is in general not generated by reflections and it does not necessarily act transitively on the chambers of $\mathfrak{a}$ --elements of $\mathcal{P}(A)$ are not necessarily conjugate. On the other hand, it is true that only the identity of $W(A)$ stabilizes a chamber.

Finally, fix a W_0-invariant inner product on $\mathfrak{a}_0$ and use this inner product to identify $\mathfrak{a}_0^*$ and $\mathfrak{a}_0$. By restriction obtain a $W(A)$-invariant inner product on $\mathfrak{a}$. To each $\alpha \in \mathfrak{a}^*$ there corresponds a unique element $i(\alpha) \in \mathfrak{a}$ such that $\langle\alpha, H(a)\rangle = (i(\alpha), H(a))$ $(a \in A)$.

Fixing a closed chamber $\bar{\mathfrak{C}} = \mathfrak{a}^+ \subset \mathfrak{a}$, we may define a corresponding dual cone $+\mathfrak{a}$ to be the set of all linear combinations of elements of $\Sigma_{\mathfrak{C}}$ with nonnegative coefficients; we define $H_M^{-1}(\mathfrak{a}^+) = M^+$ and $H_M^{-1}(+\mathfrak{a}) = +M$ of M. We also write $A^+ = A \cap M^+$ and $+A = A \cap +M$. It follows from Corollary 0.4.2 that $M^+ = {}^0M\omega A^+$, where ω is a finite subset of M^+.

§0.6. A_0-good Maximal Compact Subgroups of G.

The maximal compact subgroups of G are all open subgroups as well and they lie in finitely many conjugacy classes. A theory due to Bruhat and Tits ([4c]) gives a detailed description of these conjugacy classes.

Fix a maximal split torus A_0 of G. Write $M_0 = Z_G(A_0)$ and $M_0^+ = H_{M_0}^{-1}(\bar{\mathfrak{C}})$, where $\mathfrak{C}$ is a chamber of $\mathfrak{a}_0$. A maximal compact subgroup K of G is called an A_0-good maximal compact subgroup of G if:

(1) $G = KP = PK$ for some minimal (and, hence, every) parabolic subgroup of G.

(2) There exist representatives $y(s) \in K$ for every $s \in W(G/A_0)$.

(3) For every A_0-standard torus A and $P \in \mathcal{P}(A)$ $(P = MN)$ the group $K \cap P = (K \cap M)(K \cap N)$.

(4) The group G possesses a Cartan decomposition $G = K(M_0^+/{}^0M_0)K$.

(5) For any A_0-standard torus A, $M = Z_G(A)$, the group $K \cap M$ satisfies (1)-(4) with respect to (M, A_0). Bruhat and Tits have proved that there exist A_0-good maximal compact subgroups of G.

Note that we may also write the Cartan decomposition of G in the form $G = K(A_0^+/{}^0A_0)\omega K$, where ω is a finite subset of M_0^+ and $A_0^+ = A_0 \cap M_0^+$.

EXAMPLE:

Let D be a central division algebra of dimension ℓ^2 over Ω. For any integer $n \geq 1$ the ring $M_n(D)$ of all $n \times n$ matrices with entries in D is a central simple algebra over Ω. There is an absolute value mapping $| \ | : D \to \mathbb{Q}$ and a reduced norm mapping $\nu : M_n(D)^{\times} \to \Omega^{\times}$. Let $G = SL_n(D)$ be the group of all elements of $M_n(D)^{\times}$ which have reduced norm one. Then G is the group of Ω-points of a connected reductive Ω-group; in fact, G is simply connected and almost simple. We may choose A_0 to be the group of all diagonal matrices in G with entries in Ω; in this case, $M_0 = Z_G(A_0)$ is just the full diagonal subgroup of G. If $D \neq \Omega$, then $M_0 \neq A_0$. It is easy to see that ${}^0M_0$ is the subgroup of M_0 consisting of all diagonal matrices whose entries have absolute value one and that $[M_0 : {}^0M_0A_0] = \ell^{n-1}$.

Fix an A_0-good maximal compact subgroup K of G. Let H be a closed, hence locally compact, subgroup of G. A Haar measure dh on H will be called <u>normalized</u> (with respect to K) if $\int_{H \cap K} dh = 1$. It is of course clear that this makes sense, since $H \cap K$ is an open compact subgroup of H.

With respect to $P_0 \in \mathcal{P}(A_0)(P_0 = M_0N_0)$ and K we also have an <u>Iwasawa decomposition</u> $G = KM_0N_0 = M_0N_0K$ for G. It is also clear that, for any parabolic subgroup $P = MN$ of G, we have $G = KP = PK = MNK$.

Given a fixed A_0-good maximal compact subgroup G and a semi-standard p-pair $(P, A)(P = MN)$, we have a mapping $H_P : G \to \mathfrak{a}$ ($\mathfrak{a}$ the real Lie algebra of A) defined by setting $H_P(kmn) = H_M(m)$ $(k \in K,\ m \in M,\ n \in N)$.

Fixing a subset $S \subset M$ of representatives for $K\backslash G/N$, we obtain continuous mappings $\kappa : G \to K$ and $\eta : G \to N$ and a locally constant mapping $\mu : G \to S \hookrightarrow M$ such that $\kappa(x)\mu(x)\eta(x) = x$ and $H_P(x) = H_M(\mu(x))$ for all $x \in G$.

Chapter I. Generalities Concerning Totally Disconnected Groups and Their Representations.

The intention of this chapter, violated in the two appendices to §1.2, is to present all the ideas needed in these notes which do not depend, either for their formulations or for their proofs, upon the theory of algebraic groups.

§1.1. Functions and Distributions on Totally Disconnected Spaces.

A t.d. space or totally disconnected space is a Hausdorff space which has a countable basis consisting of open compact sets. It follows that a t.d. space is locally compact. Subordinate to any open covering of a totally disconnected space, there is a covering of the space by a countable union of disjoint open compact sets. Any closed or open subspace of a t.d. space is obviously a t.d. space with respect to the relative topology. A finite or restricted countable product of t.d. spaces is also a t.d. space.

Let X be a t.d. space and S a set. A function $f : X \to S$ will be called smooth or locally constant if every $x \in X$ lies in a neighborhood $O(x)$ whose image $f(O(x))$ is a single point of S. We write $C^{\infty}(X:S)$ for the set of all smooth S-valued functions on X. Notice that, if S is given the discrete topology, then $C^{\infty}(X:S)$ is exactly the set of all continuous mappings from X to S.

Now let S have a distinguished "zero" element. Let $f : X \to S$ be a function. By the support of f we mean the closure of the subset of X on which f assumes non-zero values. We write Supp f for the support of f. Write $C_c^{\infty}(X:S)$ for the subset of $C^{\infty}(X:S)$ which consists of those smooth f for which Supp f is compact. If $f \in C_c^{\infty}(X:S)$, then f has only finitely many distinct values and Supp f is open as well as compact.

In the event that, in addition, S has the structure of a topological space we also have the sets $C(X:S)$ and $C_c(X:S)$ of all continuous S-valued functions and, respectively, all continuous, compactly supported S-valued functions.

Now let V be a vector space over the complex numbers. In this case, the spaces $C^\infty(X:V)$, $C(X:V)$, $C_c^\infty(X:V)$, and $C_c(X:V)$ are all naturally complex vector spaces. If $V = \mathbb{C}$, the complex numbers, we generally write $C(X)$ for $C(X:\mathbb{C})$, etc. If Y is an open subset of X, then there is a natural (linear) injection $i : C_c^\infty(Y:V) \hookrightarrow C_c^\infty(X:V)$; if Z is a closed subset of X, there is a surjective "restriction" mapping $r : C_c^\infty(X:V) \to C_c^\infty(Z:V)$; i.e., r is a surjective linear mapping, because any compact open subset of Z is the intersection with Z of a compact open subset of X. There is also a monomorphism $i_V : C^\infty(X) \otimes V \hookrightarrow C^\infty(X:V)$ defined by setting $i_V(f \otimes v)(x) = f(x)v$ for any $f \in C^\infty(X)$ and $v \in V$ and extending by linearity. The mapping $i_{V,c} = i_V | C_c^\infty(X) \otimes V$ is a bijection onto $C_c^\infty(X:V)$. If V is finite dimensional, then i_V is an isomorphism.

A linear functional $T : C_c^\infty(X:V) \to \mathbb{C}$ is called a V-distribution on X. If $V = \mathbb{C}$, we call T a distribution on X. Note that these definitions include no continuity conditions. The set of V-distributions has a natural vector space structure. We write $\mathcal{D}(X:V)[\mathcal{D}(X)]$ to denote the vector space consisting of all V-distributions [distributions] on X.

Let T be a V-distribution on X. If Y is an open subset of X, then $T \circ i = T_Y$ is a V-distribution on Y, called the restriction of T to Y. The mapping $T \mapsto T_Y$ of $\mathcal{D}(X:V)$ into $\mathcal{D}(Y:V)$ is obviously linear. If the restriction T_Y of T is non-zero for every neighborhood Y of a point $x \in X$, then x is said to lie in the support of T. The support of T is obviously a closed subset of X and is denoted Supp T.

Lemma 1.1.1. Let $f \in C_c^\infty(X:V)$ and $T \in \mathcal{D}(X:V)$. If $\text{Supp } f \cap \text{Supp } T = \phi$, then $T(f) = 0$.

Proof. Since Supp f is a compact open set and $\text{Supp } f \cap \text{Supp } T = \phi$, we may write Supp f as a finite union of compact open sets Y_i $(i \in I)$ such that $T_{Y_i} = 0$. The lemma is an immediate consequence of this fact.

The set of all V-distributions which have their supports in a subset S of X comprise a subspace $\mathcal{D}_S(X:V) \subset \mathcal{D}(X:V)$. Let Z be a closed subset of X. Given $T \in \mathcal{D}(Z:V)$ we define $T^X = T \circ r$, the extension of T to X. Obviously, $T \mapsto T^X$ is a linear mapping of $\mathcal{D}(Z:V)$ into $\mathcal{D}(X:V)$.

Lemma 1.1.2. The mapping $T \mapsto T^X$ defines a bijection of $\mathcal{D}(Z:V)$ onto $\mathcal{D}_Z(X:V)$.

Proof. That the mapping is injective follows from the fact that r is surjective. It is, furthermore, clear that $\text{Supp } T^X \subset Z$ for any $T \in \mathcal{D}(Z:V)$. To prove surjectivity we shall construct an inverse mapping. Take $T \in \mathcal{D}_Z(X:V)$. For any $f \in C_c^\infty(Z:V)$ we know that there exists an element $\tilde{f} \in C_c^\infty(X:V)$ such that $i(\tilde{f}) = f$. Define $T_Z(f) = T(\tilde{f})$. To show that $T_Z \in \mathcal{D}(Z:V)$ is well defined it is sufficient to observe that, if $r(\tilde{f}) = 0$, then $T(\tilde{f}) = 0$ (Lemma 1.1.1). Finally, it is clear that $T_Z{}^X(\tilde{f}) = T_Z(f) = T(\tilde{f})$, so the mapping is surjective.

A distribution T on X is called a Radon measure on X (cf., [14a], p. 32) if:

(1) T extends to a functional defined on $C_c(X)$;

(2) $T(f)$ is a non-negative real number whenever $f \in C_c(X)$ is a non-negative real-valued function.

Let μ be a Radon measure on X and let $f \in C_c^\infty(X:V)$. Define $\int_X f d\mu = \sum_{v \in f(X)} \mu(f^{-1}(v))v$, a finite sum. If V is finite-dimensional, then we may define $\int_X f d\mu$ for $f \in C_c(X:V)$; we leave this to the reader. If $\mathrm{Supp}\,\mu$ is compact, we define $\int_X f d\mu$ for any $f \in C^\infty(X:V)[C(X:V)]$ by setting $\int_X f d\mu = \int_X f\phi\, d\mu$, where $\phi \in C_c^\infty(X)$ satisfies $\phi \equiv 1$ on $\mathrm{Supp}\,\mu$.

Let $\pi : X \to \mathrm{End}(V)$ be a mapping such that the function $x \mapsto \pi(x)v$ is smooth for any $v \in V$ (Notation: $\mathrm{End}(V)$ is the algebra of endomorphisms of V). If $f \in C_c^\infty(X)$, we set $\pi(f, \mu)v = \int_X f(x)\pi(x)v d\mu$. We write $\pi(f)$ for this operator when μ is understood.

§1.2. T. d. Groups and Relatively Invariant Measures on Homogeneous Spaces.

A t. d. group or totally disconnected group is a topological group which, as a topological space, is a t. d. space.

Let G be a t. d. group. Then every neighborhood of the identity in G contains a compact open subgroup. If G is not discrete, then a compact open subgroup is a profinite group. (For the above and other structural facts about t. d. groups the reader should consult [12].)

Let X be an open subset of G and E a complex vector space. If O is an open compact subset of X, write $Ox[xO]$ for the image of O under right [left] multiplication by $x \in G$. Write α_O for the characteristic function of O. An E-distribution T on X is called left-invariant [right-invariant] if, given any $x \in G$ and open compact subset O of $X \cap x^{-1}X[X \cap Xx^{-1}]$, we have $T(\alpha_{xO}) = T(\alpha_O)$ $[T(\alpha_{Ox}) = T(\alpha_O)]$.

Any locally compact group has defined on it a left invariant Radon measure, called a left Haar measure, which is unique up to a positive constant factor. Let $d_\ell x$ denote a left Haar measure. There is a homomorphism δ_G from G into the positive real numbers, called the modular function of G, such that $d_\ell x^{-1} = d_r x = \delta_G(x) d_\ell x$ is a right invariant measure. For any $y \in G$ the mapping $O \mapsto \int_{Oy} d_\ell x$ from compact open subsets of G to non-negative real numbers determines a unique left invariant measure which we write as $d_\ell(xy)$. We have $\delta_G(xy) d_\ell(xy) = \delta_G(x) d_\ell x$, so $\delta_G^{-1}(y) d_\ell x = d_\ell(xy)$ or $d_\ell(yxy^{-1}) = \delta_G(y) d_\ell x$.

For Γ a subgroup of G write $G/\Gamma[\Gamma\backslash G]$ for the space of left [right] cosets of Γ in G, or sometimes for a set of coset representatives. Write $G/\!\!/\Gamma$ for double cosets, i.e., for $\Gamma\backslash G/\Gamma$. If Γ is closed, then, as follows immediately from the fact that the quotient projection is both open and continuous, G/Γ and $\Gamma\backslash G$ are t.d. spaces. If Γ is open, then G/Γ is discrete, so $C^\infty(G/\Gamma) = C(G/\Gamma)$--i.e., all functions on G/Γ are both continuous and smooth. In general, we often regard $C^\infty(G/\Gamma)$ as the set of all smooth functions $f : G \to \mathbb{C}$ such that $f(g\gamma) = f(g)$ for all $g \in G$ and $\gamma \in \Gamma$. The symbol $C_c^\infty(G/\Gamma)$ can denote the subspace of $C^\infty(G/\Gamma)$ which consists of all functions on G whose support projects to a compact subset of G/Γ. Similar ideas apply to $\Gamma\backslash G$ and $G/\!\!/\Gamma$ as well.

The following lemmas will be used chiefly in §§1.7 and 1.8. The reader might consider skipping the remainder of this section and its appendices on a first reading.

Let Γ be a closed subgroup of G and E a complex vector space. Write $p(x) = \bar{x}$, where $p : G \to G/\Gamma$ is the projection mapping. For $\alpha \in C_c^\infty(G:E)$ set $\bar{\alpha}(\bar{x}) = \int_\Gamma \alpha(x\gamma) d_\ell \gamma \quad (x \in G)$.

Lemma 1.2.1. For any open compact subset X of G the mapping $\alpha \mapsto \bar{\alpha}$ sends $C_c^\infty(X:E)$ onto $C_c^\infty(\bar{X}:E)$, where $p(X) = \bar{X}$.

Proof. We prove a more general result by exactly the same methods later (Lemma 1.9.1), so, for the present, we spare the reader the details of this proof.

Corollary 1.2.2. The mapping $\alpha \mapsto \bar{\alpha}$ is a surjective mapping from $C_c(X:E)$ to $C_c(\bar{X}:E)$.

Proof omitted.

Lemma 1.2.3. For $\alpha \in C_c^\infty(G:E)$ set

$$(\lambda(y)\rho(z)\alpha)(x) = \alpha(y^{-1}xz)$$

and

$$(\bar{\lambda}(y)\bar{\alpha})(\bar{x}) = \bar{\alpha}((y^{-1}x)^-) \qquad (x, y, z \in G).$$

Then

$$(\lambda(y)\rho(\gamma)\alpha)^-(x) = \delta_\Gamma^{-1}(\gamma)(\bar{\lambda}(y)\bar{\alpha})(\bar{x}) \qquad (x, y \in G;\ \gamma \in \Gamma).$$

Proof.

$$(\lambda(y)\rho(\gamma)\alpha)^-(\bar{x}) = \int_\Gamma \alpha(y^{-1}x\gamma_1\gamma)d_\ell\gamma_1$$

$$= \delta_\Gamma(\gamma)^{-1}\int_\Gamma \alpha(y^{-1}x\gamma_2)d_\ell\gamma_2$$

$$= \delta_\Gamma(\gamma)^{-1}(\bar{\lambda}(y)\bar{\alpha})(\bar{x}).$$

Lemma 1.2.4. Let Γ be a closed subgroup of G. Then there exists a real-valued locally constant function χ on G such that:

1) $\chi(x) > 0$ for all $x \in G$ and $\chi(1) = 1$;

2) $\chi(x\gamma) = \chi(x)\chi(\gamma)$ for all $x \in G$ and $\gamma \in \Gamma$;

and

3) $\chi(\gamma) = \delta_\Gamma(\gamma)\delta_G^{-1}(\gamma)$ for all $\gamma \in \Gamma$.

Proof. Fix an open compact subgroup K of G and choose a sequence $\{x_i\}_{i=1}^\infty \subset G$ such that $x_1 = 1$ and $G = \coprod_{i=1}^\infty Kx_i\Gamma$ (disjoint union). Extend δ_Γ to a locally constant function on G by setting $\delta_\Gamma(kx_i\gamma) = \delta_G(x_i)\delta_\Gamma(\gamma)$. It is easy to see that this is unambiguous. Set $\chi(x) = \delta_\Gamma(x)\delta_G^{-1}(x)$ $(x \in G)$. All the required properties are obvious.

Lemma 1.2.5. Let Γ be a closed subgroup of G and fix a left Haar measure $d_\ell\gamma$ on Γ. Define χ as in Lemma 1.2.4. Then there exists a unique Radon measure $d\bar{x} = d_\chi\bar{x}$ on $\bar{G} = G/\Gamma$ such that

$$\int_G \alpha(x)\chi^{-1}(x)d_\ell x = \int_{\bar{G}} \bar{\alpha}(\bar{x})d\bar{x}$$

for all $\alpha \in C_c(G)$ and $\bar{\alpha}(\bar{x}) = \int_\Gamma \alpha(x\gamma)d_\ell\gamma$.

Proof. Noting that $\alpha \mapsto \bar{\alpha}$ is surjective (Corollary 1.2.2), we see that it suffices to prove the existence of $d\bar{x}$--uniqueness is then obvious. Noting, furthermore, that $\chi^{-1}(x)d_\ell x$ is a Radon measure on G and that any non-negative real-valued function in $C_c(\bar{G})$ is the image of such a function in $C_c(G)$ under the mapping $\alpha \mapsto \bar{\alpha}$, we conclude that it is sufficient to prove the following: If $\alpha \in C_c(G)$ and $\bar{\alpha} = 0$, then $\int_G \alpha(x)\chi^{-1}(x)d_\ell x = 0$, too.

For this let $\alpha \in C_c(G)$. Then

$$\int_G \alpha(x)\chi^{-1}(x)d_\ell x = \sum_{i=1}^{\infty} \sum_{\gamma \in \Gamma \cap x_i^{-1}Kx_i \backslash \Gamma} \int_{Kx_i\gamma} \alpha(x)\chi^{-1}(x)d_\ell x$$

$$= \sum_{i=1}^{\infty} \sum_{\gamma \text{ as above}} \delta_G^{-1}(x_i\gamma)\int_K \alpha(kx_i\gamma)\chi^{-1}(kx_i\gamma)dk$$

$$= \sum_{i,\, \gamma \text{ as above}} \delta_G^{-1}(x_i)\int_K \alpha(kx_i\gamma)\delta_\Gamma^{-1}(\gamma)dk .$$

Since there is a right Haar measure $d_{r,i}\gamma$ on Γ such that

$\sum_{\gamma \in \Gamma \cap x_i^{-1}Kx_i \backslash \Gamma} \psi(\gamma) = \int_\Gamma \psi(\gamma)d_{r,i}\gamma$ for all $\psi \in C_c(\Gamma \cap x_i^{-1}Kx_i \backslash \Gamma)$, we obtain

$$\int_G \alpha(x)\chi^{-1}(x)d_\ell x = \sum_{i=1}^{\infty} c_i \int_\Gamma \int_K \alpha(kx_i\gamma)dk d_\ell \gamma,$$

where c_i is a positive constant for each i. Since we may obviously change orders of summing and integrating at will, it is clear that $\bar{\alpha} = 0$ implies that $\int_G \alpha(x)\chi^{-1}(x)d_\ell x = 0$.

<u>Corollary</u> 1.2.6. $d(y\bar{x}) = \chi(y;\bar{x})d\bar{x}$, where $\chi(y;\bar{x}) = \chi(x)\chi^{-1}(yx)$.

<u>Proof.</u> Using Corollary 1.2.2, we take $\alpha \in C_c(G)$. Then, by definition, $\int_{\bar{G}} \bar{\alpha}(y^{-1}\bar{x})d\bar{x} = \int_{\bar{G}} \bar{\alpha}(\bar{x})d(y\bar{x})$, and it suffices to check that

$$\int_{\bar{G}} \bar{\alpha}(y^{-1}\bar{x})d\bar{x} = \int_G \alpha(y^{-1}x)\chi^{-1}(x)d_\ell x$$

$$= \int_G \alpha(y^{-1}x)\chi^{-1}(x)\chi(y^{-1}x)\chi^{-1}(y^{-1}x)d_\ell x$$

$$= \int_G \alpha(u)\chi^{-1}(yu)\chi(u)\chi^{-1}(u)d_\ell u$$

$$= \int_{\bar{G}} \bar{\beta}(\bar{x})d\bar{x},$$

where $\beta(x) = \alpha(x)\chi^{-1}(yx)\chi(x)$ and $\bar{\beta}(\bar{x}) = \chi^{-1}(yx)\chi(x)\bar{\alpha}(\bar{x})$.

§1.2.1. On the Haar Measure for a Parabolic Group.

Let $G = \underset{\sim}{G}(\Omega)$, where $\underset{\sim}{G}$ is a connected reductive Ω-group, Ω as in Chapter 0. Let $(P, A)(P = MN)$ be a p-pair of G. Then the subgroups M and N are each separately unimodular; however, P possesses a non-trivial modular function δ_P. We shall show that δ_P may be described in terms of the action of $\underset{\sim}{A}$ on the Lie algebra $\mathfrak{n}$ of $\underset{\sim}{N}$.

For this purpose we need a few more remarks about the structure of parabolic groups, in addition to the information presented in Chapter 0. The group $\underset{\sim}{N}$ has a finite descending chain of normal Ω-subgroups, $\underset{\sim}{N} = \underset{\sim}{N}_1 \supset \underset{\sim}{N}_2 \supset \ldots \supset \underset{\sim}{N}_r = (1)$, such that M normalizes N_i for each i and N_i/N_{i+1} is a vector group. As algebraic varieties, $\underset{\sim}{N} \cong \underset{\sim}{N}_1/\underset{\sim}{N}_2 \times \ldots \times \underset{\sim}{N}_{r-1}$. From the Lie algebra quotients $\mathfrak{n}_i/\mathfrak{n}_{i+1}$ there is a group isomorphism $\varphi_i : \mathfrak{n}_i/\mathfrak{n}_{i+1} \to N_i/N_{i+1}$ such that φ_i is both an isomorphism of Ω-vector spaces and an M-module isomorphism. We may describe the action induced by Ad_M or Ad_A on $\mathfrak{n}_i/\mathfrak{n}_{i+1}$ in terms of the A-roots in $\Sigma(P, A)$ and use this description to compute the effect of conjugation by $m \in M$ on a Haar measure of N.

We observe, first, that a product of Haar measures μ_i on the groups N_i/N_{i+1} gives a Haar measure μ for N. The action of A on $\Sigma\mathfrak{n}_i/\mathfrak{n}_{i+1}$ is the same as that on the vector space $\oplus \mathfrak{g}_\alpha$, the sum being over all $\alpha \in \Sigma(P, A)$ with multiplicities included exactly as for $\mathfrak{n}$. (In fact, this describes the action of a maximal split torus A_0 of M on $\Sigma\mathfrak{n}_i/\mathfrak{n}_{i+1}$.) Define $\rho = \rho_P$ by setting $2\rho = \Sigma\alpha$ (summed over $\alpha \in \Sigma(P, A)$, including multiplicities). It follows that, if $Y = \prod_{i=1}^{r} Y_i \subset N$, where each Y_i is a section in N_i from N_i/N_{i+1} or $\mathfrak{n}_i/\mathfrak{n}_{i+1}$, then

$$\mu(mYm^{-1}) = \prod_{i=1}^{r} \mu_i(mYm^{-1}) = q^{2\langle \rho, H(m)\rangle} \mu(Y) .$$

Lemma 1.2.1.1. Let dm and dn denote Haar measures on the groups M and N, respectively. For $f \in C_c(P)$ set $I(f) = \int_{MN} f(mn)dm\,dn$. Then $f \mapsto I(f)$ defines a left-invariant measure on P. The modular function for P is $\delta_P(p) = \delta_P(mn) = q^{2\langle \rho, H_M(m)\rangle}$ $(p = mn \in P)$.

Proof. The measure $dm\,dn$ is obviously right invariant under N and, from what has been said before, it follows that $q^{+2\langle \rho, H_M(m)\rangle} dm\,dn$ is right M-invariant. Since N, being a union of compact subgroups, lies in the kernel of the modular function and $dm\,dn$ is left M-invariant, the lemma follows.

One final remark: By choice of $m \in M^+$ and conjugation by m or m^{-1} we can "expand" or "shrink" a compact subgroup (or subset) of N arbitrarily.

§1.2.2. The Constant $\gamma(G/P)$ for a Reductive p-adic Group.

For this appendix to §1.2 let G be as in §1.2.1. Fix a maximal split torus A_0 of G and let $(P, A)(P = MN)$ and $(\bar{P}, A)(\bar{P} = M\bar{N})$ be opposite semistandard p-pairs of G. Let K be an A_0-good maximal compact subgroup of G. Since $G = KP$, the homogeneous space G/P may be identified with $K/K \cap P$. Assume that the measure $d\bar{x}$ (Lemma 1.2.5) on the compact space G/P is normalized such that $\int_{G/P} d\bar{x} = 1$. Write $d\bar{k}$ for the corresponding measure on $K/K \cap P$ and dk for the normalized measure on K. Then, normalizing $d_\ell p$ and $d_r p$ such that $\int_{K \cap P} d_\ell p = 1$, we see that $\int_{K/K\cap P} \int_{K\cap P} d\bar{k}\, d_\ell p = \int_K dk = 1$, i.e., $d\bar{k}$ is the normalized K-invariant measure on $K/K \cap P$.

Recalling that the group G is unimodular, we see that $\chi_P(g) = \delta_P(g)$, where $\delta_P(kp) = \delta_P(p)$ $(k \in K,\ p \in P)$. Let dx denote the normalized Haar measure for G, i.e., $\int_K dx = \int_K dk = 1$. Then it follows from Lemma 1.2.5 that for any $\alpha \in C_c(G)$,

$$\int_G \alpha(x)\delta_P^{-1}(x)dx = \int_{G/P} \int_P \alpha(xp)d_\ell p\, d\bar{x}$$

$$= \int_{G/P} \int_P \alpha(xp)\delta_P^{-1}(p)d_r p\, d\bar{x}\,.$$

First, it is clear that

$$\int_G \alpha(x)\delta_P^{-1}(x)dx = \int_K \int_P \alpha(kp)\delta_P^{-1}(kp)dk\, d_r p$$

$$= \int_K \int_P \alpha(kp)dk\, d_\ell p.$$

Second, using the fact that the product mapping $(\bar{n}, p) \mapsto \bar{n}p$ defines a homeomorphism of $\bar{N} \times P$ with a dense open subset of G and the existence and uniqueness of Haar measure, we see that

$$\int_G \alpha(x)\delta_P^{-1}(x)dx = \gamma^{-1}\int_{\bar{N}}\int_P \alpha(\bar{n}p)\delta_P^{-1}(\bar{n}p)d\bar{n}\, d_r p$$

$$= \gamma^{-1}\int_{\bar{N}}\int_P \alpha(\bar{n}p)\delta_P^{-1}(\bar{n})d\bar{n}\, d_\ell p,$$

where $d\bar{n}$ is the normalized measure on $\bar{N}$ $(\int_{\bar{N}\cap K} d\bar{n} = 1)$ and

$$\gamma = \gamma(G/P) = \gamma(G/P, K) = \int_{\bar{N}} \delta_P^{-1}(\bar{n})d\bar{n}.$$

In other words, the measure $d\bar{x}$ is given on the set of representatives $\bar{N}$ for G/P by $\gamma^{-1}\delta_P^{-1}(\bar{n})d\bar{n}$. It will be proved later (cf. 5.4.3) that $\gamma(G/P) = \gamma(G/M)$ does not depend upon $P \in \mathcal{P}(A)$.

Begging the reader's forgiveness for a cryptic comment, we note that the constant $\gamma = \gamma\,(G/M)$ is related to a (generalized) Poincaré polynomial for G.

§1.3. Representations of Groups.

Let G be a group. By a representation π of G in a complex vector space V (not necessarily finite dimensional!) we mean a homomorphism $\pi : G \to GL(V)$ ($GL(V)$ denotes the group of invertible linear transformations of V). To be definite we assume that $\pi(xy)v = \pi(x) \circ \pi(y)v$ ($x, y \in G$; $v \in V$), so π gives V a left G-module structure.

We say that π is a finitely generated representation (or that V is a G-module of finite type or a finitely generated G-module), if there is a finite subset $v_1, \ldots, v_r$ of V such that every $v \in V$ can be expressed as a finite sum of the form $v = \sum_{\substack{1 \leq i \leq s \\ 1 \leq j \leq r}} c_{ij}\, \pi(g_i)v_j$ ($g_1, \ldots, g_s \in G$; $c_{ij} \in \mathbb{C}$).

Let π be a representation of G in V. A subspace $W \subset V$ is called G-stable (π-stable, $\pi(G)$-stable) if W is a G-submodule; in this case we say that there is a subrepresentation of π in W. We call π irreducible (or algebraically irreducible) if V is a simple G-module, i.e., if V and (0) are the only G-submodules. We call π completely reducible if V is a direct sum of simple G-submodules. If W is G-stable, we also have a quotient representation of G on the space V/W. A quotient representation of a subrepresentation is termed a subquotient or subquotient representation. If σ is a subrepresentation [subquotient] of π we sometimes write $\sigma \subset \pi$ [$\sigma \prec \pi$].

Two representations π_1 and π_2 of G in, respectively, V_1 and V_2 are called equivalent (Notation: $\pi_1 \sim \pi_2$) if the associated modules are isomorphic, i.e., if there is a vector space isomorphism $\varphi : V_1 \to V_2$ such that $\varphi \circ \pi_1(x) = \pi_2(x) \circ \varphi$ for all $x \in G$. When no confusion seems likely to occur, we sometimes regard equivalent representations as "the same", e.g., we say that π_1 is a subquotient of π_2 if π_1 is equivalent to a subquotient of π_2, etc. The <u>class</u> of a representation π is the set of all representations equivalent to π.

Let H be a subgroup of G. Let π be a representation of G in V. By restricting the homomorphism π from G to H we define a representation of H in V. This representation is called the <u>restriction of</u> π <u>to</u> H and denoted either $\pi|H$ or π_H.

If σ is a representation of a subgroup H of G and y is an element of G, we write σ^y for the representation of $H^y = yHy^{-1}$ such that $\sigma^y(h^y) = \sigma(h)$ $(h \in H)$.

If V is a complex vector space we write V' for its <u>algebraic dual</u>, i.e., for the space of all complex-valued linear functionals defined on V. For $v \in V$ and $v' \in V'$ we write $\langle v', v\rangle$ to denote $v'(v)$, the value of v' at v. To a homomorphism of vector spaces $T : V \to W$ there corresponds the <u>transpose</u> ${}^tT : W' \to V'$, a homomorphism of the dual spaces; the transpose ${}^tT$ is defined to be the unique operator which satisfies the relation $\langle {}^tTw', v\rangle = \langle w', Tv\rangle$ for all $v \in V$ and $w' \in W'$. If π is a representation of G in V, then its <u>contragredient</u> (or algebraic contragredient) π' is a representation of G in V'. It is defined by setting $\pi'(x) = {}^t\pi(x^{-1})$ $(x \in G)$, so that $\langle v', v\rangle = \langle \pi'(x)v', \pi(x)v\rangle$ for all $x \in G$, $v \in V$, and $v' \in V'$.

In the same way we write $V^\wedge$ for the space of all anti-linear functionals and note that sending $v' \mapsto (v')^-$ = (complex conjugate)v' defines an anti-linear bijection $V' \to V^\wedge$. To a homomorphism $T : V \to W$ there corresponds the <u>adjoint</u> $T^\wedge : W^\wedge \to V^\wedge$, defined by the relation $\langle T^\wedge w^\wedge, v\rangle = \langle w^\wedge, Tv\rangle$ for all $v \in V$, $w^\wedge \in W^\wedge$. If π is a representation of G in V, then $x \mapsto \pi(x^{-1})^\wedge = \pi^\wedge(x)$

defines a representation in $V^\wedge$. We shall call this representation the algebraic adjoint representation.

A representation π of G in a complex vector space V is called unitary (or pre-unitary) if there is a hermitian positive definite form H on V such that $H(\pi(x)v_1, \pi(x)v_2) = H(v_1, v_2)$ for all $x \in G$ and $(v_1, v_2) \in V \times V$. To any non-singular hermitian form H on V there corresponds a linear injection $J_H : V \to V^\wedge$ such that $H(v_1, v_2) = \langle J_H(v_1), v_2 \rangle$ $((v_1, v_2) \in V \times V)$. If H is non-singular and $H(\pi(x)v_1, \pi(x)v_2) = H(v_1, v_2)$ for all $x \in G$, $(v_1, v_2) \in V \times V$, then $J_H(\pi(x)v) = \pi^\wedge(x)J_H(v)$ for all $x \in G$ and $v \in V$.

Let G be a group and V a vector space. Let $\mathrm{Map}(G, V)$ be the set of all V-valued functions defined on G. Then $\mathrm{Map}(G, V)$ is itself a vector space. We set $(\rho(x)f)(y) = f(yx)$ and $(\lambda(x)f)(y) = f(x^{-1}y)$ $(x, y \in G,\ f \in \mathrm{Map}(G, V))$ and call ρ and λ, respectively, the right and left regular representations of G on $\mathrm{Map}(G, V)$. In fact, when no confusion seems likely, we use ρ and λ to denote the corresponding representations obtained by restriction on an arbitrary G-stable subspace of $\mathrm{Map}(G, V)$. A function of the form $\rho(x)f[\lambda(x)f]$ $(x \in G,\ f \in \mathrm{Map}(G, V))$ is called a right [left] translate of f.

Let G be a group and H a subgroup. A function f on G with values in a vector space is called right (left) [double] H-finite if its translates on the right (left) [both sides] by elements of H generate a finite-dimensional space of functions.

Write $\mathbb{Z}$ for the set of rational integers regarded as a group under addition.

Lemma 1.3.1. Let V be a $\mathbb{C}$-vector space. Let $f : \mathbb{Z} \to V$ and assume that f is $\mathbb{Z}$-finite. If $f(x) = 0$ for all $x \geq c$ $(c \in \mathbb{Z})$, then f is identically zero.

The proof is left to the reader.

By a <u>double representation</u> τ of G on V we mean a pair (τ_1, τ_2) where τ_1 is a representation of G in V and τ_2 gives V a right G-module structure with the additional condition that, for $v \in V$ and $x_1, x_2 \in G$, $(\tau_1(x_1)v)\tau_2(x_2) = \tau_1(x_1)(v\tau_2(x_2))$. We usually write $\tau(x_1)v\tau(x_2)$ for $\tau_1(x_1)v\tau_2(x_2)$. We generally write (V, τ) to denote a vector space V with a double representation τ defined on it.

We say that a double representation $\tau = (\tau_1, \tau_2)$ of G is <u>unitary</u> on V if there is a pre-hilbert space structure on V with respect to which both τ_1 and τ_2^{-1} are unitary representations of G.

§1.4. <u>Smooth Representations of t. d. Groups.</u>

Let G be a t. d. group and π a representation of G in a vector space V. An element $v \in V$ is called <u>smooth</u> (or π-<u>smooth</u>) if the mapping $x \mapsto \pi(x)v$ is a smooth mapping from G to V. For an element $v \in V$ to be smooth it is necessary and sufficient that there exists an open compact subgroup K of G such that $\pi(k)v = v$ for all $k \in K$. In this case we say that v is K-<u>fixed</u>; we write V_K for the subspace consisting of all K-fixed elements of V. Let $V_\infty = \cup V_K$, K running over all open compact subgroups of G, so V_∞ is the space of all π-smooth elements of V. Then V_∞ is G-stable; we write π_∞ for the subrepresentation of π on V_∞. A representation π of G is called <u>smooth</u> if $\pi = \pi_\infty$, i.e., $V = V_\infty$.

Let G be a t. d. group. A smooth one-dimensional representation of G is called a <u>quasi-character of</u> G. A unitary quasi-character of G is called a <u>character</u> or <u>one-dimensional character</u> of G. Write $\mathfrak{X}(G)$ for the set of quasi-characters of G and $\hat{G}$ for the set of characters of G. There is an obvious multiplication on $\mathfrak{X}(G)$ with respect to which $\mathfrak{X}(G)$ is a group and $\hat{G}$ a subgroup.

To say that a double representation $\tau = (\tau_1, \tau_2)$ is a <u>smooth double representation</u> will mean that both τ_1 and τ_2 are smooth.

The following spaces are stable under the left [right] regular representation $\lambda[\rho]$ on $\mathrm{Map}(G:\mathbb{C})$ or $\mathrm{Map}(G:U)$ (U a complex vector space): $C^\infty(G)$, $C^\infty(G:U)$, $C(G/K)[C(K\backslash G)]$, $C_c(G/K)[C_c(K\backslash G)]$, $C_c^\infty(G)$, etc. Notice that $C^\infty(G)$ is not a smooth representation space under either ρ or λ, unless G is compact or discrete. However, $C_c^\infty(G)$ is always smooth under either regular representation. In order that $f \in C^\infty(G)$ be a ρ-smooth vector it is necessary and sufficient that $f \in C(G/K)$ for some open compact subgroup K.

Note that $C_c^\infty(G)$ is an associative algebra under the convolution product

$$(f*g)(x) = \int_G f(y)g(y^{-1}x)d_\ell y$$
$$= \int_G f(xy^{-1})g(y)d_r y.$$

If π is a smooth representation of G in V we set

$$\pi(f) = \int_G f(x)\pi(x)\delta_G(x)^{\frac{1}{2}}d_\ell x$$
$$= \int_G f(x)\pi(x)\delta_G(x)^{-\frac{1}{2}}d_r x.$$

It follows that $\pi(f*g) = \pi(f)\pi(g)$, so we obtain a representation, which we also denote as π, of $C_c^\infty(G)$ in V.

Write $\lambda' = \delta_G^{-\frac{1}{2}}\lambda$ and $\rho' = \delta_G^{-\frac{1}{2}}\rho$, where λ and ρ are, respectively, the left and right regular representations on $C_c^\infty(G)$.

<u>Lemma</u> 1.4.1. Let π be a smooth representation of G. Then, for all $y \in G$ and $f \in C_c^\infty(G)$,

$$\pi(y)\pi(f) = \pi(\lambda'(y)f) \text{ and}$$

$$\pi(f)\pi(y^{-1}) = \pi(\rho'(y)f).$$

<u>Proof.</u> Omitted.

For each open compact subgroup K of G the space $C_c(G/\!\!/K)$ is a subalgebra of $C_c^\infty(G)$. Indeed $C_c^\infty(G) = \bigcup_{K'} C_c(G/\!\!/K')$, where K' ranges over all open compact subgroups of G. Each algebra $C_c(G/\!\!/K)$ contains an identity element which we write as E_K. The function E_K is a multiple of the characteristic function of K, the factor being one over the Haar measure of K. For π a smooth representation of G, $\pi(E_K)$ is an idempotent operator in V whose image is the space V_K of all K-fixed vectors in V.

If σ is a representation of $C_c^\infty(G)$ in a space W, we may use the relations of Lemma 1.4.1 to define a smooth representation σ_1 of G in the subspace $W_1 = \sigma(C_c^\infty(G))W$ as follows: To define $\sigma_1(g)w$ for $w \in W_1$ and $g \in G$ observe that $\sigma(E_K)w = w$ for some K; set $\sigma_1(g)w = \sigma(\lambda'(g)E_K)w$. It is easy to see that σ_1 is a smooth representation of G *in* W_1 *and that the subrepresentation of* $C_c^\infty(G)$ on W_1 is σ_1. Indeed, it is clear that $W_1 = W_\infty$.

Let π' be the algebraic contragredient of π in the space V'. We write $\tilde{\pi}$ for $(\pi')_\infty$ and $\tilde{V}$ for $(V')_\infty$. We call $\tilde{V}$ the space of <u>smooth linear functionals</u> on V. We call $\tilde{\pi}$ the <u>smooth contragredient</u> of π, or, simply, the contragredient when no confusion is likely to arise. Clearly, $v' \in \tilde{V}$ if and only if ${}^t\pi(E_K)v' = v'$ for some open compact subgroup K in G, i.e., if and only if

$\langle v', v\rangle = \langle v', {}^t\pi(E_K)v\rangle$ for every $v \in V$ and some open compact subgroup K of G.

Let π be a smooth representation of G and $\tilde{\pi}$ its contragredient. A function of the form $f(x) = \langle \tilde{v}, \pi(x)v\rangle$ $(x \in G;\ v \in V$ and $\tilde{v} \in \tilde{V})$ is called a <u>matrix coefficient</u> for π.

In the same way we define the space V^* of <u>smooth anti-linear functionals</u> on V and observe that complex conjugation defines a bijection between $\tilde{V}$ and V^*, i.e., between $(\tilde{V})_K$ and $(V^*)_K$ for every open compact subgroup K. We also define the <u>smooth adjoint representation</u> π^* of G on V^* by setting $\pi^*(x) = (\pi^\wedge)_\infty(x)$ $(x \in G)$.

If π is an irreducible smooth representation of G and ω denotes the class of π, we usually write $\tilde{\omega}\,[\omega^*]$ for the class of $\tilde{\pi}\,[\pi^*]$. If a class ω is unitary, then $\omega = (\tilde{\omega})^- = \omega^*$.

<u>Lemma</u> 1.4.2. Let π be a smooth representation of G in a vector space V. Let $\tilde{V}\,[V^*]$ be the smooth dual space [smooth antilinear dual space] of V.

(1) The bilinear [sesquilinear] form $\langle \tilde{v}, v\rangle$ $[\langle v^*, v\rangle]$ on $\tilde{V} \times V$ $[V^* \times V]$ is non-degenerate.

(2) There is a natural linear injection $V \hookrightarrow \tilde{\tilde{V}}$ $[V \hookrightarrow V^{**}]$.

The proof is easy and omitted.

Let Z be a central subgroup of G and χ a quasi-character of Z. Let U be a complex vector space. We write $C^\infty(G, Z:U)$ for the space of Z-finite elements of $C^\infty(G:U)$, $C_c^\infty(G, Z:U)$ *for the subspace which consists of functions* which are compactly supported modulo Z. (Notice that, if Z is not compact, then a function cannot be both Z-finite and compactly supported on G.) We write

$C^\infty(G:U)_\chi[C_c^\infty(G:U)_\chi]$ for the space which consists of all $f \in C^\infty(G, Z:U)$ $[C_c^\infty(G, Z:U)]$ such that, for all $z \in Z$, $(\rho(z)-\chi(z))^n f = 0$ with $n = n(f) \geq 0$. One can show that $C^\infty(G, Z:U) = \oplus\, C^\infty(G:U)_\chi$ and $C_c^\infty(G, Z:U) = \oplus\, C_c^\infty(G:U)_\chi$ --in each direct sum χ runs over all elements of $\mathfrak{X}(Z)$. For any $\chi \in \mathfrak{X}(Z)$ we reserve the notation $C^\infty(G,\chi:U)[C_c^\infty(G,\chi:U)]$ for the subspace of $C^\infty(G:U)_\chi[C_c^\infty(G:U)_\chi]$ which consists of functions satisfying $(\rho(z)-\chi(z))f = 0$ for all $z \in Z$. Notice that $C_c^\infty(G, Z:U)$ is a smooth G-module under either λ or ρ.

Remark. For the following simple lemma, due to an apparent excess of Gründlichkeit, we have given two proofs, the first due to Harish-Chandra and the second to the note taker. Howe has remarked: "The point of course is that the elements in W are already determined by their values on a compact set; being locally constant, their values on this compact set are invariant by some open compact subgroup. Perhaps it would be simpler to just say this. The reader has a long way to go, and we should try to make the way as easy as possible."

Lemma 1.4.3. Let G be a t.d. group, H a closed subgroup, and U a vector space. Let W be a finite-dimensional ρ_H-stable subspace of $C^\infty(G:U)$, where ρ_H denotes the restriction to H of the right regular representation ρ of G on $C^\infty(G:U)$. Then W is a smooth H-module under ρ_H.

Proof. Let $\dim W = r$. For $x \in G$ and $u' \in U'$ define $\varphi_{x,u'} \in W'$ by setting $\varphi_{x,u'}(\psi) = \langle u', \psi(x)\rangle$ $(\psi \in W)$. Obviously, we may choose $(x_1, u_1'), \ldots, (x_r, u_r')$ such that $\varphi_{x_1,u_1'}, \ldots, \varphi_{x_r,u_r'}$ is a basis for W'. Let $(x_1, u_1'), \ldots, (x_r, u_r')$ map to a basis for W'. Let $\psi_1, \ldots, \psi_r$ be the dual basis in W, so that

$$\det(\langle u_j', \psi_i(x_j)\rangle) \neq 0.$$

Then, for $i = 1, \ldots, r$ and $h \in H$, we have

$$\rho_H(h)\psi_i = \sum_{\ell=1}^{r} c_{i\ell}(h)\psi_\ell$$

with $c_{i\ell}(h) \in \mathbb{C}$ $(1 \leq i, \ \ell \leq r)$; thus,

$$\langle u_j', \psi_i(x_j h)\rangle = \sum_{\ell=1}^{r} c_{i\ell}(h)\langle u_j', \psi_\ell(x_j)\rangle,$$

$1 \leq i, \ j \leq r$. We have the matrix product

$$(\langle u_j', \psi_i(x_j h)\rangle)(\langle u_j', \psi_\ell(x_j)\rangle)^{-1} = (c_{i\ell}(h)).$$

To complete the proof we need only observe that $\psi_i \in C^\infty(G:U)$ implies that $\langle u_j', \psi_i(x_j h)\rangle$ is a smooth function on H. It then follows that the left side of our matrix product is the product of a matrix of smooth functions on H by a constant matrix, so $c_{ij}(h) \in C^\infty(H)$ for $1 \leq i, \ j \leq r$.

Proof (2). Let $\{H_n\}_{n=1}^\infty$ be a sequence of open compact subgroups of H such that $\bigcap_{n=1}^\infty H_n = 1$. For each $n \geq 1$ define $\rho_n : W \to W$ by setting $\rho_n\psi = \rho_H(E_{H_n})\psi$, $\psi \in W$. Let $W_n = \ker \rho_n$. Then $\{\dim W_n\}$ is a decreasing sequence of non-negative integers. Let $W_0 = W_{n_0}$ have minimal dimension. Then $W_0 = \{\psi \in W \mid \rho_n\psi = 0 \text{ for all } n\} = (0)$, since, given $0 \neq \psi \in W$, there exists $x \in G$ such that $\psi(xh) = \psi(x) \neq 0$ for all $h \in H_n$ for some n. It follows that W is a smooth H-module.

We shall conclude this section with a summary, for reference purposes, of some elementary results concerning finite dimensional smooth representations of abelian t.d. groups.

Let A be an abelian t.d. group. Note that $\mathfrak{X}(A) \subset C^\infty(A)$. For $\alpha \in C_c^\infty(A)$ and $\chi \in \mathfrak{X}(A)$ write

$$\chi(\alpha) = \hat{\alpha}(\chi) = \int_A \alpha(a)\chi(a)da.$$

Lemma 1.4.4. Let A be an abelian t.d. group and let $\chi_1, \ldots, \chi_n$ be a finite set of distinct quasi-characters of A. Then there exist elements $\alpha_1, \ldots, \alpha_n$ of $C_c^\infty(A)$ such that $\chi_i(\alpha_j) = \delta_{ij}$ $(1 \leq i, j \leq n)$.

Proof. Write $\mathfrak{a} = C_c^\infty(A)$ and let $\mathfrak{m}_i = \ker \chi_i$ $(i = 1, \ldots, n)$. Let $\mathfrak{m} = \bigcap_{i=1}^{n} \mathfrak{m}_i$ and set $\mathfrak{B} = \mathfrak{a}/\mathfrak{m}$. Then $\dim(\mathfrak{B}) \leq n$, since

$$\alpha \mapsto \begin{pmatrix} \chi_1(\alpha) & & 0 \\ & \ddots & \\ 0 & & \chi_n(\alpha) \end{pmatrix}$$

defines a faithful representation of $\mathfrak{B}$. It follows that $\mathfrak{B}$ is a finite-dimensional commutative semisimple algebra whose distinct maximal ideals are $\mathfrak{m}_i/\mathfrak{m}$ $(i = 1, \ldots, n)$, i.e., $\dim \mathfrak{B} = n$. The lemma is now an immediate consequence of a well-known theorem of Dedekind.

Let σ be a smooth representation of A in a vector space W. For any $\chi \in \mathfrak{X}(A)$ write W_χ for the set of all $w \in W$ such that $(\chi(t) - \sigma(t))^d w = 0$ for all $t \in A$ and some integer $d = d(w) \geq 0$. If σ is finite-dimensional, then σ is equivalent to a representation by (upper) triangular matrices, which means that $W = \oplus W_{\chi_i}$ for some $\chi_1, \ldots, \chi_n$ in $\mathfrak{X}(A)$. For any $w \in W_{\chi_i}$ we have $d(w) \leq \dim W_{\chi_i}$. Whether W is finite-dimensional or not, one sees that $w \in W_\chi$ if and only if there exists $d = d(w) > 0$ such that $\prod_{j=1}^{d} (\sigma(\alpha_j) - \chi(\alpha_j))w = 0$ for all $\alpha_1, \ldots, \alpha_d$ in $C_c^\infty(A)$.

Corollary 1.4.5. Let σ be a finite-dimensional smooth representation of A in a complex vector space V. Let $\chi_1, \ldots, \chi_r$ be the distinct quasi-characters of A such that $V = \bigoplus_{i=1}^{r} V_{\chi_i}$. Let $P_i : V \rightarrow V_{\chi_i}$ denote the corresponding projections, $i = 1, \ldots, r$. There exist elements $\beta_1, \ldots, \beta_r \in C_c^\infty(A)$ such that $\sigma(\beta_i)v = P_i v$

for any $v \in V$.

We assume that a finite-dimensional complex vector space V has a norm $\| \ \|$ defined on it.

Lemma 1.4.6. Let T be a unipotent transformation of a finite-dimensional complex vector space V. Then there exist positive constants c_1, c_2, and $r \leq \dim V$ such that, for any $v \in V$,

$$c_1 \|v\| \leq \|T^n v\| \leq c_2 n^r \|v\|$$

for all positive integers n.

Proof. Omitted.

Corollary 1.4.7. Let T be as above. Then $T^n v \to 0, n \to \infty$, if and only if $v = 0$.

Corollary 1.4.8. Let the notations be as in Corollary 1.4.5. Let $x \in A$ and $v \in V$. The following statements are equivalent:

(1) $\sigma(x^n)v \to 0$, $n \to \infty$.

(2) $\sigma(x^n)P_i v \to 0$, $n \to \infty$ for all $i = 1, \ldots, n$.

(3) For all $i = 1, \ldots, n$ such that $P_i v \neq 0$, $\chi_i(x^n) \to 0, n \to \infty$.

Proof. Omitted.

§1.5. Admissible Representations of t.d. Groups.

Let G be a t.d. group. A representation π of G in a vector space V is called admissible if:

(1) *π is smooth, and*

(2) for every open (compact) subgroup K of G, the space of K-fixed vectors, V_K, is finite-dimensional.

As remarked earlier, π is smooth if $V = \bigcup_K V_K$, where K ranges over all open compact subgroups of G. If, in addition, $\dim V_K < \infty$ for all K, then π is admissible.

Let π be irreducible and admissible. Then, clearly, every equivalent representation has the same properties. We write $\mathcal{E}_{\mathbb{C}}(G)$ for the set of classes of irreducible admissible representations of G, $\mathcal{E}(G)$ for the subset of $\mathcal{E}_{\mathbb{C}}(G)$ which consists of all the unitary classes. Recall that if K is any compact group, then all continuous irreducible representations are finite-dimensional and unitary; in fact, the same is true for all measurable = continuous, irreducible representations of K. Consequently, if K is t.d. and compact, then $\mathcal{E}_{\mathbb{C}}(K) = \mathcal{E}(K)$.

Now return to G an arbitrary t.d. group. A representation π of G is admissible, as one sees easily, if and only if, for some (and, hence, for every) open compact subgroup K of G, $\pi|K$ decomposes with finite multiplicities, i.e., $\pi|K \sim \oplus m_{\underline{d}}\underline{d}$, where $\underline{d} \in \mathcal{E}(K)$ and $m_{\underline{d}} < \infty$ for all $\underline{d}$.

Let π be a representation of G in V. A subspace $W \subset V$ is called admissible if W is G-stable and the subrepresentation of π in W is admissible. The representation π is called <u>quasi-admissible</u> if every $v \in V$ lies in an admissible subspace; equivalently, $V = \bigcup_{\{W\}} W$, where $\{W\}$ is the set of admissible subspaces.

<u>Lemma</u> 1.5.1. Let $0 \to U \to V \to W \to 0$ be a short exact sequence of G-modules.

(1) If V is smooth [admissible], then so are U and W.

(2) If U and W are admissible and V is smooth, then V is admissible.

The proof is easy and omitted.

Remark. It is not known to the writer whether U and W smooth implies V smooth.

Lemma 1.5.2. Let π be an admissible representation of G in a vector space V and let $\tilde{\pi}[\pi^*]$ denote the smooth contragredient [adjoint] of π. Then:

(1) $\tilde{\pi}[\pi^*]$ is admissible;

(2) $\dim(V_K) = \dim(\tilde{V})_K = \dim(V^*)_K$;

in fact, the spaces V_K and $(\tilde{V})_K[(V^*)_K]$ stand naturally in duality.

(3) There is a natural bijection $V \approx \tilde{\tilde{V}}[\approx V^{**}]$ and $(\tilde{\pi})\tilde{} \sim \pi[\pi^{**} \sim \pi]$.

Again the proof is easy and omitted.

Let π be a smooth representation of G in V. In §1.4 we pointed out that π extends to a representation of $C_c^\infty(G)$. A vector $v \in V$ is called Hecke-finite if, for each open compact subgroup K of G, the subspace $\pi(C_c(G/\!/K)) \cdot v \subset V$ is finite-dimensional. It is easy to see that $v \in V$ is Hecke-finite if and only if the cyclic G-module $\pi(G) \cdot v$ is admissible. In an admissible G-module every element is Hecke-finite.

We close the section with the following easy lemmas, whose proofs are left to the reader.

Lemma 1.5.3. Let V be a finitely generated admissible G-module. Then V has irreducible admissible quotient modules.

Lemma 1.5.4. Let V be an admissible G-module. Let Z be a central subgroup of G. Then every element of V is Z-finite.

§1.6. Intertwining Operators and Forms.

Let π_1 and π_2 be representations of a group G on V_1 and V_2, respectively. By an intertwining form for π_1 and π_2 we mean a bilinear form B on $V_1 \times V_2$ such that $B(\pi_1(x)v_1, \pi_2(x)v_2) = B(v_1, v_2)$ for all $x \in G$, $v_1 \in V_1$, and $v_2 \in V_2$. Write $\mathcal{B}(\pi_1, \pi_2)$ for the space of all intertwining forms for π_1 and π_2, $I(\pi_1, \pi_2)$ for the dimension of $\mathcal{B}(\pi_1, \pi_2)$.

An intertwining operator from π_1 to π_2 is a linear mapping $T : V_1 \to V_2$ such that $T(\pi_1(g)v) = \pi_2(g) \cdot T(v)$ for all $v \in V_1$ and $g \in G$. Write $\mathcal{T}(\pi_2 | \pi_1)$ for the space of intertwining operators from π_1 to π_2, $J(\pi_2 | \pi_1)$ for its dimension.

Now assume that G is a t. d. group and that π_1 and π_2 are smooth. Let $\tilde{\pi}_2$ be the smooth contragredient of π_2. For $T \in \mathcal{T}(\tilde{\pi}_2 | \pi_1)$ set $B_T(v_1, v_2) = \langle Tv_1, v_2 \rangle$ $(v_1 \in V_1, v_2 \in V_2)$. Then $B_T \in \mathcal{B}(\pi_1, \pi_2)$ and the mapping $T \mapsto B_T$ is injective. It follows that $I(\pi_1, \pi_2) \geq J(\tilde{\pi}_2 | \pi_1)$ and $I(\pi_1, \tilde{\pi}_2) \geq J(\tilde{\tilde{\pi}}_2 | \pi_1) \geq J(\pi_2 | \pi_1)$.

As usual, a vector space V equipped with a hermitian positive definite form is termed a pre-hilbert space or, if V is complete with respect to the associated norm, then V is called a Hilbert space. We denote the value of a hermitian form H on V at $(v_1, v_2) \in V \times V$ by (v_1, v_2), $(v_1, v_2)_H$, or $H(v_1, v_2)$. The norm $(v, v)^{\frac{1}{2}}$ of an element $v \in V$ is written as $\|v\|$ or $\|v\|_H$. We write $S^{\perp}$ for the subspace of V which is orthogonal to a subset S of V.

Lemma 1.6.1. Let π_0 be an admissible unitary representation of a t. d. group G in a vector space V. Let W be a π_0-stable subspace. Then $W^{\perp}$ is also π_0-stable and $V = W \oplus W^{\perp}$.

Proof. It is obvious that $W^\perp$ is π_0-stable and $W \cap W^\perp = (0)$. We shall show only that, if $v \in V$, then $v \in W + W^\perp$. There exists an open compact subgroup $K \subset G$ such that $v \in V_K$; moreover, $\dim V_K < \infty$. Let U denote the orthogonal complement of $W \cap V_K$ in V_K. Then $V_K = W \cap V_K + U$, since $\dim V_K < \infty$. It is sufficient to show that $U \subset W^\perp$. However, this is immediate: Given $w \in W$ and $u \in U$, we have $(u, w) = (\pi(E_K)u, w) = (u, \pi(E_K)w) = 0$, since $\pi(E_K)w \in W \cap V_K$.

Lemma 1.6.2.

(1) If π_1 and π_2 are admissible, then $J(\pi_2 | \pi_1) = I(\tilde{\pi}_2, \pi_1)$.

(2) If π is admissible and irreducible, then $I(\pi, \tilde{\pi}) = 1$.

(3) If π is admissible and unitary and if $I(\pi, \tilde{\pi}) = 1$, then π is irreducible.

Proof.

(1) We have to show that, for π_1 and π_2 admissible, $I(\tilde{\pi}_2, \pi_1) \leq J(\pi_2 | \pi_1)$. Let $B \in \mathcal{B}(\pi_1, \tilde{\pi}_2)$. For $v \in V_1$ let $T_B(v)$ be the linear functional on $\tilde{V}_2$ such that $T_B(v)(\tilde{v}_2) = B(v, \tilde{v}_2)$ for all $\tilde{v}_2 \in \tilde{V}_2$. Since V_2 is admissible, we may identify $\tilde{\tilde{V}}_2$ with V_2 and regard $v \mapsto T_B(v)$ as a map from V_1 to V_2. For all $v \in V_1$, $\tilde{v}_2 \in \tilde{V}_2$, and $x \in G$ we have $\langle T_B(\pi_1(x)v), \tilde{v}_2 \rangle = B(\pi_1(x)v, \tilde{v}_2) = B(v, \tilde{\pi}_2(x^{-1})\tilde{v}_2) = \langle T_B(v), \tilde{\pi}_2(x^{-1})\tilde{v}_2 \rangle = \langle \pi_2(x) T_B(v), \tilde{v}_2 \rangle$. It is thus clear that the mapping $B \mapsto T_B$ is an injective linear mapping of $\mathcal{B}(\pi_1, \tilde{\pi}_2)$ to $\mathcal{T}(\pi_2 | \pi_1)$, so (1) is proved.

(2) By (1) it suffices to show that $J(\pi | \pi) = 1$. Let π act in the space V. We observe that, for any $T \in \mathcal{T}(\pi | \pi)$ and any compact open subgroup K, $T(V_K) = V_K$. For any K we have a mapping $T \mapsto \lambda_{T,K} \in \mathbb{C}$, since the image of T in $\mathrm{End}(V_K)$ lies in a skew field, finite dimensional over $\mathbb{C}$, i.e., $\mathbb{C}$ itself. Obviously, $\lambda_{T,K} = \lambda_T$, a constant independent of K. This proves (2).

(3) If π is not irreducible, then there is a subspace $W \subsetneq V$ and its orthogonal

complement $W^{\perp}$, both stable under π (Lemma 1.6.1). The orthogonal projections of V on W and $W^{\perp}$ are obviously linearly independent elements of $\mathcal{T}(\pi|\pi)$, so $J(\pi|\pi) \geq 2$.

Lemma 1.6.3. Let π_1 and π_2 be admissible representations of G. In order that π_1 and π_2 be equivalent it is necessary and sufficient that there exist a non-degenerate $B \in \mathcal{B}(\pi_1, \tilde{\pi}_2)$.

The proof is omitted.

Let π be a representation of a group G on a vector space V. A bilinear or hermitian form B on V is called invariant or π-invariant if $B(\pi(x)v_1, \pi(x)v_2) = B(v_1, v_2)$ for all $x \in G$ and all $(v_1, v_2) \in V^2$. Paraphrasing an earlier definition, we may say that π is unitary with respect to a positive definite hermitian form H on V if and only if H is π-invariant. In this case the operator $J_H : V \to V^{\wedge}$ (cf., §1.4) intertwines π and $\pi^{\wedge}$.

Lemma 1.6.4. Let π be an admissible representation of a t.d. group G. Then π is unitary if and only if $\tilde{\pi}$ is unitary, in which case $\pi \sim \pi^*$.

The proof is omitted.

Lemma 1.6.5. Let π be a unitary admissible representation of a t.d. group G in a vector space V. Assume that V is a G-module of finite type. Then V is a finite direct sum of mutually orthogonal irreducible unitary submodules.

Proof. Let the finite set F generate V and assume that $F \subset V_K$, where K is an open compact subgroup of G. Let us first observe that, if W is a G-submodule and $W \perp V_K$, then $W = (0)$. To see this note that, given $v \in V$, there exist constants $c_1, \ldots, c_r$ and elements $g_1, \ldots, g_r$ of G such that

$\sum_{i=1}^{r} c_i \pi(g_i) v_i = v$, where $v_1, \ldots, v_r \in V_K$. We have, for any $w \in W$,

$(v, w) = \sum_{i=1}^{r} (c_i \pi(g_i) v_i, w) = \sum_{i=1}^{r} (v_i, \bar{c}_i \pi(g_i^{-1}) w) = 0$, so $W \perp V$. Since every $w \in W$ has norm zero, $W = (0)$. To conclude the proof apply Lemma 1.6.1 and mathematical induction to show that the cardinality of the set of mutually orthogonal irreducible subspaces cannot exceed $\dim V_K$.

Corollary 1.6.6. A unitary quasi-admissible representation of a t.d. group G is a direct sum of irreducible unitary admissible representations of G.

§1.7. Smooth Induced Representations of t.d. Groups.

Let G be a t.d. group and H a closed subgroup. Let σ be a representation of H on V. In this section we shall define the "big" and "little" smooth induced representations, to be denoted σ^G and $\mathrm{Ind}_H^G \sigma$, and study their most elementary properties.

First, the "big" representations. Let $\mathcal{V}_\ell(H, \sigma)$ denote the space of all functions $\beta : G \to V$ which satisfy:

1) $\beta(xh) = \sigma(h^{-1})\beta(x)$ $(x \in G,\ h \in H)$, and

2) $\beta \in C(K_0 \backslash G : V)$, for some open subgroup $K_0 = K(\beta)$ depending upon β.

Define $(\sigma_\ell^G(y)\beta)(x) = \beta(y^{-1}x)$ $(x \in G,\ y \in G)$. It is clear that σ_ℓ^G is a smooth representation of G, which we call the big left representation.

Similarly, we define the big right representation, σ_r^G. It will be easy to show that $\sigma_\ell^G \sim \sigma_r^G$, so most of the time we shall not have to distinguish these representations, i.e., we refer to either representation as σ^G, and it is a matter of taste or convenience as to which explicit representation we choose to work with.

We let $\mathcal{V}_r(H,\sigma)$ denote the space of functions $\beta : G \to V$ which satisfy:

1) $\beta(hg) = \sigma(h)\beta(g)$ $(h \in H,\ g \in G)$, and

2) $\beta \in C(G/K_0 : V)$ for some open subgroup $K_0 = K(\beta)$ depending on β.

We define $(\sigma_r^G(y)\beta)(x) = \beta(xy)$ $(x \in G,\ y \in G)$. We call σ_r^G the <u>big right representation.</u> It is also obviously smooth.

To define the <u>little left</u> (<u>little right</u>) induced representation, we introduce the subspace $\mathcal{H}_\ell(H,\sigma) \subset \mathcal{V}_\ell(H,\sigma)$ $(\mathcal{H}_r(H,\sigma) \subset \mathcal{V}_r(H,\sigma))$, where $\beta \in \mathcal{H}_\ell(\mathcal{H}_r)$ provided β is compactly supported mod H on the right (left). It is clear that $\mathcal{H}_\ell(\mathcal{H}_r)$ is $\sigma_\ell^G(\sigma_r^G)$ stable. We write $\lambda = \pi_\ell = \mathrm{Ind}_H^G(\sigma,\ell)$ $(\rho = \pi_r = \mathrm{Ind}_H^G(\sigma, r))$ to denote the subrepresentation of $\sigma_\ell^G(\sigma_r^G)$ on $\mathcal{H}_\ell(\mathcal{H}_r)$.

We define $T : \mathcal{V}_\ell(H,\sigma) \to \mathcal{V}_r(H,\sigma)$ (or vice versa) by setting $(T\beta)(x) = \beta(x^{-1})$. It is easy to see that T intertwines σ_ℓ^G and σ_r^G, also π_ℓ and π_r.

If G/H is compact, then $\sigma^G = \mathrm{Ind}_H^G\sigma$, so, in this case, there is no distinction between "big" and "little" induced representations; we speak of the <u>smooth induced representation.</u> Notice that σ^G and, hence, $\mathrm{Ind}_H^G\sigma$ are smooth even when σ is not, i.e., $\sigma^G = (\sigma_\infty)^G$ and $\mathrm{Ind}_H^G\sigma = \mathrm{Ind}_H^G(\sigma_\infty)$.

<u>Lemma</u> 1.7.1. Let η be a quasi-character of G and write η_H for $\eta|H$. Then $\eta\sigma^G \sim (\eta_H\sigma)^G$ and $\eta\,\mathrm{Ind}_H^G(\sigma) \sim \mathrm{Ind}_H^G(\eta_H\sigma)$.

<u>Proof.</u> The mapping $\beta \mapsto \eta\beta$ maps $\mathcal{V}_r(H,\sigma)$ bijectively on $\mathcal{V}_r(H,\eta_H\sigma)$, $\mathcal{H}_r(H,\sigma)$ on $\mathcal{H}_r(H,\eta_H\sigma)$. The respective equivalences are clear.

We omit the proofs of the next two easy lemmas.

Lemma 1.7.2. Let L be a closed subgroup of G and assume that $H \subset L$. Then $(\sigma^L)^G \sim \sigma^G$ and $\mathrm{Ind}_L^G(\mathrm{Ind}_H^L \sigma) \sim \mathrm{Ind}_H^G(\sigma)$.

Lemma 1.7.3. Fix $y \in G$ and recall that σ^y denotes the representation of $H^y = yHy^{-1}$ which satisfies $\sigma^y(h^y) = \sigma(h)$ $(h \in H,\ h^y = yhy^{-1} \in H^y)$. Then $(\sigma^y)^G \sim \sigma^G$ and $\mathrm{Ind}_{H^y}^G(\sigma^y) \sim \mathrm{Ind}_H^G(\sigma)$.

Lemma 1.7.4. Let H be a cocompact closed subgroup of G. If σ is an admissible representation of H, then $\mathrm{Ind}_H^G \sigma$ is an admissible representation of G.

Proof. Let σ act in a space V. Let $\pi = \mathrm{Ind}_H^G(\sigma, r)$ act in the space $\mathcal{H}$ of all smooth functions $\beta : G \to V$ such that $\beta(hg) = \sigma(h)\beta(g)$ $(h \in H,\ g \in G)$. Let K be an open compact subgroup of G. Then $H\backslash G/K$ is a finite set. To prove the lemma it suffices to show that, for any $x \in G$, the space of K-invariant elements of $\mathcal{H}$ with support in HxK is finite-dimensional.

Given such a function β, we must have $\beta(x) = \pi(k)\beta(x) = \beta(xkx^{-1}x) = \sigma(xkx^{-1})\beta(x)$ for all $k \in K \cap x^{-1}Hx$. Thus, $\beta(x)$ must lie in the finite-dimensional subspace of V comprised of $xKx^{-1} \cap H$-fixed vectors. Since the value of β at x determines β, the lemma follows.

In the following lemma let χ denote the product $\delta_H \delta_G^{-1}$ regarded as a quasi-character of H.

Lemma 1.7.5. For $\alpha \in C_c^\infty(G : V)$ set

$$P\alpha(x) = \int_H \chi(h)^{\frac{1}{2}} \sigma(h)\alpha(xh)\,d_\ell h,$$

$d_\ell h$ a left Haar measure on H. Then $P\alpha \in \mathcal{H}$ and $P(\lambda(g)\alpha) = \pi(g)P(\alpha)$ for any $g \in G$. Moreover, the mapping $P : C_c^\infty(G : V) \to \mathcal{H}$ is surjective.

Proof. The first part is clear. To prove the surjectivity it suffices to show that, for C an arbitrarily small open compact subset of G and $\psi \in \mathcal{H}$ with $\mathrm{Supp}\psi \subset CH$, there exists $\alpha \in C_c^\infty(G : V)$ such that $P\alpha = \psi$. Take an open compact subgroup $K \subset H$ and assume that $C^{-1}C \cap H \subset K$. Set $\alpha(x) = \psi(x)$ for $x \in CK$ and $\alpha(x) = 0$, $x \notin CK$. Then $\alpha \in C_c^\infty(G : V)$ and

$$P\alpha(x) = \int_H \chi(h)^{\frac{1}{2}} \sigma(h)\alpha(xh) d_\ell h$$

$$= \int_K \chi(h)^{\frac{1}{2}} \sigma(h)\psi(xh) d_\ell h$$

$$= \int_K d_\ell h \cdot \psi(x), \quad \text{for } x \in C;$$

$$P\alpha(x) = 0, \quad x \notin CH.$$

This implies the lemma.

Lemma 1.7.6. For every $h \in H$ and $\alpha \in C_c^\infty(G : V)$

$$P(\rho(h)\alpha) = \delta^{\frac{1}{2}}(h)\delta_G(h)^{\frac{1}{2}} P(\sigma(h^{-1})\alpha).$$

Proof.
$$P(\rho(h)\alpha)(x) = \int_H \chi(h')^{\frac{1}{2}} \sigma(h')(\rho(h)\alpha)(xh') d_\ell h'$$

$$= \int_H \chi(h')^{\frac{1}{2}} \sigma(h')\alpha(xh'h) d_\ell h'$$

$$= \int_H \chi(h')^{\frac{1}{2}} \sigma(h')\alpha(xh'h)\delta^{-1}(h') d_r h'$$

$$= \chi(h)^{-\frac{1}{2}} \delta(h) \int_H \chi(h'')^{\frac{1}{2}} \sigma(h'')\sigma(h^{-1})\alpha(xh'') d_\ell h''$$

$$= \chi(h)^{-\frac{1}{2}} \delta(h) P(\sigma(h^{-1})\alpha).$$

Theorem 1.7.7. Let G, H and σ be as before. Let $\tilde{\sigma}$ be the smooth contragredient of σ, acting in the space $\tilde{V}$. Let $\chi = \delta_H^{\frac{1}{2}} \delta_G^{-\frac{1}{2}}$, a quasi-character of H. Form the representations $\pi = \pi_\ell = \mathrm{Ind}_H^G(\chi^{\frac{1}{2}}\sigma)$ and $\mu = (\chi^{\frac{1}{2}}\tilde{\sigma})_\ell^G$. Then $\mu \sim \tilde{\pi}$, where $\tilde{\pi}$ denotes the smooth contragredient of π.

<u>Proof.</u> Let $\mathcal{H}_1 = C_c^\infty(G:V)$ and $\mathcal{V}_1' = \bigcup_{K_0} C(K_0\backslash G:V')$, where V' denotes the algebraic dual of V (cf. §1.3) and K_0 ranges over a fundamental sequence of open compact subgroups of G. Clearly, each space is, under λ, a smooth G-module.

Given $f \in C_c^\infty(G:V)$ and $\varphi \in C(K\backslash G:V')$, we set $\langle f, \varphi\rangle = \int_G \langle f(x), \varphi(x)\rangle d_\ell x$. Since obviously the bilinear form $\langle f,\varphi\rangle$ is both $\lambda(G)$-invariant (§1.6) and nondegenerate, it is clear that we may regard $\mathcal{V}_1'$ as at least a subspace of the smooth dual space $\mathcal{H}_1^\sim$ of $\mathcal{H}_1$. On the other hand, since $\mathcal{H}_1 = \bigcup_{K_0} C_c(K_0\backslash G:V)$, it is sufficient, in order to specify any linear functional on $\mathcal{H}_1$, to specify the linear functional on each space $C_c(K_0\backslash G:V)$. But $C_c(K_0\backslash G:V)$ has a basis consisting of elements $E_x \otimes v_i$, where E_x is the characteristic function of a coset and v_i belongs to a fixed basis of V. It follows by standard arguments (from the fact that the dual of a free module is a product module) that $C(K_0\backslash G:V')$ is the dual of $C_c(K_0\backslash G:V)$. Since $(\mathcal{H}_1)'_K = (\mathcal{V}_1')_K$ and $\bigcup_K (\mathcal{H}_1)'_K = \widetilde{\mathcal{H}}_1$, we see that $\mathcal{V}_1' = \bigcup (\mathcal{V}_1')_K = \widetilde{\mathcal{H}}_1$. Thus, in the special case in which, in effect, we induce from $H = (1)$, the theorem is true.

To extend the argument to the case of general H and σ, we first use Lemma 1.7.5 to see--immediately--that $(\mathcal{H}_\ell(H,\sigma), \pi)$ is a quotient representation of $(\mathcal{H}_1, \lambda)$. Thus, $\widetilde{\pi}$ may be regarded as a subrepresentation of $(\mathcal{V}_1', \lambda)$. A function $\varphi \in \mathcal{V}_1'$ corresponds to an element of $\mathcal{H}_\ell(H,\sigma)^\sim$ if and only if $\langle\alpha,\varphi\rangle = 0$ for all $\alpha \in \mathcal{H}_1$ such that $P(\alpha) = 0$ (P as in Lemma 1.7.5). By Lemma 1.7.6, $P(\alpha) = P(\delta_H^{-\frac{1}{2}}(h)\delta_G^{-\frac{1}{2}}(h)\sigma(h)\rho(h)\alpha)$ for all $\alpha \in \mathcal{H}_1$, which implies that if $\varphi \in \mathcal{H}_\ell(H,\sigma)^\sim \subset \mathcal{V}_1'$, then $\langle\alpha,\varphi\rangle = \int_G \langle\alpha(x), \varphi(x)\rangle d_\ell x =$
$= \int_G \langle\delta_H^{-\frac{1}{2}}(h)\delta_G^{-\frac{1}{2}}(h)\sigma(h)\alpha(xh), \varphi(x)\rangle d_\ell x = \int_G \langle\alpha(x), \delta_H^{-\frac{1}{2}}(h)\delta_G^{\frac{1}{2}}(h)\,{}^t\sigma(h)\varphi(xh^{-1})\rangle d_\ell x$. Thus,

$\varphi(xh^{-1}) = \delta_H^{\frac{1}{2}}(h)\delta_G^{-\frac{1}{2}}(h)^t\sigma(h^{-1})\varphi(x)$ for all $x \in G$. Moreover, since $\varphi(xh) = \varphi(xhx^{-1}\cdot x) = \varphi(x)$ for all $h \in H \cap x^{-1}K_0 x$, it follows that $\varphi(x) \in \tilde{V} \subset V'$ for each $x \in G$. We have proved that the subspace $\mathcal{V}_\ell(H,\tilde{\sigma}) \supset \mathcal{H}_\ell(H,\sigma)^\sim$.

To show that $\mathcal{V}_\ell(H,\tilde{\sigma}) = \mathcal{H}_\ell(H,\sigma)^\sim$ it now suffices to show that $P(\alpha) = 0$ implies $<\alpha,\varphi> = 0$, $\varphi \in \mathcal{V}_\ell(H,\tilde{\sigma})$. We now let $\chi(x)$ denote the function of Lemma 1.2.4. Using Lemma 1.2.5, we compute:

$$\int_G <\alpha(x),\varphi(x)> d_\ell x = \int_{G/H}\int_H <\alpha(xh),\varphi(xh)>\chi(xh)d_\ell h d\bar{x}$$

$$= \int_{G/H}\int_H \frac{\delta_H(h)}{\delta_G(h)} <\delta_H^{-\frac{1}{2}}(h)\delta_G^{\frac{1}{2}}(h)\sigma(h)\alpha(xh),\varphi(x)>\chi(x)d_\ell h d\bar{x}$$

$$= \int_{G/H} <P(\alpha)(\bar{x}),\varphi(\bar{x})>\chi(\bar{x})d\bar{x} \quad \text{(Lemma 1.7.5)}.$$

Consequently, if $P(\alpha) = 0$, then $<\alpha,\varphi> = 0$ for all $\varphi \in \mathcal{V}_\ell(H,\tilde{\sigma})$; thus, $\mathcal{V}_\ell(H,\tilde{\sigma}) = \mathcal{H}_\ell(H,\sigma)^\sim$.

<u>Corollary</u> 1.7.8. Let H be cocompact in G. Let $\pi_1 = \mathrm{Ind}_H^G(\delta_H^{\frac{1}{2}}\delta_G^{-\frac{1}{2}}\sigma)$ and $\pi_2 = \mathrm{Ind}_H^G(\delta_H^{\frac{1}{2}}\delta_G^{-\frac{1}{2}}\sigma^\sim)$. Then $\pi_1 \sim \tilde{\pi}_2$ and $\pi_2 \sim \tilde{\pi}_1$.

<u>Corollary</u> 1.7.9. Let H be cocompact and σ unitary and smooth. Then $\pi = \mathrm{Ind}_H^G(\delta_H^{\frac{1}{2}}\delta_G^{-\frac{1}{2}}\sigma)$ is unitary and smooth.

<u>Proof.</u> The map which assigns $v \in V$ the element $v' \in V'$ such that $<v',w> = H_V(w,v)$ defines an antilinear isomorphism of V on $\tilde{V} \subset V'$. For $\beta_i \in \mathcal{H}_\ell(H,\sigma)$ $(i = 1,2)$ define $H(\beta_1,\beta_2) = \int_{G/H} H_V(\beta_1(\bar{x}),\beta_2(\bar{x}))\chi(\bar{x})d\bar{x}$, χ and $d\bar{x}$ as in Lemmas 1.2.4 and 1.2.5. It is clear that H is G-invariant and positive definite, so π is unitary.

We close this section with a version of Frobenius' reciprocity theorem for smooth representations and an obvious consequence.

Theorem 1.7.10. Let π be a smooth representation of G. Then there is a canonical bijection between $\mathcal{J}(\sigma|\pi_H)$ and $\mathcal{J}(\sigma^G|\pi)$.

Proof. The point is that the proof in the case of finite groups carries over to the present context. Let π act in $\mathcal{H}$. Given $T \in \mathcal{J}(\sigma|\pi_H)$, set $T^G(h)(x) = T(\pi(x)h)$ $(x \in G, h \in \mathcal{H})$. It is easy to see that $T^G: \mathcal{H} \to \mathcal{V}(H,\sigma)$ and intertwines π and σ^G. If $S \in \mathcal{J}(\sigma^G|\pi)$, $S_H = \Delta \circ S$, where $\Delta\beta = \beta(1)$, intertwines π_H and σ. To complete the proof the reader should check that $(T^G)_H = T$ and $(S_H)^G = S$.

Corollary 1.7.11. Let π and σ be irreducible. Then $\pi \subset \sigma^G$ if and only if σ occurs as a quotient of π_H.

§1.8. Some Results Concerning Distributions Defined on a t.d. Group.

Theorem 1.8.1. Let T be a left invariant distribution (cf. §1.2) on a nonempty open subset X of G. Then there exists $c \in \mathbb{C}$ such that, for every $f \in C_c^\infty(X)$,

$$T(f) = c\int_G f(x)d_\ell x.$$

Proof. The proof proceeds in two steps: (1) We show that T can be represented by a locally constant function. (2) We show that the locally constant function is actually constant.

(1) Fix $x_0 \in X$ and choose an open subgroup $K \subset G$ such that $Kx_0 = X_0 \subset X$. Choose $\alpha \in C_c^\infty(K)$ such that $\int_K \alpha(x)dx = 1$. Let $f \in C_c^\infty(X_0)$ and,

for $x \in X_0$, define

$$g(x) = \int_K \alpha(y) f(y^{-1}x)dy.$$

Observe that $g \in C_c^\infty(X_0)$ and, as follows from the left invariance of T, $T(g) = T(f)$. Moreover, writing $\delta = \delta_G$, we have

$$g(x) = \int_K \alpha(y^{-1}) f(yx)dy$$

$$= \int_{Kx} \alpha(xy^{-1}) f(y) d_r y$$

$$= \int_{Kx} \alpha(xy^{-1}) f(y) \delta(y) d_\ell y$$

$$= \int_{X_0} (\rho(y^{-1})\alpha)(x) f(y) \delta(y) d_\ell y.$$ (If $x \in X_0$, $Kx = X_0$; if $x \notin X_0$, the integral is zero.)

Thus $T(f) = \int_{X_0} T((\rho(y^{-1})\alpha)(x)) \delta(y) f(y) d_\ell y$

$$= \int_{X_0} \beta(y) f(y) d_\ell y,$$

where $\beta(y) = T((\rho(y^{-1})\alpha)(x))\delta(y)$. The function β is independent of f and locally constant. It follows easily that there exists a locally constant function F defined on X such that $T(f) = \int_X fF d_\ell x$ for all $f \in C_c^\infty(X)$.

(2) To show that F is constant on X we take $x_1, x_2 \in X$ and set $y = x_2 x_1^{-1}$; then $x_1 = y^{-1}x_2 \in X \cap y^{-1}X$. By hypothesis $T(f) = T(\lambda(y)f)$ for all $f \in C_c^\infty(X \cap y^{-1}X)$. Thus,

$$\int f(x) F(x) d_\ell x = \int f(y^{-1}x) F(x) d_\ell x$$

$$= \int f(x) F(yx) d_\ell x,$$

so

$$0 = \int f(x)[F(yx) - F(x)] d_\ell x$$

for all $f \in C_c^\infty(X \cap y^{-1}X)$. Therefore, $F(x) - F(yx) \equiv 0$ on $X \cap y^{-1}X$, which implies that $F(x_1) = F(x_2)$.

q. e. d.

Corollary 1.8.2. Let T be a left invariant E-distribution on a non-empty open set $X \subset G$. Then there exists a unique linear functional φ on E such that

$$T(f) = \int_G \langle \varphi, f(x) \rangle d_\ell x$$

for all $f \in C_c^\infty(X : E)$.

Proof. Identifying $C_c^\infty(X : E)$ with $C_c^\infty(X) \otimes E$, we define for $v \in E$ the distribution T_v on $C_c^\infty(X)$ by setting $T_v(\alpha) = T(\alpha \otimes v)$, $\alpha \in C_c^\infty(X)$. Then T_v is left invariant, so there exists $\varphi(v) \in \mathbb{C}$ such that $T_v(\alpha) = \varphi(v) \int_X \alpha d_\ell x$ for all $\alpha \in C_c^\infty(X)$. The mapping $v \mapsto \varphi(v)$ is a linear functional, obviously uniquely determined by T, such that, for any $f = \sum_{i=1}^r \alpha_i \otimes v_i \in C_c^\infty(X) \otimes E$, we have

$$T(f) = \sum_{i=1}^r T_{v_i}(\alpha_i) = \sum_{i=1}^r \varphi(v_i) \int_U \alpha_i d_\ell x .$$

This proves the corollary.

Corollary 1.8.3. Let σ be a smooth right representation of G on E (i.e., σ gives E a right G-module structure). Let X be a non-empty open subset of G and T an E-distribution on X such that $T(\lambda(x)f) = T(f\sigma(x))$ for all $x \in G$ and $f \in C_c^\infty(X \cap x^{-1}X : E)$. Then there exists a unique linear functional φ on E such that $T(f) = \int_G \langle \varphi, f(x)\sigma(x) \rangle d_\ell x$ for all $f \in C_c^\infty(X : E)$.

Proof. Let $f \in C_c^\infty(X : E)$ and set $f'(x) = f(x)\sigma(x)$. Then $f \mapsto f'$ maps $C_c^\infty(X : E)$ bijectively to itself. Define an E-distribution T' on $C_c^\infty(X : E)$ by setting $T'(f') = T(f)$, f as before. Then

$$(\lambda(y)f')(x) = f'(y^{-1}x)$$
$$= f(y^{-1}x)\sigma(y^{-1}x)$$
$$= f(y^{-1}x)\sigma(y^{-1})\sigma(x)$$
$$= g'(x),$$

where $g(x) = (\lambda(y)f)(x)\sigma(y^{-1})$. It follows that, if $\mathrm{Supp}\, f \subset X \cap y^{-1}X$, $T'(\lambda(y)f') = T'(g') = T(g) = T(f) = T'(f')$, so T' is a left invariant E-distribution. By Corollary 1.8.2 there exists a unique linear functional φ on E such that

$$T(f) = T'(f') = \int_X < \varphi, f(x)\sigma(x) > d_\ell x.$$

Let $\chi(x)$ be as in Lemma 1.2.4. Let Γ be a closed subgroup of G. Let $p : G \to G/\Gamma$ be the projection mapping; write $p(x) = \bar{x}$. For $\alpha \in C_c^\infty(G : E)$ define $\bar{\alpha}(\bar{x}) = \int_\Gamma \alpha(x\gamma)d_\ell\gamma$. For $y \in G$ let $\bar{\lambda}(y)\bar{\alpha} = (\lambda(y)\alpha)^-$ (cf., Lemma 1.2.3).

Theorem 1.8.4. Let σ be a smooth right representation of G in E. Let $\bar{X}$ be a non-empty open subset of $\bar{G} = G/\Gamma$. Let T be an E-distribution on $\bar{X}$ such that $T(\bar{\lambda}(x)f) = T(f\sigma(x))$ for all $x \in G$ and $f \in C_c^\infty(\bar{X} \cap x^{-1}\bar{X} : E)$. Then there exists a unique linear functional φ on E such that:

1) $< \varphi, e\sigma(\gamma) > \chi^{-1}(\gamma) = < \varphi, e >$ for all $e \in E$ and $\gamma \in \Gamma$;

2) $T(f) = \int_{\bar{G}} < \varphi, f(\bar{x})\sigma(x) > \chi(x)d\bar{x}$ for all $f \in C_c^\infty(\bar{X} : E)$.

Proof. Let $X = p^{-1}(\bar{X})$. Define an E-distribution T' on X by setting $T'(\alpha) = T(\bar{\alpha})$ for $\alpha \in C_c^\infty(X : E)$. Then, if $\mathrm{Supp}\,\alpha \in X \cap y^{-1}X$, we have

$$T'(\lambda(y)\alpha) = T(\bar{\lambda}(y)\bar{\alpha}) = T(\overline{\alpha\sigma(y)}) = T'(\alpha\sigma(y)),$$

so T' satisfies the hypotheses of Lemma 1.8.3. Consequently, there exists a unique linear functional φ on E such that

$$T'(\alpha) = \int_G \langle \varphi, \alpha(x)\sigma(x)\rangle d_\ell x.$$

By Lemma 1.2.5 this is equivalent to 2), provided 1) is true.

Let us verify 1). By Lemma 1.2.3

$$T'(\rho(\gamma)\alpha) = T((\rho(\gamma)\alpha)^-) = \delta_\Gamma(\gamma)^{-1}T(\bar{\alpha}) = \delta_\Gamma(\gamma)^{-1}T'(\alpha) \quad (\gamma \in \Gamma).$$

Therefore, for $\gamma \in \Gamma$,

$$\begin{aligned} \delta_\Gamma(\gamma)^{-1}T'(\alpha) &= T'(\rho(\gamma)\alpha) \\ &= \int_G \langle \varphi, \alpha(x\gamma)\sigma(x)\rangle d_\ell x \\ &= \int_G \langle \varphi\,{}^t\sigma(\gamma^{-1}), \alpha(x\gamma)\sigma(x\gamma)\rangle d_\ell x \\ &= \delta_G(\gamma)^{-1}\int_G \langle \varphi\,{}^t\sigma(\gamma^{-1}), \alpha(x)\sigma(x)\rangle d_\ell x, \end{aligned}$$

so
$$T'(\alpha) = \frac{\delta_\Gamma(\gamma)}{\delta_G(\gamma)}\int_G \langle \varphi\,{}^t\sigma(\gamma^{-1}), \alpha(x)\sigma(x)\rangle d_\ell x.$$

Since φ is unique,
$$\varphi = \varphi\,{}^t\sigma(\gamma) \cdot \left(\frac{\delta_\Gamma(\gamma)}{\delta_G(\gamma)}\right)^{-1},$$

which verifies 1).

In §1.9 we shall apply the following version of Theorem 1.8.4. The two statements are of course equivalent because of the involution $x \mapsto x^{-1}$ on G. Define a measure $d\bar{x}$ on $\Gamma\backslash G$ by means of this involution.

Theorem 1.8.4R. Let σ be a smooth right representation of G on E. Let $\bar{X}$ be a non-empty open subset of $\bar{G} = \Gamma\backslash G$ ($x \mapsto \bar{x}$ the projection mapping; $\bar{\rho}$ the representation by right translations on $C_c^\infty(\Gamma\backslash G : E)$). Let T^0 be an E-distribution on $\bar{X}$ such that $T^0(\bar{\rho}(x)f) = T^0(f\sigma(x))$ for all $x \in G$ and $f \in C_c^\infty(\bar{X} \cap \bar{X}x : E)$. Then there exists a unique linear functional $\varphi = \varphi_{T^0}$ on E such that:

1) $<\varphi, e\sigma(\gamma)>\chi(\gamma) = <\varphi, e>$ for all $e \in E$ and $\gamma \in \Gamma$;

2) $T^0(f) = \int_{\Gamma\backslash G} <\varphi, f(\bar{x})\sigma(x^{-1})>\chi^{-1}(x)d\bar{x}$ for all $f \in C_c^\infty(\bar{X} : E)$.

§1.9. Invariant Distributions and Intertwining Forms for Induced Representations.

The essential ideas for this section come from Bruhat's thesis ([4a, b]). The most important result is stated as Theorem 1.9.5, which reduces the study of intertwining forms for induced representations to the study of certain spaces of linear functionals. "Bruhat's Theory" will continue in § 2.5.

Let G be a t.d. group and H_i a closed subgroup ($i = 1, 2$). Set $\chi_i = \delta_{H_i}\delta_G^{-1} = \delta_i\delta_G^{-1}$. Let σ_i be a smooth representation of H_i in a vector space V_i. Let $\pi_i = \mathrm{Ind}_{H_i}^G(\chi_i^{\frac{1}{2}}\sigma_i, \ell)$ and recall that the representation space $\mathcal{H}_i$ for π_i consists of all smooth functions $\psi : G \to V_i$ which are compactly supported mod H_i and satisfy the relation $\psi(xh^{-1}) = \chi_i^{\frac{1}{2}}(h)\sigma_i(h)\psi(x)$ ($x \in G$, $h \in H_i$). We have $(\pi_i(y)\psi)(x) = \psi(y^{-1}x)$ ($x, y \in G$). As in Lemma 1.7.5, define, for $\alpha \in C_c^\infty(G : V_i)$, $P_i\alpha(x) = \int_{H_i}\chi_i(h)^{\frac{1}{2}}\sigma_i(h)\alpha(xh)d_\ell h$, where $d_\ell h$ denotes a left Haar measure on H_i. Recall that $P_i : C_c^\infty(G : V_i) \to \mathcal{H}_i$ is an epimorphism of G-modules, intertwining λ and π_i, $i = 1, 2$.

Write $E = V_1 \otimes V_2$, $\mathcal{H} = H_1 \times H_2$, and $\mathcal{G} = G \times G$. In the following we frequently identify $C_c^\infty(\mathcal{G} : E)$ with $C_c^\infty(G : V_1) \otimes C_c^\infty(G : V_2)$.

Let $B \in \mathcal{B}(\pi_1, \pi_2)$ (§1.6) and set $T(\alpha_1 \otimes \alpha_2) = B(P_1\alpha_1, P_2\alpha_2)$ for $\alpha_i \in C_c^\infty(G : V_i)$. Then $T(\lambda(x)\alpha_1 \otimes \lambda(x)\alpha_2) = T(\alpha_1 \otimes \alpha_2)$ for any $x \in G$. Via the identification mapping we have $T(\lambda(x, x)\beta) = T(\beta)$ for any $\beta \in C_c^\infty(\mathcal{G} : E)$.

Define $\beta'(x, y) = \beta(x, xy)$. Then $\beta \mapsto \beta'$ maps $C_c^\infty(\mathcal{G} : E)$ bijectively to itself and

$$
\begin{aligned}
(\lambda(x,x)\beta)'(y,z) &= (\lambda(x,x)\beta)(y,yz) \\
&= \beta(x^{-1}y, x^{-1}yz) \\
&= \beta'(x^{-1}y, z) \\
&= \lambda(x,1)\beta'(y,z).
\end{aligned}
$$

We define another E-distribution T' on $\mathfrak{G}$ by setting $T'(\beta') = T(\beta)$. It follows that $T'(\lambda(x,1)\beta') = T'(\beta')$ $(x \in G)$.

Interpret T' as a functional on $C_c^\infty(G) \otimes C_c^\infty(G : E)$ which is left (i.e., $\lambda(G)$-) invariant with respect to the first factor and obtain:

<u>Lemma</u> 1.9.1. To $B \in \mathfrak{B}(\pi_1, \pi_2)$ there corresponds a linear mapping from $C_c^\infty(G : E)$ to the space of left-invariant distributions on G. The mapping is defined as follows: Take $\alpha \in C_c^\infty(G)$, $\beta \in C_c^\infty(G : E)$, and T' as above; define $\tau_\beta(\alpha) = T'(\alpha \otimes \beta)$. The left invariance is expressed by the relation $\tau_\beta(\lambda(x)\alpha) = \tau_\beta(\alpha)$ for any $x \in G$.

Observe next that, by Theorem 1.8.1, $\tau_\beta(\alpha) = c(\beta)\int_G \alpha(x)d_\ell x$ with $c(\beta)$ a complex number. Set $T^0(\beta) = c(\beta)$ and note that T^0 is an E-distribution on G. Thus, $T'(\alpha \otimes \beta) = T^0(\beta)\int_G \alpha(x)d_\ell x$ for $\alpha \in C_c^\infty(G)$ and $\beta \in C_c^\infty(G : E)$; equivalently, $T(\gamma) = T'(\gamma') = T^0(j(\gamma'))$, where the map $j : C_c^\infty(\mathfrak{G} : E) \to C_c^\infty(G : E)$ is defined by setting $j(\gamma)(y) = \int_G \gamma(x,y)d_\ell x$. Setting $\gamma_0 = j(\gamma')$, we have a surjection $\gamma \mapsto \gamma_0$ of $C_c^\infty(\mathfrak{G} : E)$ to $C_c^\infty(G : E)$. If $T^0 = 0$, then $T' = 0$, $T = 0$, and $B = 0$, since $B(P_1\alpha_1, P_2\alpha_2) = T(\alpha_1 \otimes \alpha_2)$ and the maps P_i are surjective to (Lemma 1.7.5). We summarize with the following theorem.

<u>Theorem</u> 1.9.2. There is an injective linear mapping $B \mapsto T_B^0$ from $\mathfrak{B}(\pi_1, \pi_2)$ into the space of E-distributions on G such that $B(P_1\alpha_1, P_2\alpha_2) = T_B^0(\gamma_0)$, where

$\gamma_0 \in C_c^\infty(G : E)$ is the function defined by the formula

$$\gamma_0(y) = \int_G \alpha_1(x) \otimes \alpha_2(xy) d_\ell x$$

for $\alpha_i \in C_c^\infty(G : V_i)$ $(i = 1, 2)$.

For our applications we shall require a slightly different formulation of the above theorem. We proceed to develop this reformulation.

Let $h = (h_1, h_2) \in \mathcal{H}$. Note that the modular functions satisfy $\delta_{\mathcal{G}}(h) = \delta_G(h_1)\delta_G(h_2)$ and $\delta_{\mathcal{H}}(h) = \delta_{H_1}(h_1)\delta_{H_2}(h_2)$. Write $\chi_{\mathcal{H}} = \delta_{\mathcal{H}}\delta_{\mathcal{G}}^{-1}$ and $\sigma(h) = \sigma_1(h_1) \otimes \sigma_2(h_2)$, so that σ is a representation on $E = V_1 \otimes V_2$.

<u>Lemma</u> 1.9.3. Let $\gamma \in C_c^\infty(\mathcal{G} : E)$ and set $\gamma_0(y) = \int_G \gamma(x, xy) d_\ell x$. Then, for $h = (h_1, h_2) \in \mathcal{H}$, we have $T_B^0(\delta_G(h_1)\lambda(h_1)\rho(h_2)\gamma_0) = T(\delta_{\mathcal{H}}(h)^{\frac{1}{2}}\delta_{\mathcal{G}}(h)^{\frac{1}{2}}\sigma(h^{-1})\gamma)$.

<u>Proof.</u> First we shall show that

$$(\rho(h)\gamma)_0 = \delta_G(h_1)\lambda(h_1)\rho(h_2)\gamma_0 :$$

$$\begin{aligned}(\rho(h)\gamma)_0(y) &= \int_G (\rho(h_1, h_2)\gamma)(x, xy) d_\ell x \\ &= \int_G \gamma(xh_1, xyh_2) d_\ell x \\ &= \int_G \gamma(xh_1, xyh_2)\delta_G^{-1}(x) d_r x \\ &= \int_G \gamma(x, xh_1^{-1}yh_2)\delta_G^{-1}(xh_1^{-1}) d_r x \\ &= \delta_G(h_1)\gamma_0(h_1^{-1}yh_2).\end{aligned}$$

Next, using Lemma 1.7.6 and the first part of this lemma, we have

$$\delta_{\mathcal{H}}^{\frac{1}{2}}(h)\delta_{\mathcal{Y}}(h)^{\frac{1}{2}}T(\sigma(h^{-1})\gamma) = T(\rho(h)\gamma)$$
$$= T_B^0((\rho(h)\gamma)_0) = T_B^0(\delta_G(h_1)\lambda(h_1)\rho(h_2)\gamma_0).$$

Our reformulation follows immediately from Theorem 1.9.2 and Lemma 1.9.3 together with the obvious fact that $\gamma \mapsto \gamma_0$ is surjective on $C_c^\infty(G : E)$.

Theorem 1.9.4. Let $\mathcal{J}$ be the space of all E-distributions T^0 on G such that

$$T^0(\delta_G(h_1)\lambda(h_1)\rho(h_2)\gamma_0) = T^0(\delta_{\mathcal{H}}(h)^{\frac{1}{2}}\delta_{\mathcal{Y}}(h)^{\frac{1}{2}}\sigma(h^{-1})\gamma_0)$$

for all $h = (h_1, h_2) \in \mathcal{H}$ and $\gamma_0 \in C_c^\infty(G : E)$. There exists an injective linear mapping of $\mathcal{B}(\pi_1, \pi_2)$ into $\mathcal{J}$.

Now let $\mathcal{H}$ operate on G by setting $x \cdot h = h_1^{-1} x h_2$ for $h = (h_1, h_2) \in \mathcal{H}$. Fix an orbit $\mathcal{O}$ and $y \in \mathcal{O}$. Write $\mathcal{H}(y)$ for the isotropy subgroup of y in $\mathcal{H}$. Then $\mathcal{H}(y)$ is the set of pairs $(h_1, h_1^{y^{-1}})$ with $h_1 \in H_1 \cap H_2^y$. There is a mapping from $\mathcal{H}(y) \backslash \mathcal{H}$ onto $\mathcal{O}$. Let $\mathcal{T}(\mathcal{O})$ denote the subspace of $\mathcal{T}$ consisting of those T^0 with support in the closure $\bar{\mathcal{O}}$ of $\mathcal{O}$.

Using Lemma 1.1.2, we may identify elements of $\mathcal{T}(\mathcal{O})$ with their restrictions to $\bar{\mathcal{O}}$.

Theorem 1.9.5. Let $\mathcal{O}$ be an orbit of $\mathcal{H} = H_1 \times H_2$ in G and assume that $\mathcal{O}$ is homeomorphic to $\mathcal{H}(y) \backslash \mathcal{H}$, where $\mathcal{H}(y)$ is the isotropy subgroup of $y \in \mathcal{O}$. Let $E'(y)$ denote the vector space of all linear functionals φ on E which satisfy the relation

$$\delta_{\mathcal{H}(y)}(h)\langle \varphi, \sigma(h^{-1})e\rangle = \delta_{\mathcal{H}}^{\frac{1}{2}}(h)\langle \varphi, e\rangle$$

for all $h = (h_1, h_1^{y^{-1}}) \in \mathcal{H}(y)$ and $e \in E$. There exists a linear mapping $T^0 \mapsto \varphi_{T^0}$

from $\mathcal{T}(\mathcal{O})$ to $E'(y)$ such that $\varphi_{T^0} = 0$ if and only if $(\text{Supp } T^0) \cap \mathcal{O} = \phi$.

Remark. See Montgomery-Zippin [11], p. 65, for a theorem implying that $\mathcal{H}(y)\backslash\mathcal{H}$ and $\mathcal{O}$ are homeomorphic.

Proof. Since $\mathcal{O}$ is homeomorphic to $\mathcal{H}(y)\backslash\mathcal{H}$, $\mathcal{O}$ is locally compact and, hence, open in its closure $\bar{\mathcal{O}}$.

Let $T^0 \in \mathcal{J}(\mathcal{O})$. It is an immediate consequence of Theorem 1.9.4 that either $\mathcal{O} \subset \text{Supp}(T^0)$ or $\mathcal{O} \cap \text{Supp}(T^0) = \phi$. In the latter case, set $\varphi_{T^0} = 0$. Assume that $\mathcal{O} \subset \text{Supp}(T^0)$. To define φ_{T^0}, we first identify T^0 with a distribution on $\bar{\mathcal{O}}$, then consider the restriction of this distribution to $\mathcal{O}$. Using the fact that $\mathcal{O}$ and $\mathcal{H}(y)\backslash\mathcal{H}$ are homeomorphic, we obtain a distribution, also denoted T^0, on $\mathcal{H}(y)\backslash\mathcal{H}$. Under the identification of $C_c^\infty(\mathcal{O} : E)$ with $C_c^\infty(\mathcal{H}(y)\backslash\mathcal{H} : E)$ we see that $\lambda(h_1)\rho(h_2)\gamma_0$ becomes $\rho_{\mathcal{H}}(h)\gamma_0$ for any $h = (h_1, h_2) \in \mathcal{H}$, where $\rho_{\mathcal{H}}$ is the representation by right translations on $C_c^\infty(\mathcal{H}(y)\backslash\mathcal{H} : E)$ and $\gamma_0 \in C_c^\infty(\mathcal{H}(y)\backslash\mathcal{H} : E)$.

Since $\delta_{\mathcal{Y}}(h)^{\frac{1}{2}} = \delta_G(h_1)$ for $h = (h_1, h_1^{y^{-1}}) \in \mathcal{H}(y)$, we have $T^0(\rho_{\mathcal{H}}(h)\gamma_0) = T^0(\delta_{\mathcal{H}}^{\frac{1}{2}}(h)\sigma(h^{-1})\gamma_0)$ $(h \in \mathcal{H}(y))$ for all $\gamma_0 \in C_c^\infty(\mathcal{H}(y)\backslash\mathcal{H} : E)$, by Theorem 1.9.4. We may, therefore, apply Theorem 1.8.4R, part 1), and conclude that there exists a unique linear functional $\varphi_{T^0} \in E'(y)$ satisfying

$$<\varphi_{T^0}, \delta_{\mathcal{H}}(h)^{\frac{1}{2}}\sigma(h^{-1})e>\chi_{\mathcal{H}(y)}(h) = <\varphi_{T^0}, e>$$

for all $h \in \mathcal{H}(y)$ and $e \in E$. Since $\chi_{\mathcal{H}(y)} = \delta_{\mathcal{H}(y)}\delta_{\mathcal{H}}^{-1}$, we have obtained exactly the required relation. Finally, Theorem 1.8.4R, part 2), implies that if $\varphi_{T^0} = 0$, then $T^0 = 0$ on $\mathcal{O}$.

§1.10. Automorphic Forms on a t. d. Group.

Let G be a t. d. group and U a complex vector space. We shall define here a space $\mathcal{A}(G : U)$ of U-valued "automorphic forms" on G. For $U = \mathbb{C}$ the space $\mathcal{A}(G : U) = \mathcal{A}(G)$ consists simply of all matrix coefficients for admissible representations of G. The space $\mathcal{A}(G : U)$ in general turns out to be identifiable with $\mathcal{A}(G) \otimes U$.

Analogous notions occur in the representation theory of semisimple real groups ([7d]) and in the theory of automorphic forms associated to $G_{\mathbb{R}}/\Gamma$ (G s. s. and Γ arithmetic; cf. [7a]).

Let K be an open compact subgroup of G. The algebra $H_K = C_c(G /\!/ K)$ is sometimes called the *Hecke algebra with respect to* K. A function $f \in C^\infty(G : U)$ is called right (left) [double] H_K-finite if the vector space generated by all functions of the form $f * \alpha_1 (\alpha_1 * f)[\alpha_1 * f * \alpha_2]$, with $\alpha_1, \alpha_2 \in H_K$, is finite-dimensional. If f is right (left) [double] H_K-finite for all K, then f is called right (left) [double] *Hecke-finite*. To say that f is left (right) Hecke-finite is equivalent, as we shall see, to saying that f generates under $\lambda(\rho)$ an admissible G-module, i. e., f is a Hecke-finite vector in the sense of §1. 5.

In general, if $f \in C^\infty(G : U)$ is arbitrary, there is no compact open subgroup K_0 of G such that $f \in C(G/K_0 : U)$. However, if f is right K-finite for some open compact K, then it follows from Lemma 1. 4. 3 that $f \in C(G/K_0 : U)$ for some K_0. The converse is of course clear: If $f \in C(G/K_0 : U)$, then f is right K-finite for any K.

Define the space $\mathcal{A}_r(G : U)[\mathcal{A}_\ell(G : U)]$ to be the set of all $f \in C^\infty(G : U)$ such that:

1) f is double K-finite;

2) f is right [left] Hecke-finite.

Soon we shall see that $\mathcal{A}_r(G : U) = \mathcal{A}_\ell(G : U)$, so we may write simply $\mathcal{A}(G : U)$. If $U = \mathbb{C}$, we employ the notations $\mathcal{A}_r(G)$, $\mathcal{A}_\ell(G)$, and $\mathcal{A}(G)$.

Denote by ρ the right regular representation of G on $\mathcal{A}_r(G : U)$, so $\rho(x)f(y) = f(yx)$.

Lemma 1.10.1. ρ is quasi-admissible on $\mathcal{A}_r(G : U)$.

Proof. That ρ is smooth follows immediately from condition 1) in the definition of $\mathcal{A}_r(G : U)$. Fix $f \in \mathcal{A}_r(G : U)$ and write V_f for the subspace of $\mathcal{A}_r(G : U)$ spanned by the right translates of f. Let $V_f(K_1)$ be the subspace of V_f spanned by the K_1-fixed vectors (K_1 a compact and open subgroup of G). We must show that $\dim V_f(K_1) < \infty$.

Clearly, $\rho(C_c(K_1\backslash G))f = V_f(K_1)$; in fact, assuming that $f \in C^\infty(G/K_0 : U)$, we have $\rho(C_c(K_1\backslash G/K_0))f = V_f(K_1)$. Let $K_2 \subset K_0 \cap K_1$. Then $\rho(H_{K_2})f \supset V_f(K_1)$. Since f is right Hecke-finite, $\dim \rho(H_{K_2})f < \infty$, which implies the lemma.

Now let Z be any central subgroup of G.

Corollary 1.10.2. If $f \in \mathcal{A}_r(G : U)$, then f is Z-finite.

Proof. By the lemma V_f is admissible under ρ. Furthermore, $f \in V_f(K_0)$ for some K_0. Since Z is central, the finite-dimensional space $V_f(K_0)$ is $\rho(Z)$-stable.

Remark. It is not asserted that the representation of Z on $V_f(K_0)$ is diagonalizable. In interesting cases this will not be true.

Now we shall show that $\mathcal{A}_r(G : U) = \mathcal{A}_\ell(G : U)$. For this let π be an admissible representation of G on a vector space W and write $\tilde{\pi}$ for the contragredient representation of G on $\tilde{W}$. We know that $\tilde{\pi}$ is admissible too. Consider the matrix coefficient $f(x) = <\tilde{w}, \pi(x)w>$ $(\tilde{w} \in \tilde{W},\ w \in W)$.

Lemma 1.10.3. $f \in \mathcal{A}(G)$.

Proof. Obviously, $f \in C(G/\!/K_0)$ for some K_0, i.e., f is double K-finite. For $\alpha \in H_{K_1}$ we have

$$\begin{aligned}(f * \alpha)(x) &= \int_G f(xy^{-1})\alpha(y)d_r y \\ &= \int_G <\tilde{w}, \pi(xy^{-1})w>\alpha(y)d_r y \\ &= <\tilde{w}, \pi(x)\pi(\alpha')w>,\end{aligned}$$

where $\alpha'(x) = \delta_G^{\frac{1}{2}}(x)\alpha(x^{-1})$. From the fact that $\pi(\alpha')w \in W(K_1)$, the finite-dimensional space spanned by the K_1-fixed vectors in W, one sees that f is right H_{K_1}-finite. The proof that f is left H_{K_1}-finite, for any K_1, is similar.

Lemma 1.10.3 implies that, in order to show that $\mathcal{A}_r(G) = \mathcal{A}_\ell(G)$, it suffices to show that any $f \in \mathcal{A}_r(G)$ is a matrix coefficient of an admissible representation.

Lemma 1.10.4. $\mathcal{A}_r(G) \subset \mathcal{A}_\ell(G)$.

Proof. If $f_0 \in \mathcal{A}_r(G)$, then V_{f_0}, the space generated by all right translates of f_0, is ρ-admissible. We shall show that there exists $\delta_0 \in \tilde{V}_{f_0}$ such that $<\delta_0, \rho(x)f_0> = f_0(x)$. Let $f_0 \in C(G/\!/K_0)$. Let δ be the "Dirac measure" at the identity, so $\delta(\rho(x)f) = f(x)$ for any $f \in V_{f_0}$; however, $\delta \in V' - \tilde{V}$. To define δ_0 we "smooth" δ.

Let $E_0 = \int_{K_0} \rho(k)dk$. (Assume $\int_{K_0} dk = 1$.) Set $\delta_0 = \delta \circ E_0$. Then, clearly, $\langle \delta_0, f\rangle = \langle \delta_0, E_0 f\rangle = \langle {}^tE_0\delta_0, f\rangle$, so $\delta_0 \in \widetilde{V}_{f_0}$. Since every $f \in V_{f_0}$, being a linear combination of right translates of f_0, is left K_0-invariant, we have $\langle \delta_0, f\rangle = \int_{K_0} f(k)dk = f(1) = \delta(f)$ for all $f \in V_{f_0}$. Therefore, $\langle \delta_0, \rho(x)f_0\rangle = f_0(x)$, too.

The proof that $\mathcal{A}_\ell(G) \subset \mathcal{A}_r(G)$ is similar, so $\mathcal{A}_\ell(G) = \mathcal{A}_r(G) = \mathcal{A}(G)$. We introduce the notation $\mathcal{A}(\pi)$ to denote the vector space generated by the matrix coefficients of an admissible representation π of G. We know that $\mathcal{A}(G) = \bigcup_\pi \mathcal{A}(\pi)$; more precisely, every element of $\mathcal{A}(G)$ <u>is</u> a matrix coefficient of some admissible representation of G.

Now assume $\dim U > 1$.

In the following we regard $C^\infty(G) \otimes U$ as a subspace of $C^\infty(G : U)$.

<u>Lemma</u> 1.10.5. $\mathcal{A}_r(G : U) \subset C^\infty(G) \otimes U$.

<u>Proof.</u> Let $f_0 \in \mathcal{A}_r(G : U)$. We shall show that there is a finite linearly independent set $\{u_1, \ldots, u_r\} \subset U$ such that $f_0(x) = \sum_{i=1}^{r} a_i(x)u_i$ with $a_i \in C^\infty(G)$.

Let $V = V_{f_0}$ be the cyclic submodule of $\mathcal{A}_r(G : U)$ generated by f_0 under ρ. Let U' be the algebraic dual of U. For $\varphi \in U'$ define $\delta_\varphi \in V'$ (the algebraic dual of V) by setting $\langle \delta_\varphi, f\rangle = \langle \varphi, f(1)\rangle$. As in the proof of Lemma 1.10.4 we "smooth" δ_φ to construct an element $\delta_{\varphi,K} \in \widetilde{V}$. For this let K be an open compact subgroup of G. Let E_K be the identity element in the algebra $C_c(G/\!/K)$ and set $E_0 = \rho(E_K)$. Then E_0 is an idempotent operator on V whose image is V_K, the space of K-fixed elements of V. Define $\delta_{\varphi,K} = \delta_\varphi \circ E_0$.

Clearly, $\delta_{\varphi,K} \in \widetilde{V}$; in fact, $\delta_{\varphi,K} \in (\widetilde{V})_K = (V_K)'$.

By Lemma 1.10.1 V_{f_0} is admissible, so V_K is finite-dimensional. It follows that the mapping $\varphi \mapsto \delta_{\varphi,K}$ from U' to $(V_K)'$ has a kernel U_0' which is of finite codimension in U'. Let $U_1 = \{u \in U \mid f(x) = u$ for some $f \in V_K$ and $x \in G\}$. Then, if $\varphi \in U_0'$ and $f \in V_K$, we see that $<\delta_{\varphi,K}, \rho(x)f> =$ $<\delta_\varphi, E_0(\rho(x)f)> = \varphi((E_0 f)(x)) = <\varphi, f(x)> = 0$, so $U_1 \perp U_0'$. Since any linear functional on U_1 extends to a linear functional on U, we have a surjection $U'/U_0' \to U_1'$, so U_1' is a finite-dimensional space. We may conclude that U_1 is a finite-dimensional vector space.

Let $u_1, \ldots, u_r$ be a base for U_1. Then, if $f \in V_K$, we may write $f(x) = \sum_{i=1}^{r} b_i(x) u_i$ with $b_i \in C(G/K)$. In particular, $f_0 \in V_K$ for some K, so the lemma is proved.

Corollary 1.10.6. Let $f \in \mathcal{A}(G)$ and let $V = V_f$ be the G-module spanned by $\{\rho(g)f \mid g \in G\}$. Write π for the representation ρ on V. Then:

1) There exist elements $v_0 \in V$ and $\widetilde{v}_0 \in \widetilde{V}$ such that $f(x) =$ $<\widetilde{v}_0, \pi(x)v_0>$ $(x \in G)$.

2) $\widetilde{V}$ is spanned by $\widetilde{\pi}(g)\widetilde{v}_0$.

3) $\mathcal{A}(\pi)$ is spanned by $\lambda(x)\rho(y)f$ $(x, y \in G)$.

Proof. Part 1) follows from the proof of Lemma 1.10.4. For 2) we note first that the mapping $\pi(g)v_0 \mapsto \rho(g)f$ extends to a G-module epimorphism of $\pi(G)v_0$ to V (i.e., if $\sum c_i \pi(g_i)v_0 = 0$, then $\sum c_i f(xg_i) = 0$ for all x). To prove 2) observe that $<\widetilde{\pi}(x)\widetilde{v}_0, \sum_{i=1}^{r} c_i \pi(g_i)v_0> = 0$ for all $x \in G$ $(c_i \in \mathbb{C}$, $g_i \in G)$, if and only if $\sum_{i=1}^{r} c_i \rho(g_i) f = 0$. Complete the proof by making use of the admissibility of V. Part 3) follows from the fact that v_0 and $\tilde{v}_0$ are cyclic vectors in, respectively, V and $\tilde{V}$.

Corollary 1.10.7. $\mathcal{A}_r(G : U) = \mathcal{A}(G) \otimes U$.

Proof. Let $f \in \mathcal{A}_r(G : U)$. Then, according to Lemma 1.10.5, there exist $u_1, \ldots, u_r$ linearly independent and $a_1, \ldots, a_r \in C^\infty(G)$ such that $f(x) = \sum_{i=1}^{r} a_i(x) u_i$. Since $f \in C(G /\!/ K : U)$ for some open compact subgroup K, it is clear that $a_i \in C(G /\!/ K)$, too. Similarly f right Hecke finite implies that the same is true for each a_i.

Corollary 1.10.8. $\mathcal{A}_r(G : U) = \mathcal{A}_\ell(G : U)$.

Thus, if π is an admissible representation of G and U is a $\mathbb{C}$-vector space, we may--and often do--identify $\mathcal{A}(\pi) \otimes U$ and $\mathcal{A}(\pi : U)$.

Lemma 1.10.9. If $\alpha \in C_c^\infty(G)$ and $f \in \mathcal{A}(G)[\mathcal{A}(\pi)]$, then both $\alpha * f$ and $f * \alpha$ belong to $\mathcal{A}(G)[\mathcal{A}(\pi)]$.

Proof. Clearly, $\alpha * f \in C(G /\!/ K_0)$ for some open compact subgroup K_0 of G. Moreover, by the associativity of convolution, it is clear that $\alpha * f$ is right Hecke-finite. The proof that $f * \alpha \in \mathcal{A}(G)$ goes similarly. We leave to the reader the proof that, if $f \in \mathcal{A}(\pi)$, $\alpha * f$ and $f * \alpha$ are too.

§1.11. The Space of Finite Operators for an Admissible Module.

The space $\mathrm{End}^0(V)$, to be introduced below, plays much the same role, relative to admissible representations of t.d. groups, as the space of Hilbert-Schmidt operators plays, relative to unitary representations of Lie groups.

Let G be a t.d. group and π an admissible representation of G in V. Then π induces in a natural way a $G \times G$-module structure on the space $\mathrm{End}(V)$ of all endomorphisms of V: $(x, y) \cdot T = \pi(x)T\pi(y^{-1})$ $(x, y \in G;\ T \in \mathrm{End}(V))$.

Let $\mathcal{T} = \mathrm{End}^0(V)$ denote the space consisting of all $T \in \mathrm{End}(V)$ such that the mappings $x \mapsto \pi(x)T$ and $x \mapsto T\pi(x)$ $(x \in G)$ are both smooth, i.e., $\mathrm{End}^0(V) = (\mathrm{End}(V))_\infty$ with respect to the above $G \times G$-module structure. Clearly, $\mathcal{T}$ is a subalgebra as well as a $G \times G$-submodule of $\mathrm{End}(V)$. We write π_2 for the representation of $G \times G$ on $\mathcal{T}$ and observe that π_2 is obviously admissible. Indeed, if $T \in \mathcal{T}$, then there exists an open compact subgroup K of G such that the identity element $E_K \in C_c(G/\!/K)$ satisfies $\pi(E_K)T\pi(E_K) = T$. Writing $V_K = \pi(E_K)V$, we have a natural injection $i : \mathrm{End}(V_K) \hookrightarrow \mathcal{T}$. Denoting $i(\mathrm{End}(V_K)) = \mathcal{T}_K$ we see that $\mathcal{T}_K$ is the space of $K \times K$-fixed elements in $\mathcal{T}$ (in fact in $\mathrm{End}(V)$) and $\mathcal{T} = \cup\, \mathcal{T}_K$. In particular, this means that every $T \in \mathcal{T}$ is representable by a matrix with only finitely many non-zero entries. Since each $\mathcal{T}_K$ is a simple algebra, it is easy to see that $\mathcal{T}$ is a simple algebra too.

It is sometimes useful to note that $V_K \otimes V'_K \cong \mathrm{End}(V_K)$ and $V \otimes \widetilde{V} = \bigcup_K V_K \otimes V'_K$; there is a canonical bijection $i : V \otimes \widetilde{V} \hookrightarrow \mathcal{T}$ defined by setting $i(v \otimes \widetilde{v})(u) = \widetilde{v}(u)v$ for $v, u \in V$ and $\widetilde{v} \in \widetilde{V}$. Identifying $V \otimes \widetilde{V}$ with $\mathcal{T}$ via i, one sees that $\pi_2(x, y)(v \otimes \widetilde{v}) = \pi(x)v \otimes \widetilde{\pi}(y)\widetilde{v}$. Observe also that the representation of $C_c^\infty(G)$ associated with π maps $C_c(G/\!/K)$ into $\mathcal{T}_K$, so $C_c^\infty(G)$ maps into $\mathcal{T}$.

Before proceeding further we define the <u>trace</u> for $T \in \mathrm{End}(V)$ of finite rank. Let V_0 be a subspace of V of finite codimension such that $V_0 \subset \ker T$. Let $\bar{V} = V/V_0$ and let $T_{V_0} \in \mathrm{End}(\bar{V})$ be defined from T by projection. Set $\mathrm{tr}(T) = \mathrm{tr}(T_{V_0})$. Obviously, $\mathrm{tr}(T)$ does not depend upon the choice of V_0 and this notion of trace agrees with the usual one when $\dim(V) < \infty$. One sees also that, if T_1 and T_2 are endomorphisms of V such that T_1 is of finite rank, then the composition in either order is of finite rank and $\mathrm{tr}(T_2T_1) = \mathrm{tr}(T_1T_2)$.

Returning to an admissible representation π in a vector space V, we see that, for any $T \in \mathcal{T}$ and $x \in G$, both $\mathrm{tr}(\pi(x)T)$ and $\mathrm{tr}(T\pi(x))$ are defined and have the same value. We set $f_T(x) = \mathrm{tr}(\pi(x)T)$. Taking $T = v \otimes \tilde{v} \in V \otimes \tilde{V}$, we note that $f_T(x) = \langle \tilde{v}, \pi(x)v \rangle$. It follows that $T \mapsto f_T$ maps $\mathcal{T}$ surjectively to $\mathcal{A}(\pi)$.

<u>Lemma</u> 1.11.1. The admissible representation π is irreducible if and only if the mapping $T \mapsto f_T$ of $\mathcal{T}$ to $\mathcal{A}(\pi)$ is injective.

<u>Proof.</u> Let W be a proper G-submodule of V. Then, since $V = \bigcup V_K$, we have $V_K \neq W_K$ for some open compact subgroup K of G. It follows that there exists a non-zero element $\tilde{v} \in (\tilde{V})_K$ such that $\langle \tilde{v}, w \rangle = 0$ for all $w \in W_K$. Since W is a G-module, it follows that, for any $w \in W$, $\langle \tilde{v}, w \rangle = \langle \tilde{\pi}(E_K)\tilde{v}, w \rangle = \langle \tilde{v}, \pi(E_K)w \rangle = 0$. Let $T = \tilde{v} \otimes w$ with $0 \neq w \in W$. Then $f_T(x) = \langle \tilde{v}, \pi(x)w \rangle = 0$, so $T \mapsto f_T$ is not injective.

Conversely, if $T \mapsto f_T$ is injective, then $0 \neq f_{v \otimes \tilde{v}}(x) = \langle \tilde{v}, \pi(x)v \rangle$ for every non-zero element $v \otimes \tilde{v} \in V \otimes \tilde{V}$, so V contains no proper G-submodule.

Lemma 1.11.2. Let $\alpha \in C_c^\infty(G)$ and $T \in \mathcal{J}$. Define $\alpha'(x) = \alpha(x^{-1})$. Then:

(1) $\alpha * f_T = f_{T\pi(\delta_G^{\frac{1}{2}}\alpha')}$,

(2) $f_T * \alpha = f_{\pi(\delta_G^{-\frac{1}{2}}\alpha')T}$,

(3) $\lambda(\alpha)f_T = f_{T\pi(\alpha')}$, and

(4) $\rho(\alpha)f_T = f_{\pi(\alpha)T}$.

Proof. We check only (1):

$$\alpha * f_T(x) = \int_G \alpha(y) f_T(y^{-1}x) d_\ell y$$

$$= \int_G \alpha'(y)\delta_G^{\frac{1}{2}}(y) f_T(yx)\delta_G^{-\frac{1}{2}}(y) d_r y$$

$$= \int_G \operatorname{tr}(\alpha'(y)\delta_G^{\frac{1}{2}}(y)\pi(y) \cdot \pi(x)T)\delta_G^{-\frac{1}{2}}(y) d_r y$$

$$= \operatorname{tr}(\pi(x)T\pi(\delta_G^{\frac{1}{2}}\alpha'))$$

$$= f_{T\pi(\delta_G^{\frac{1}{2}}\alpha')}.$$

Lemma 1.11.3. Let π be an admissible representation of G in V. Then the following five conditions are equivalent:

1) π is irreducible.

2) π_2 is irreducible on $\mathcal{T} = V \otimes \tilde{V}$.

3) $\pi(C_c^\infty(G)) = \mathcal{T}$.

4) As a double module over $C_c^\infty(G)$ $\mathcal{A}(\pi)$ is simple.

5) For every open compact subgroup K_0 of G, the space $\mathcal{A}(\pi) \cap C(G/\!/K_0)$ is a simple double module over $C_c(G/\!/K_0)$.

Proof. In view of Lemmas 1.11.1 and 1.11.2, statements 2) and 4) are equivalent. Assertions 4) and 5) are obviously equivalent. Since $\mathcal{T}$ is a simple algebra and π is an algebra homomorphism of $C_c^\infty(G)$ into $\mathcal{T}$, 2) and 3) are also obviously equivalent. Since π is irreducible if and only if $\tilde{\pi}$ is irreducible, 1) implies 2); on the other hand, clearly, 3) implies 1).

Lemma 1.11.4. Let π be an irreducible admissible representation of G in a vector space V. Let $\mathcal{B}$ be a non-zero ρ-stable subspace of $\mathcal{A}(\pi)$. Let ρ_0 denote ρ acting in the space $\mathcal{B}$. Then $J(\pi|\rho_0) \neq 0$.

Proof. Recall that $j : T \mapsto f_T$ is a bijection of $\mathcal{T}$ on $\mathcal{A}(\pi)$ (Lemma 1.11.1); moreover, $\rho(\alpha)f_T = f_{\pi(\alpha)T}$ for any $\alpha \in C_c^\infty(G)$ (Lemma 1.11.2). It follows that $j^{-1}(\rho(\alpha)f) = \pi(\alpha)j^{-1}(f)$ for any $f \in \mathcal{A}(\pi)$. Put $\mathcal{T}_0 = j^{-1}(\mathcal{B}) \neq (0)$.

Fix $v_0 \in V$ such that $T_0v_0 \neq 0$ for some $T_0 \in \mathcal{T}_0$. Define a non-zero mapping $\varphi : \mathcal{B} \to$ by setting $\varphi(f) = j^{-1}(f)v_0$. Then $\varphi(\rho_0(\alpha)f) = j^{-1}(\rho_0(\alpha)f)v_0 = \pi(\alpha)j^{-1}(f)v_0 = \pi(\alpha)\varphi(f)$, so $\varphi \in \mathcal{T}(\pi|\rho_0)$.

Lemma 1.11.5. Let π be as before. Let π_1 be an irreducible and admissible representation of G in a vector space V_1. If there is a subspace $\mathcal{B} \subset \mathcal{A}(\pi)$ and a non-zero G-module homomorphism $\mu : \mathcal{B} \to V_1$, then $\pi_1 \sim \pi$.

Proof. For $\tilde{v} \in \tilde{V}$ and $v \in V$ let $f_{\tilde{v},v}(x) = \langle\tilde{v}, \pi(x)v\rangle$ $(x \in G)$. Then $f_{\tilde{v},\pi(g)v}(x) = f_{\tilde{v},v}(xg) = \rho(g)f_{\tilde{v},v}(x)$ $(x, g \in G)$, so, for any non-zero $\tilde{v} \in \tilde{V}$ there is an injection $t_{\tilde{v}} : V \to \mathcal{A}(\pi)$ which intertwines π and ρ. Let $\sum_{i=1}^{r} f_{\tilde{v}_i,v_i} \in \mathcal{B}$ and assume that $\mu(\sum_{i=1}^{r} f_{\tilde{v}_i,v_i}) \neq 0$. Then $\mu \circ \Sigma t_{\tilde{v}_i}$ is a non-zero G-module mapping of V^r into V_1. Let $i_j : V \hookrightarrow V^r$ be the injection of V to the j-th component. Obviously, for some j, $\mu \circ \Sigma t_{\tilde{v}_i} \circ i_j$ is a non-zero intertwining operator between π and π_1.

Corollary 1.11.6. Let $\pi_1, \ldots, \pi_n$ be a finite set of non-zero inequivalent irreducible admissible representations of G. Then the sum $\sum_{i=1}^{n} \mathcal{A}(\pi_i)$ is direct.

Proof. If $n = 1$, there is nothing to prove, so assume the Corollary for $\mathcal{A}(\pi_2)+\ldots+\mathcal{A}(\pi_n) = \mathcal{B}$. Assume that $\mathcal{A}(\pi_1) \cap \mathcal{B} \neq (0)$. Then, since $\mathcal{A}(\pi_1)$ is a simple double module, we must have $\mathcal{A}(\pi_1) \subset \mathcal{B}$. Let $E_2, \ldots, E_n$ be the projections on $\mathcal{A}(\pi_2), \ldots, \mathcal{A}(\pi_n)$, i.e., such that $E_j \mathcal{B} = \mathcal{A}(\pi_j)$ and $\sum_{j=2}^{n} E_j = I$ on $\mathcal{B}$. Clearly, $E_j \mathcal{A}(\pi_1) \neq (0)$ for some j and, since E_j is obviously a double module morphism, this implies $\mathcal{A}(\pi_1) = \mathcal{A}(\pi_j)$. By Lemma 1.11.5 this contradicts $\pi_1 \not\sim \pi_j$.

We have already identified $V \otimes \tilde{V}$ with $\mathrm{End}^0(V)$. It follows that, by identification, $\mathrm{End}^0(\tilde{V}) = \tilde{V} \otimes \tilde{\tilde{V}} = \tilde{V} \otimes V$. Thus, the transpose maps $V \otimes \tilde{V}$ to $\tilde{V} \otimes V$, equivalently $\mathrm{End}^0(V)$ to $\mathrm{End}^0(\tilde{V})$. There is a non-degenerate symmetric bilinear form on $\mathrm{End}^0(V)$: For $T_1, T_2 \in \mathrm{End}^0(V)$ we have $B(T_1, T_2) = \mathrm{tr}(T_1 \circ T_2)$. For $S \in \mathrm{End}^0(\tilde{V})$ and $T \in \mathrm{End}^0(V)$ define $\langle S, T\rangle = B({}^tS, T)$. Then $\langle\ ,\ \rangle$ is also a non-degenerate bilinear form. If $S = \tilde{v}_1 \otimes v_1$ and $T = v_2 \otimes \tilde{v}_2$, then it is easy to see that $\langle S, T\rangle = \tilde{v}_1(v_2) \cdot \tilde{v}_2(v_1)$. This implies that $\langle(\tilde{\pi})_2(x,y)S, \pi_2(x,y)T\rangle = \langle S, T\rangle$, so we may also regard $(\tilde{\pi})_2$ as $(\pi_2)^{\sim}$ and $\tilde{V} \otimes V = (V \otimes \tilde{V})^{\sim}$.

Now assume--as we may without loss of generality--that there is a positive definite hermitian form H defined on V and that there exists an open compact subgroup K_0 such that $\pi | K_0$ is unitary with respect to H. Given $T \in \mathrm{End}(V)$, define $T^* = T_H^*$ so that $H(T^* v_1, v_2) = H(v_1, Tv_2)$ $((v_1, v_2) \in V \times V)$. Since $H(\pi(k)v_1, v_2) = H(v_1, \pi(k)v_2)$ $(k \in K_0)$, we conclude easily that, if $T \in \mathcal{T}_K$, then $T^* \in \mathcal{T}_K$, too. One sees that $T \mapsto T^*$ defines an anti-linear involution on $\mathcal{T}$. It is clear that $(v_1, v_2) \mapsto H(TT^* v_1, v_2)$ defines a non-negative hermitian form on V, non-zero for $T \neq 0$.

This, essentially, implies the following lemma.

Lemma 1.11.7. Set $\|T\|_H^2 = \mathrm{tr}(TT_H^*)$. Then $\| \ \|_H$ defines a pre-hilbert space norm on $\mathcal{T}$. If π is unitary with respect to H, then π_2 is unitary with respect to $\| \ \|_H$ on $\mathcal{T}$.

Write $L_r^2(G)$ for the space of all measurable functions f on G which satisfy

$$\|f\|_r^2 = \int_G |f(x)|^2 d_r x < \infty.$$

Let Z be a closed central subgroup of G and $\chi \in \hat{Z}$. Write $L_r^2(G, \chi)$ for the set of all measurable f which satisfy $f(xz) = \chi(z)f(x)$ $(x \in G,\ z \in Z)$ and

$$\|f\|_{G/Z, r}^2 = \int_{G/Z} |f(x^*)|^2 d_r x^* < \infty.$$

Both $L_r^2(G)$ and $L_r^2(G, \chi)$ are unitary representation spaces for the right regular representation ρ.

Corollary 1.11.8. Let π be an irreducible admissible representation of G in a space V and let χ be a character of a closed central subgroup Z of G. Assume that $\mathcal{A}(\pi) \subset L_r^2(G)[L_r^2(G, \chi)]$. Then there exists a scalar product for V with respect to which π is unitary.

Proof. We check only the case $\mathcal{A}(\pi) \subset L_r^2(G)$, as the other case goes similarly. With respect to the $L_r^2(G)$-norm $\mathcal{A}(\pi)$ is a pre-hilbert space and a (pre-)unitary representation space for ρ. Consider $0 \neq f \in \mathcal{A}(\pi)$ and let V_f be the cyclic admissible G-module generated by f under ρ. Define a G-module epimorphism $\varphi : V_f \to V$ (use Lemmas 1.5.3, 1.11.4 and 1.11.5). Let $\mathcal{H} = (\ker\varphi)^{\perp} \subset V_f$. Lemma 1.6.1 implies that $\mathcal{H}$ is an admissible unitary representation space. Noting that $\varphi|\mathcal{H}$ is an isomorphism of G-modules, we transport the metric on $\mathcal{H}$ to V.

Remark. It is obviously sufficient that there exists one non-zero $f \in \mathcal{A}(\pi) \cap L^2_r(G)[\mathcal{A}(\pi) \cap L^2_r(G, \chi)]$ for the conclusion of Corollary 1.11.8 to hold.

Let π be an irreducible admissible representation of G. If there exists a closed central subgroup Z of G with respect to which $\mathcal{A}(\pi) \subset L^2_r(G, \chi)$ for some $\chi \in \hat{Z}$, then we say that π belongs to the right discrete series of G. We write $\mathcal{E}_{2,r}(G)$ for the set of such classes.

If the group G is unimodular, we speak simply of discrete series and write $\mathcal{E}_2(G) = \mathcal{E}_{2,r}(G)$.

§1.12. Double Representations of K and Automorphic Forms.

Let K be an open compact subgroup of a t.d. group G and let τ be a smooth double representation of K on a vector space V. We define the space $C(G, \tau)$ of all functions $\psi : G \to V$ such that $\psi(k_1 x k_2) = \tau(k_1)\psi(x)\tau(k_2)$ (all $x \in G$, all $k_1, k_2 \in K$) and the subspace $C_c(G, \tau) = C(G, \tau) \cap C_c(G : V)$. We write $\mathcal{A}(G, \tau)$ for the space $(\mathcal{A}(G) \otimes V) \cap C(G, \tau)$ and $\mathcal{A}(\pi, \tau)$ for the subspace $\mathcal{A}(G, \tau) \cap (\mathcal{A}(\pi) \otimes V)$, when π is an admissible representation of G.

Let Z be a closed central subgroup of G and $\chi \in \hat{Z}$. We write $C(G, Z, \tau)$ for the subspace of $C(G, \tau)$ which consists of Z-finite functions, $C(G, \tau)_\chi[C(G, \chi, \tau)]$ for the subspace of $C(G, Z, \tau)$ which consists of functions annihilated by some finite [the first] power of $\chi(z)-\rho(z)$ (all $z \in Z$). Similarly, $\mathcal{A}(G, Z, \tau)$, $\mathcal{A}(G, \tau)_\chi$, and $\mathcal{A}(G, \chi, \tau)$ denote the intersection of $\mathcal{A}(G : V)$ with, respectively, $C(G, Z, \tau)$, $C(G, \tau)_\chi$, and $C(G, \chi, \tau)$.

If V is an algebra, we shall say that the pair (V, τ) satisfies "<u>associativity conditions</u>" if:

$$(\tau(k)v_1) \cdot v_2 = \tau(k)(v_1 \cdot v_2),$$

$$v_1 \cdot (v_2 \tau(k)) = (v_1 \cdot v_2)\tau(k),$$

and

$$(v_1 \tau(k)) \cdot v_2 = v_1 \cdot (\tau(k)v_2)$$

for all $k \in K$ and $v_1, v_2 \in V$. In this case we may convolve elements of $C(G, \tau)$ with elements of $C_c(G, \tau)$ and regard $C(G, \tau)$ as a double module over the ring $C_c(G, \tau)$. It is clear that $\mathcal{A}(G, \tau)$ may be regarded as a submodule.

Let V be a pre-hilbert space with respect to the scalar product $(\, , \,)$ and let τ be a unitary double representation in V. In this context we shall sometimes assume that there is an involution $* : V \to V$ and a trace function $\mathrm{tr} : V \to \mathbf{C}$ which satisfy the conditions:

1) $(cv)^* = \bar{c}v^*$;

2) $(u \cdot v)^* = v^* \cdot u^*$;

3) $(u, v) = \mathrm{tr}(u^* \cdot v)$;

4) $\mathrm{tr}(u \cdot v) = \mathrm{tr}(v \cdot u)$;

5) $\mathrm{tr}(\tau(k)u) = \mathrm{tr}(u\tau(k))$;

and 6) $(\tau(k_1)u\tau(k_2))^* = \tau(k_2^{-1})u^*\tau(k_1^{-1})$ $(k, k_1, k_2 \in K;\ u, v \in V;$ and $c \in \mathbf{C})$.

We shall now give the two most important examples of double representations (V, τ). The reader may check that they possess all the additional structure defined above. The remainder of the section will develop the properties of example 2).

1) Let τ be an admissible representation of K on a vector space U. Let $V = \mathrm{End}^0(U)$ (cf. §1.11). For $k_1, k_2 \in K$ and $v \in V$ the operator $\tau(k_1)v\tau(k_2) \in V$, so we may regard τ as a double representation on V. It is trivial that the pair (V, τ) satisfies "associativity conditions".

Fixing a pre-Hilbert space structure on U with respect to which τ is unitary, we may let v^* be the adjoint of $v \in V$ and $\mathrm{tr}\, v$ the ordinary trace (cf. §1.11). The scalar product $\mathrm{tr}(u^*v) = (u, v)$ is that associated to the Hilbert-Schmidt norm on V. The six conditions are all well known and obvious.

2) Let $V = C^\infty(K \times K)$ and set $(v_1 \cdot v_2)(k_1 : k_2) = \int_K v_1(k_1 : k^{-1}) v_2(k : k_2) dk$. Under this product V is an associative algebra. Set $\tau(k)v(k_1 : k_2) = v(k_1 k : k_2)$ and $v(k_1 : k_2)\tau(k) = v(k_1 : k k_2)$. One checks that (V, τ) satisfies the associativity conditions.

Define $v(k_1 : k_2)^* = \bar{v}(k_2^{-1} : k_1^{-1})$ and $\mathrm{tr}\, v = \int_K v(k : k^{-1}) dk$. The associated scalar product is

$$(u, v) = \int_{K \times K} \bar{u}(k_1 : k_2) v(k_1 : k_2) d(k_1, k_2).$$

All the above assumptions hold for this example too.

If K_0 is an open subgroup of K we define a subspace $V(K_0) \subset V$ by letting $V(K_0) = C(K_0 \backslash K \times K / K_0)$. Obviously, $\dim V(K_0) < \infty$ and $V = \bigcup V_{K_0}$, the union being over all open compact subgroups K_0. Clearly, for each K_0, $V(K_0)$ is a $K \times K$-submodule of V. Moreover, if K_1 is an open normal subgroup of K then $C(K_0 \backslash K / K_1, K_1 \backslash K / K_0)$ is also a submodule.

A finite-dimensional smooth double representation V of K is called irreducible if the representation $(k_1, k_2) \cdot v = \tau(k_1)v\tau(k_2^{-1})$ $(v \in V)$ is irreducible as a representation of $K \times K$ on V. There is a similarly obvious definition of completely reducible. Every (finite-dimensional) smooth double representation of K is completely reducible. Given an irreducible double representation of K, we may always regard it, up to equivalence, as a subrepresentation of $C(K_0 \backslash K/K_1, K_1 \backslash K/K_0)$ for some choice of K_0 and K_1, i.e., any irreducible double representation occurs, up to equivalence, as a subrepresentation of $C^\infty(K \times K)$.

With τ as in 2) above, we define a mapping $j: C^\infty(G) \to C(G, \tau)$, $j(f) = \underset{\sim}{f}$, by setting $\underset{\sim}{f}(x) : (k_1, k_2) \mapsto f(k_1 x k_2)$; we also write $\underset{\sim}{f}(k_1 : x : k_2) = f(k_1 x k_2)$. Clearly, j is bijective, $f(x) = \underset{\sim}{f}(1 : x : 1)$, and $j(C_c^\infty(G)) = C_c(G, \tau)$. It is also clear that $j(\alpha * f) = \underset{\sim}{\alpha} * \underset{\sim}{f}$ and $j(f * \alpha) = \underset{\sim}{f} * \underset{\sim}{\alpha}$ for all $\alpha \in C_c^\infty(G)$ and $f \in C^\infty(G)$ (the convolution $\underset{\sim}{\alpha} * \underset{\sim}{f}$ includes the multiplication in $V = C^\infty(K \times K)$).

Lemma 1.12.1. $j(\mathcal{A}(G)) = \mathcal{A}(G, \tau)$. If π is an admissible representation of G, then $j(\mathcal{A}(\pi)) = \mathcal{A}(\pi, \tau)$.

Proof. First, we shall show that $j^{-1}(\mathcal{A}(G, \tau)) \subset \mathcal{A}(G)$. Since $\mathcal{A}(G, \tau) \subset \mathcal{A}(G) \otimes V$, we have $\underset{\sim}{f} = \sum_{i=1}^{r} f_i v_i$ where $f_i \in \mathcal{A}(G)$ and $v_i \in V$ $(i = 1, \ldots, r)$. Clearly,
$j^{-1}(\underset{\sim}{f})(x) = \underset{\sim}{f}(1 : x : 1) = \sum_{i=1}^{r} f_i(x) v(1 : 1) \in \mathcal{A}(G)$.

To prove the reverse inclusion consider $f \in \mathcal{A}(G)$. Clearly, $\underset{\sim}{f} \in C(G, \tau)$; indeed $\underset{\sim}{f}(x) \in V(K_0)$ for some open compact subgroup K_0, so $\underset{\sim}{f} \in C(G /\!/ K_1, \tau)$ for some open compact group K_1 (since $\dim V(K_0) < \infty$); in other words, $\underset{\sim}{f} \in (C(G/\!/K_1) \otimes V(K_0)) \cap C(G, \tau)$. It suffices to show that $\underset{\sim}{f}$ is right Hecke-finite. Consider $h \in H_{K_2}$ where K_2 is an open normal subgroup of K. We have:

$$(\underset{\sim}{f} * h)(k_1 : x : k_2) = \int_G f(k_1 x y k_2) h(y^{-1}) d_\ell y$$

$$= \int_G f(k_1 x y) h(k_2 y^{-1}) d_\ell y$$

$$= (\underset{\sim}{f} * h_{k_2})(k_1 : x : 1)$$

$$= (f * h_{k_2})(k_1 x),$$

where $h_{k_2} \in H_{K_2}$. Now let $f_1, \ldots, f_N$ be a basis for the finite-dimensional space $f*H_{K_2}$. Then $\underset{\sim}{f} * h(k_1 : x : k_2) = \sum_{i=1}^{N} c_i(k_2 : h) f_i(k_1 x)$ with $c_i(k_2 : h) \in \mathbb{C}$. We may let k_1 run over a set of representatives for $K_0 \backslash K$ and conclude that $N[K : K_0] \geq \dim(\underset{\sim}{f} * H_{K_2})$.

For π an irreducible admissible representation of G it is clear that $\mathcal{A}(\pi, \tau) = j(\mathcal{A}(\pi))$.

<u>Lemma</u> 1.12.2. For any irreducible admissible representation π of G, the space $\mathcal{A}(\pi, \tau)$ is a simple double module over $C_c(G, \tau)$.

<u>Proof.</u> By Lemma 1.11.3 $\mathcal{A}(\pi)$ is a simple double module over $C_c^\infty(G)$. Since j commutes with convolution, j is an isomorphism of simple modules.

<u>Lemma</u> 1.12.3. Let π be an admissible representation of G. Let (V, τ) be a smooth double representation of K such that V is an algebra and associativity conditions are satisfied. Then $\mathcal{A}(G, \tau)$ and $\mathcal{A}(\pi, \tau)$ are double modules over $C_c(G, \tau)$.

<u>Proof.</u> Let $\alpha = \sum_{i=1}^{r} \alpha_i \otimes v_i$ and $f = \sum_{j=1}^{s} f_j \otimes w_j$ be elements of, respectively, $C_c(G, \tau)$ and $\mathcal{A}(G, \tau)$. Then, certainly, $\alpha * f \in C(G, \tau)$. Moreover, $\alpha * f =$

$\Sigma\alpha_i * f_j \otimes v_i \cdot w_j$. To check that $\alpha * f \in \mathcal{A}(G) \otimes V$, i.e., $\in \mathcal{A}(G, \tau)$, it suffices to observe that $\alpha_i * f_j \in \mathcal{A}(G)$, which follows from Lemma 1.10.9. The other statements needed in order to give a complete proof of this lemma follow quite similarly.

§1.13. On the Characters of Admissible Representations.

Let π be an admissible representation of G in a vector space V. For any $f \in C_c^\infty(G)$ we know that $\pi(f) \in \mathrm{End}^0(V)$ (§1.11), so $\mathrm{tr}(\pi(f))$ is defined. The mapping $f \mapsto \Theta_\pi(f) = \mathrm{tr}\,\pi(f)$ is a distribution on G, called the character of π. For f as above and fixed $x \in G$, let $f^x(g) = f(x^{-1}gx)$ $(g \in G)$. One sees that $\Theta_\pi(f^x) = \delta_G(x)\Theta_\pi(f)$. This fact is expressed by saying that Θ_π is an invariant distribution.

It is clear that, if π_1 and π_2 are equivalent representations, then $\Theta_{\pi_1} = \Theta_{\pi_2}$. The converse holds under the additional hypothesis that each representation is completely reducible. We prove the following more precise statement.

Lemma 1.13.1. Let $\pi_1, \ldots, \pi_n$ be inequivalent irreducible admissible representations of G. Then the characters $\Theta_{\pi_1}, \ldots, \Theta_{\pi_n}$ are linearly independent over $\mathbb{C}$.

Proof. Let $f \in C_c^\infty(G)$ and let π_i act in a space $\mathcal{H}_i$, $i = 1, \ldots, n$. Lemma 1.4.1 implies that, for any $x \in G$, $\mathrm{tr}(\pi_i(\lambda'(x)f)) = \mathrm{tr}(\pi_i(x)\pi_i(f)) = f_{T_i}(x)$, where $T_i = \pi_i(f) \in \mathrm{End}^0(\mathcal{H}_i)$, so $f_{T_i} \in \mathcal{A}(\pi_i)$. If $\sum_{i=1}^n c_i\Theta_{\pi_i} = 0$, then $\sum_{i=1}^n c_i f_{T_i} = 0$, which contradicts Corollary 1.11.6, since one can choose f such that not all $f_{T_i} = 0$.

Let $\tilde{\pi}$ $[\pi^*]$ be the contragredient [adjoint] of the admissible representation π. Given $\alpha \in C_c^\infty(G)$, set $\alpha'(x) = \alpha(x^{-1})$ and $\alpha^\sim(x) = \bar{\alpha}(x^{-1})$ $(x \in G)$. Then, as one sees easily, $\Theta_{\tilde{\pi}}(\alpha) = \Theta_\pi(\alpha')$ and $\Theta_{\pi^*}(\tilde{\alpha}) = \Theta_\pi(\alpha)$. If π is unitary, then $\Theta_\pi(\tilde{\alpha}) = \bar{\Theta}_\pi(\alpha)$, since, in this case, $\pi(\tilde{\alpha}) = \pi(\alpha)^*$, so $\Theta_\pi(\tilde{\alpha} * \alpha) = \operatorname{tr}(\pi(\alpha)^* \pi(\alpha)) \geq 0$. This is usually expressed by saying that the character of a unitary admissible representation is a <u>distribution of positive type</u>. Although we shall not use this fact, the converse holds too: If π is irreducible and admissible and Θ_π is a distribution of positive type, then π is unitary.

Now let Γ be a cocompact closed subgroup of G. Fix right Haar measures $d_r x$ and $d_r \gamma$ on G and Γ, respectively. Let χ be the function constructed in Lemma 1.2.4; in particular, extend δ_Γ to a function defined on all of G as in the same lemma. Then it follows from Lemma 1.2.5 and the existence of the involution $x \mapsto x^{-1}$ on G that there is a unique Radon measure $d\bar{y}$ on $\Gamma\backslash G$ such that $\chi^{-1}(y^{-1})d_r y = d_r \gamma d\bar{y}$.

In the following theorem we abuse notation to write χ for the quasi-character $\delta_\Gamma \delta_G^{-1}$ of Γ obtained by restriction of the function χ to Γ and α^y for the function $\alpha^y | \Gamma$.

<u>Theorem</u> 1.13.2. Let Γ be as above and let σ be an admissible representation of Γ in a vector space U. Then $\pi = \operatorname{Ind}_\Gamma^G(\chi^{\frac{1}{2}}\sigma)$ is an admissible representation of G (Lemma 1.7.4). For any $\alpha \in C_c^\infty(G)$ the function $\Theta_\sigma(\alpha^y \delta_\Gamma(y^{-1}))$ is a function on $\Gamma\backslash G$; moreover, $\Theta_\pi(\alpha) = \int_{\Gamma\backslash G} \Theta_\sigma(\delta_\Gamma(y^{-1})\alpha^y) d\bar{y}$.

<u>Proof.</u> Let $\mathcal{H}[\tilde{\mathcal{H}}]$ be the space of all smooth functions $\beta : G \to U[\tilde{U}]$ such that $\beta(\gamma x) = \chi^{\frac{1}{2}}(\gamma)\sigma(\gamma)\beta(x)[\beta(\gamma x) = \chi^{\frac{1}{2}}(\gamma)\sigma^\sim(\gamma)\beta(x)]$ for all $\gamma \in \Gamma$ and $x \in G$. We may re-

gard $\pi[\pi^\sim]$ as the representation by right translations in $\mathcal{H}[\mathcal{H}^\sim]$: $(\pi(y)\beta)(x) = \beta(xy)$ $(x, y \in G)$. In the following we use notations which were explained in §§1.3-1.4.

Let $\alpha \in C_c^\infty(G)$. Then:

$$
\begin{aligned}
(\pi(\alpha)\beta)(x) &= \int_G \alpha(y)\beta(xy)\delta_G^{\frac{1}{2}}(y)d_\ell y \\
&= \int_G \alpha(x^{-1}y)\beta(y)\delta_G^{-\frac{1}{2}}(xy)\chi(y^{-1})\chi^{-1}(y^{-1})d_r y \\
&= \int_{\Gamma\backslash G}\int_\Gamma \alpha(x^{-1}\gamma y)\beta(\gamma y)\delta_G^{-\frac{1}{2}}(x\gamma y)\chi(y^{-1}\gamma^{-1})d_r\gamma d\bar{y} \\
&= \int_{\Gamma\backslash G}\sigma(\lambda'(x)\rho'(y)\alpha)\chi(y^{-1})\beta(y)d\bar{y}.
\end{aligned}
$$

One observes that the kernel $\kappa_\alpha(x:y) = \sigma(\lambda'(x)\rho'(y)\alpha)$ is a smooth function of $(x, y) \in G \times G$ and that it satisfies the relation $\kappa_\alpha(\gamma_1 x : \gamma_2 y) = \chi^{\frac{1}{2}}(\gamma_1)\sigma(\gamma_1)\kappa_\alpha(x:y)\chi^{\frac{1}{2}}(\gamma_2)\sigma(\gamma_2^{-1})$. Since, for any (x, y), $\kappa_\alpha(x:y) \in \mathrm{End}^0(U) = U \otimes \widetilde{U}$, it is clear that we may identify this kernel with an element of $\mathcal{H}\otimes\mathcal{H}^\sim$. The space $\mathcal{H}\otimes\mathcal{H}^\sim$ acts on $\mathcal{H}$: $\Sigma(h_i \otimes \widetilde{h}_i)h = \Sigma\langle \widetilde{h}_i, h\rangle h_i$, where $\langle \widetilde{h}_i, h\rangle = \int_{\Gamma\backslash G}\langle \widetilde{h}_i(y), h(y)\rangle\chi(y^{-1})d\bar{y}$ $(h \in \mathcal{H};\ h_i \otimes \widetilde{h}_i \in \mathcal{H}\otimes\mathcal{H}^\sim)$. *This is the action given by the above integral formula.*

The trace of an operator $\Sigma h_i \otimes \widetilde{h}_i$ is $\Sigma\widetilde{h}_i(h_i)$, so

$$
\begin{aligned}
\operatorname{trace}\pi(\alpha) &= \int_{\Gamma\backslash G}\chi(y^{-1})\Theta_\sigma(\lambda'(y)\rho'(y)\alpha)d\bar{y} \\
&= \int_{\Gamma\backslash G}\Theta_\sigma(\delta_\Gamma(y^{-1})\alpha^y)d\bar{y},
\end{aligned}
$$

as claimed.

Of course, if $G = K\Gamma$, where K is compact and $\Gamma\backslash G$ is taken as $K\cap\Gamma\backslash K$, then δ_Γ disappears from this formula and, assuming normalized measures, we obtain

$$\operatorname{trace} \pi(\alpha) = \int_K \Theta_\sigma(\alpha^k)dk$$

$$= \Theta_\sigma(\alpha_K),$$

where $\alpha_K(\gamma) = \int_K \alpha(k^{-1}\gamma k)dk.$

Chapter 2. Jacquet's Theory, Bruhat's Theory, the Elementary Theory of the Constant Term.

In this chapter and all remaining chapters of these notes we assume that G is the set of Ω-points of a connected reductive Ω-group $\underset{\sim}{G}$, Ω being a non-archimedean local field. For the properties of $\underset{\sim}{G}$ which we shall need the reader should consult Chapter 0.

The symbol Z will always denote the split component of G, i.e., the set of Ω-points of the largest Ω-split torus $\underset{\sim}{Z}$ lying in the center of $\underset{\sim}{G}$.

The chapter opens (§2.1) with a structure theorem for "infinitesimal" neighborhoods of the identity in G. In many applications this result replaces the differential operators and universal enveloping algebra, which play a ubiquitous role in the theory of harmonic analysis for real groups.

§2.2 brings in Jacquet's definition of supercuspidal representation along with equivalent statements. §2.3 proves a "going-down" theorem: from an admissible but not supercuspidal representation of G, one can construct admissible representations of Levi factors of parabolic subgroups. §2.4, besides proving Jacquet's subrepresentation theorem, introduces certain concepts needed for the Bruhat theory, concepts to be enlarged upon later. The "Bruhat theory" (§2.5) provides sufficient conditions for the irreducibility of induced representations and, more generally, bounds for their intertwining numbers.

The last three sections of the chapter introduce the "constant term" for an automorphic form on G and develop those properties of the constant term which do not involve "exponents". The notion of "constant term" did not, of course, originate in the present context; the reader should consult [7a] and [7d] for analogous theories.

§2.1. A Decomposition Theorem for Certain Compact Open Subgroups of G.

The structure theory of Bruhat and Tits [4c] gives stronger versions of Theorem 2.1.1. The present proof depends upon the most elementary properties of Ω-imbeddings in GL_n.

We remind the reader that, if (P, A) is a p-pair of G with Levi decomposition $P = MN$, then $\overline{P} = M\overline{N}$ is the Levi decomposition of the unique p-pair $(\overline{P}, A)$ such that $\Sigma(G, A) = \Sigma(P, A) \cup \Sigma(\overline{P}, A)$.

Notation: Let X be a group. During the proof of the following theorem we shall employ the notation $\langle A, B, C, D, \text{etc.}\rangle$ to denote the subgroup of X generated by the subsets A, B, C, D, etc., of X.

Theorem 2.1.1. Let (P_0, A_0) be a minimal p-pair of G. Then there exists a sequence of open compact subgroups of G, $\{K_j\}_{j=1}^{\infty}$, with the following properties:
(1) $\{K_j\}_{j=1}^{\infty}$ is a fundamental sequence of neighborhoods of the identity in G.
(2) Let (P, A) be a p-pair of G with the Levi decomposition $P = MN$. If (P, A) is standard with respect to (P_0, A_0), then $K_j = \overline{N}_j M_j N_j = N_j M_j \overline{N}_j$ with the notations $N_j = N \cap K_j$, $M_j = M \cap K_j$, and $\overline{N}_j = \overline{N} \cap K_j$.

Proof. Recall that in any affine algebraic group defined over an algebraically closed field a Borel subgroup is a maximal connected solvable subgroup. Borel subgroups constitute a single conjugacy class.

Without loss of generality we may (and shall) regard $\underset{\sim}{G}$ as an Ω-subgroup of $\underset{\sim}{L} = \underset{\sim}{GL}_n$. (As usual, we use "$\sim$" to denote the points in an algebraic closure of Ω.) Let $\underset{\sim}{T}_0$ be a maximal torus of $\underset{\sim}{L}$ such that $\underset{\sim}{A}_0 \subset \underset{\sim}{T}_0$. Let $\underset{\sim}{Q}_0$ be a Borel subgroup of $\underset{\sim}{L}$ which contains both $\underset{\sim}{T}_0$ and a Borel subgroup

$\underset{\sim}{B}_0$ of $\underset{\sim}{G}$ (clearly, $\underset{\sim}{A}_0 \subset \underset{\sim}{B}_0$). A minimal Ω-subgroup of $\underset{\sim}{G}$ which contains $\underset{\sim}{B}_0$ is, by definition, a minimal parabolic Ω-subgroup of $\underset{\sim}{G}$. Since minimal parabolic Ω-subgroups are G-conjugate, we may assume that $\underset{\sim}{P}_0$ is the minimal Ω-subgroup of $\underset{\sim}{G}$ which contains $\underset{\sim}{B}_0$.

Now let $\Phi = \Sigma^0(\underset{\sim}{Q}_0, \underset{\sim}{T}_0)$ be the set of simple roots associated to the pair $(\underset{\sim}{Q}_0, \underset{\sim}{T}_0)$. Let $F = \{\alpha \in \Phi \mid \alpha \mid \underset{\sim}{A} = 1\}$. Let $(\underset{\sim}{Q}, \underset{\sim}{T}) = (\underset{\sim}{Q}_0, \underset{\sim}{T}_0)_F$, so $\underset{\sim}{T} = (\bigcap_{\alpha \in F} \operatorname{Ker} \alpha)^0$ and $\underset{\sim}{Q} = Z_{\underset{\sim}{L}}(\underset{\sim}{T})\underset{\sim}{Q}_0$. Observe that both $\underset{\sim}{T}$ and $\underset{\sim}{Q}$ are Ω-subgroups of $\underset{\sim}{L}$. Obviously $\underset{\sim}{A} = (\underset{\sim}{T} \cap \underset{\sim}{G})^0$. Clearly, $\underset{\sim}{Q}$ is an Ω-subgroup and $\underset{\sim}{T}$ is an Ω-split torus, which we may assume diagonalized. Since $\underset{\sim}{A} = (\underset{\sim}{T} \cap \underset{\sim}{G})^0$ and $(Q \cap G, A)$ is a p-pair of G, we see that $Q \cap G = P$.

Let (Q, T) have the Levi decomposition $Q = M_Q N_Q$. Then the Levi decomposition for (P, A) is $P = MN$, with $M = M_Q \cap G$ and $N = N_Q \cap G$. If $\bar{N}_Q$ denotes the unipotent radical of $\bar{Q}$ in L, then we have, as the unipotent radical for $\bar{P}$ in G, the group $\bar{N} = \bar{N}_Q \cap G$. Recall that $NM\bar{N}$ and $\bar{N}MN$ are open subsets of G, that $NM\bar{N} \cong N \times M \times \bar{N} \cong \bar{N}MN$. Similar statements of course hold for $N_Q M_Q \bar{N}_Q$ and $\bar{N}_Q M_Q N_Q$ with respect to L.

Let $L_j = \{(x_{ik}) \in L \mid \operatorname{ord}(x_{ik} - \delta_{ik}) \geq j\}$, where j is a positive integer and $\operatorname{ord}: \Omega^\times \to \mathbb{Z}$ is the valuation map. Set $K_j = L_j \cap G$. Then the sequence $\{K_j\}$ obviously satisfies (1). For j large $K_j \subset K \cap NM\bar{N} \cap \bar{N}MN$. Clearly, $K_j \supset N_j M_j \bar{N}_j$ and $\bar{N}_j M_j N_j$. Since $L_j = \bar{N}_{Q,j} M_{Q,j} N_{Q,j} = N_{Q,j} M_{Q,j} \bar{N}_{Q,j}$ with obvious notations, it follows from the uniqueness of the decomposition that, if $x \in K_j$ and $x = \bar{n}mn = n'm'\bar{n}'$, each factor must lie in K_j.

<u>Corollary</u> 2.1.2. Let $K_j = \bar{N}_j M_j N_j$ as in Theorem 2.1.1. The Haar measure on

the compact group K_j is the product measure with respect to the Haar measures of the three factors.

Proof omitted.

Corollary 2.1.3. Let K_j be as above and let P be a standard parabolic subgroup of G. Set $P_j = P \cap K_j$. For any $\phi \in C_c^{\infty}(G/P_j)$ the integral $\psi(x) = \int_{\bar{N}_j} \phi(x\bar{n})d\bar{n}$ defines a function $\psi \in C_c(G/K_j)$.

Proof. Assuming normalized measures, we have

$$\int_{K_j} \phi(xk)dk = \int_{\bar{N}_j P_j} \phi(x\bar{n}p)d\bar{n}dp$$

$$= \int_{\bar{N}_j} \phi(x\bar{n})d\bar{n}$$

$$= \psi(x).$$

The result is clearly true.

§2.2. J-supercuspidal Representations.

In his Montecatini lectures on $GL(n,\Omega)$ ([9], see also [6]) Jacquet defines "pointué" representations and proves that their matrix coefficients are locally constant functions of compact support modulo the center. Here we give Jacquet's definition for general reductive groups and prove that every supercuspidal representation is "J-supercuspidal". We postpone the proof that J-supercuspidal implies supercuspidal until §2.8, as the proof depends either explicitly or implicitly upon properties of the constant term.

Let π be a smooth representation of G on a complex vector space V. Let P be a parabolic subgroup of G with N its unipotent radical. Write $V(N)$ or $V(P)$ for the subspace of V generated by all elements of the form $\pi(n)v-v$ $(n \in N,\ v \in V)$.

Lemma 2.2.1. Let $v \in V$. Then $v \in V(N)$ if and only if there exists a compact open subgroup $N(v)$ of N such that

$$\int_{N(v)} \pi(n)v\,dn = 0.$$

Proof. If $v' = \pi(n_0)v-v$ with $n_0 \in N$, then take $N(v')$ to be any open compact subgroup of N such that $n_0 \in N(v')$ (recall that N is a union of open compact subgroups). Then $\int_{N(v')} \pi(n)v'\,dn = \int_{N(v')} \pi(n)(\pi(n_0)v-v)dn$

$$= \int_{N(v')} (\pi(nn_0)v-\pi(n)v)dn$$

$$= 0 \quad (dn \text{ being a Haar measure}).$$

The extension to the case in which v'' is a linear combination of elements of the form v' is clear.

Conversely, suppose that there exists $N(v)$ such that

$$\int_{N(v)} \pi(n)v\,dn = 0.$$

Normalizing the measure on $N(v)$, we have

$$v = \int_{N(v)} v\,dn = \int_{N(v)} (v-\pi(n)v)dn.$$

Since π is smooth and $N(v)$ is compact, it follows that v is a (finite) sum of elements of the form $\lambda_i(v-\pi(n_i)v)$ $(\lambda_i > 0,\ n_i \in N(v))$.

We say that π satisfies Jacquet's condition or call π J-supercuspidal if $V(P) = V$ for all $P \neq G$.

Remarks:

1) Let P_1 and P_2 be parabolic subgroups of G. If $P_1 \supset P_2$, then $N_2 \supset N_1$, so $V(P_2) \supset V(P_1)$. In order to check that π is J-supercuspidal it suffices to check Jacquet's condition with respect to maximal parabolic subgroups.

2) Fix a minimal parabolic subgroup P_0. Then, given any parabolic subgroup P, we can find $x \in G$ such that $P^x \supset P_0$. Clearly $V(P^x) = \pi(x)V(P)$, so $V(P^x) = V$ if and only if $V(P) = V$. This means that it suffices to check Jacquet's condition for standard maximal parabolic subgroups.

Let U be a complex vector space. The space ${}^0C_c^\infty(G, Z : U)$ is, by definition, the set of all $f \in C_c^\infty(G, Z : U)$ (cf. §1.4) which satisfy $\int_N f(xn)dn = 0$ for every parabolic subgroup $P = MN$ of $G(N \neq (1))$. It is proved in [7c], pages 20-25, that ${}^0C_c^\infty(G, Z : U) \subset \mathcal{A}(G : U)$. We frequently write ${}^0\mathcal{A}(G : U)$ for ${}^0C_c^\infty(G, Z : U)$, ${}^0\mathcal{A}(G)$ for ${}^0C_c^\infty(G, Z)$.

An admissible representation π of G is called <u>supercuspidal</u> if the space of matrix coefficients (cf., §1.10) $\mathcal{A}(\pi) \subset {}^0\mathcal{A}(G)$. We write ${}^0\mathcal{E}_{\mathbb{C}}(G)[{}^0\mathcal{E}(G)]$ *for the set of classes of irreducible [unitary] supercuspidal representations of* G. It is clear that ${}^0\mathcal{E}(G) \subset \mathcal{E}_2(G)$.

<u>Lemma</u> 2.2.2. Let π be an admissible supercuspidal representation of G on a vector space V. Then π is J-supercuspidal.

<u>Proof.</u> First assume that π is irreducible. Let (P, A) be a p-pair with $P = MN$ its Levi decomposition. We want to show that $V = V(N)$. Let $\mathcal{A}(\pi) \subset {}^0\mathcal{A}(G)$ be the space of automorphic forms generated by the matrix coefficients of π. For $0 \neq \tilde{v} \in \tilde{V}$ we define $i_{\tilde{v}} : V \hookrightarrow \mathcal{A}(\pi)$ by setting $i_{\tilde{v}}(v) = f_{\tilde{v}, v}$, where

$f_{\tilde{v},v}(x) = \langle \tilde{v}, \pi(x)v \rangle$. Since $f_{\tilde{v},\pi(g)v}(x) = f_{\tilde{v},v}(xg) = \rho(g)f_{\tilde{v},v}(x)$, it is clear that $i_{\tilde{v}}$ intertwines π and the representation ρ on $\mathcal{A}(\pi)$. Thus, it suffices to show that, for any $f \in \mathcal{A}(\pi)$, there is an open compact subgroup $N(f) \subset N$ such that $\int_{N(f)} f(xn)dn = 0$ for all $x \in G$. Clearly, for each $x \in G$, there is an open compact subgroup $N(x)$ such that $\int_{N(x)} f(xn)dn = 0$. Since $f \in C(G/\!/K_0)$ for some K_0 and $N(x) = N(kx)$ for all $k \in K_0$, $x \mapsto N(x)$ is a locally constant function. Since $\operatorname{Supp} f$ is compact mod Z and $N(x) = N(zx)$ $(z \in Z,\ x \in G)$, it is clear that we may choose $N(f)$ independent of $x \in G$. This shows that $\mathcal{A}(\pi)(N) = \mathcal{A}(\pi)$ with respect to ρ and completes the proof of Lemma 2.2.2 for the case in which π is irreducible.

To extend the argument to the case in which π is not completely reducible we argue as follows. Let ${}^0G = \bigcap \operatorname{Ker} |\chi|$, χ ranging over all rational characters of G. We know that the center of ${}^0G$ is compact. Although ${}^0G$ is not generally a reductive algebraic group, it is clear that ${}^0G$ retains sufficient structure for our purposes: (1) Every compact subgroup of G is also a subgroup of ${}^0G$, so $\pi|{}^0G$ is admissible. (2) Every unipotent radical N also lies in ${}^0G$, so the matrix coefficients of $\pi|{}^0G$ are supercusp forms--they are compactly supported and satisfy the cusp form condition $\int_N f(xn)dn = 0$ for all $x \in {}^0G$ and for every N.

Let $v_0 \in V$ and set $V_0 = \pi({}^0G)v_0$. The representation π_0 of ${}^0G$ on V_0 is admissible. We want to show that $V_0(N) = V_0$. For this it suffices to prove that π_0 is unitarizable and, hence, completely reducible (Corollary 1.6.6). The argument given for π irreducible then applies without essential change.

To show that π_0 is unitarizable it suffices to show that the contragredient representation $\tilde{\pi}_0$ of ${}^0G$ in the space $\tilde{V}_0$ is unitarizable. For this define $\tilde{v} \mapsto f_{\tilde{v}}(x) = \langle \tilde{v}, \pi_0(x)v_0 \rangle$. Clearly, $0 \neq f_{\tilde{v}} \in {}^0C_c^\infty({}^0G)$ and $f_{\tilde{\pi}_0(y)\tilde{v}}(x) = \langle \tilde{\pi}_0(y)\tilde{v}, \pi_0(x)v_0 \rangle = \langle \tilde{v}, \pi_0(y^{-1}x)v_0 \rangle = f_{\tilde{v}}(y^{-1}x) = \lambda(y)f_{\tilde{v}}(x)$ (x and $y \in G$). Thus, we have an injection $\tilde{V}_0 \hookrightarrow {}^0\mathcal{A}({}^0G)$ which intertwines $\tilde{\pi}_0$ and λ. Define $\|\tilde{v}\|^2 = \|f_{\tilde{v}}\|^2 = \int_{{}^0G} |f_{\tilde{v}}(x)|^2 dx$. This gives a $\tilde{\pi}_0$-invariant pre-Hilbert space structure on $\tilde{V}_0$ and completes the proof of Lemma 2.2.2.

<u>Lemma</u> 2.2.3. Let π be an admissible representation of G in a vector space V. Let V' denote the algebraic dual space of V. There exists a non-zero element $v' \in V'$ and a proper parabolic subgroup $P = MN$ of G such that ${}^t\pi(n)v' = v'$ for all $n \in N$ if and only if π is not J-supercuspidal.

<u>Proof.</u> Let $v' = {}^t\pi(n)v'$ for all $n \in N$ and assume that $v' \neq 0$. Then there exists $v \in V$ such that $\langle v', v \rangle \neq 0$. If π is J-supercuspidal, then there exists an open compact subgroup $N(v)$ of N such that $\int_{N(v)} \pi(n)dn \cdot v = 0$. However, $\int_{N(v)} \langle v', \pi(n)v \rangle dn = \int_{N(v)} \langle {}^t\pi(n)v', v \rangle dn = c\langle v', v \rangle$, where $c > 0$. This is a contradiction.

Conversely, if π is not J-supercuspidal, then there exists $P = MN$ ($N \neq \{1\}$) for which $V \neq V(N)$. Let $0 \neq \varphi \in V'$ be such that $\varphi | V(N) = 0$. Since $\varphi(v - \pi(n)v) = 0$ for any $v \in V$ and $n \in N$, it is clear that ${}^t\pi(n)\varphi = \varphi$ for all $n \in N$.

<u>Corollary</u> 2.2.4. Let σ be any admissible representation of P trivial on N. If π is a supercuspidal representation of G, then π is not a subrepresentation of $\mathrm{Ind}_P^G \sigma$.

<u>Proof.</u> Let σ act in a vector space U. Let $\rho = \mathrm{Ind}_P^G \sigma$ and let ρ act in $\mathcal{H}$, the

space of all smooth functions $\beta : G \to U$ which satisfy $\beta(pg) = \sigma(p)\beta(g)$ ($p \in P$, $g \in G$). Let $\Delta\beta = \beta(1)$. Let $\mathcal{H}_0$ be a G-submodule of $\mathcal{H}$ and assume that $\mathcal{H}_0$ transforms as π. There is an element $\beta_0 \in \mathcal{H}_0$ such that $\beta_0(1) \neq 0$. Let $u' \in U'$ satisfy $\langle u', \beta_0(1)\rangle \neq 0$. Then $u' \circ \Delta = h_0'$ is a nonzero ${}^t\pi(N)$-invariant linear functional on $\mathcal{H}_0$, so π is not supercuspidal.

Remark. For a sharper result, cf., §5.4.1.

§2.3. Jacquet's Quotient Theorem.

Let (P, A) be a p-pair of G with $P = MN$ its Levi decomposition. Let π be a smooth representation of G on a vector space V. Clearly, the subspace $V(P)$ (cf., §2.2) is both stable and smooth under $\pi|P$, so there is a smooth representation σ of P on $V/V(P)$ (Lemma 1.5.1). Since $\sigma|N = 1$, we may regard σ as a representation of M. One of the major goals of this section is Corollary 2.3.6, which asserts that, if V is an admissible G-module, then $V/V(P)$ is an admissible M-module.

Let τ be a smooth representation of P which is trivial on N. Set $\rho = \mathrm{Ind}_P^G \tau$. Let τ act in a vector space U and ρ in $\mathcal{H}$, so that $\mathcal{H}$ is the space of all smooth U-valued functions β on G which satisfy $\beta(pg) = \tau(p)\beta(g)$ ($p \in P$, $g \in G$).

Define $\Delta : \mathcal{H} \to U$ by setting $\Delta\beta = \beta(1)$. Jacquet's main theorem is the following:

Theorem 2.3.1. Let $\mathcal{H}_0$ be a nonzero admissible G-submodule of $\mathcal{H}$. Then $\Delta(\mathcal{H}_0)$ is a nonzero admissible M-submodule of U.

The next three lemmas prove Jacquet's theorem. Consequences of Jacquet's theorem will occur as corollaries at the end of the section.

For the remainder of the section fix a minimal p-pair (P_0, A_0) such that $(P, A) \succeq (P_0, A_0)$ and let $\{K_j\}_{j=1}^{\infty}$ be a sequence of compact open subgroups of G which satisfy the conditions of Theorem 2.1.1.

Lemma 2.3.2.

(1) Δ intertwines $\rho|P$ with τ.

(2) If $\mathcal{H}_0$ is a nonzero G-submodule of $\mathcal{H}$, then $\Delta(\mathcal{H}_0) \neq (0)$.

(3) $\mathcal{H}(P) \subset \operatorname{Ker} \Delta$.

(4) Δ is surjective.

Proof.

(1) $\Delta(\rho(p)\beta) = \beta(p) = \tau(p)\beta(1) = \tau(p)\Delta(\beta)$ $(\beta \in K,\ p \in P)$.

(2) If $\beta(x) \neq 0$, then $\Delta(\rho(x)\beta) = \beta(x) \neq 0$.

(3) Let $\beta \in \mathcal{H}(P)$. Let $\int_{N(\beta)} \rho(n)dn\beta = 0$. Then

$$\mu(N(\beta)\ )\cdot\Delta(\beta) = \int_{N(\beta)} \tau(n)dn \cdot \Delta(\beta) = \Delta(\int_{N(\beta)} \rho(n)dn\beta) = 0.$$

(4) Let F be a compact open subgroup of M which fixes $u \in U$. Take K_j as in Theorem 2.1.1 such that $M_j \subset F$. For $x = pk$ with $p \in P$ and $k \in K_j$ define $\beta_u(x) = \tau(p)u$; for $x \notin PK_j$ set $\beta_u(x) = 0$. Then β_u is an element of $\mathcal{H}$ and $\Delta(\beta_u) = u$.

Now let $\mathcal{H}_0$ be a nonzero admissible submodule of $\mathcal{H}$. Then $\Delta(\mathcal{H}_0) = U_0$ is a nonzero M-submodule of U and, to prove Jacquet's theorem, it is sufficient to establish that U_0 is admissible.

Actually, we shall prove a more precise statement. Given an integer $j > 0$, we set $\mathcal{H}_0(j) = \{\beta \in \mathcal{H}_0 | \rho(k)\beta = \beta \ (k \in K_j)\}$ and $U_0(j) = \{u \in U_0 | \tau(m)u = u \ (m \in M_j)\}$. Clearly, $\Delta\mathcal{H}_0(j) \subset U_0(j)$. In the next two lemmas we shall prove that $\Delta\mathcal{H}_0(j) = U_0(j)$. Since $\Delta\mathcal{H}_0(j)$ is finite-dimensional, this will imply the admissibility of U_0.

For $t \geq 1$ let $A^+(t) = \{a \in A | \ |\alpha(a)| \geq t \ (\alpha \in \Sigma(P, A))\}$.

Lemma 2.3.3. For any $j > 0$ the space $\Delta\mathcal{H}_0(j)$ is stable under $\tau | A$.

Proof. Since $\Delta\mathcal{H}_0(j)$ has finite dimension, the set $S = \{a \in A | \tau(a)\Delta\mathcal{H}_0(j) \subset \Delta\mathcal{H}_0(j)\}$ is a group. Since A is the group generated by $A^+(t)$ for any $t > 0$ it is sufficient to show that there exists $t > 0$ such that $a^{-1} \in S$ for all $a \in A^+(t)$.

Choose t such that $a\bar{N}_j a^{-1} \subset \bar{N}_j$ provided that $a \in A^+(t)$. For $a \in A^+(t)$ and $\beta \in \mathcal{H}_0(j)$ set $\beta_{(a, j)} = \int_{K_j} \rho(ka^{-1})dk \ \beta$. Then $\beta_{(a, j)} \in \mathcal{H}_0(j)$ and it is sufficient to show that $\Delta\beta_{(a, j)} = \tau(a^{-1})\Delta\beta$ (assuming normalized measures). For this we have

$$\beta_{(a, j)}(1) = \int_{N_j M_j \bar{N}_j} \rho(nm\bar{n}a^{-1})dndmd\bar{n}\beta(1)$$

$$= \int_{\bar{N}_j} d\bar{n}\beta(\bar{n}a^{-1})$$

$$= \beta(a^{-1}), \text{ since } \bar{N}_j^a \subset \bar{N}_j.$$

Lemma 2.3.4. $U_0(j) \subset \Delta\mathcal{H}_0(j)$.

Proof. We must show that, if $u_0 \in U_0(j)$, then there exists $\beta_0 \in \mathcal{H}_0(j)$ such that $\Delta\beta_0 = u_0$. Let $u_0 = \Delta\beta \in U_0(j)$. Then, if $\phi = E_{M_j} \in C_c(M /\!/ M_j)$, we have

$$u_0 = \int_{M_j} \phi(m)\tau(m)dmu_0 = \Delta(\int_{M_j} \phi(m)\rho(m)dm \ \beta) = \int_{M_j} \phi(m)\beta(m)dm.$$

To find β_0 we proceed as follows. Assume that $\beta \in \mathcal{H}_0(j')$ and choose $t > 0$ so that, if $a \in A^+(t)$, then $\bar{N}_j^a \subset \bar{N}_{j'}$. Fix $a \in A^+(t)$. Set $\phi_a(m) = \phi(ma)$ $(m \in M)$ and $\phi'_a(x) = \phi_a(m)$ if $x = km \in K_jM$, 0 if $x \notin K_jM$. Then $\phi'_a \in C_c(K_j\backslash G)$, so $\rho(\phi'_a)\beta \in \mathcal{H}_0(j)$. Set $\beta_1 = \rho(\phi'_a)\beta$. We shall show that $\Delta\beta_1 = \tau(a)u_0 \in \Delta\mathcal{H}_0(j)$. It then follows from Lemma 2.3.3 that we may choose $\beta_0 \in \Delta^{-1}(u_0)$.

Concluding the proof, we have (assuming normalized measures)

$$\begin{aligned}(\rho(\phi'_a)\beta)(1) &= \int_G \phi'_a(x)\beta(x)dx \\ &= \int_{K_j} \beta(xa^{-1})dx \\ &= \tau(a)\int_{\bar{N}_j} \beta(\bar{n}^a)d\bar{n} \\ &= \tau(a)\beta(1),\end{aligned}$$

since $\bar{N}_j^a \subset \bar{N}_{j'}$.

This proves Theorem 2.3.1.

Corollary 2.3.5. If ρ is admissible, then so is τ.

Proof. Use the surjectivity of Δ (Lemma 2.3.2(4)).

Corollary 2.3.6. If V is an admissible G-module, then $V/V(P)$ is an admissible M-module.

Proof. Let $v \mapsto \bar{v}$ denote the canonical surjection of V on $V/V(P)$. Let π be the representation of G in V and σ the representation of M (or P) in $V/V(P)$. Let $\rho' = \mathrm{Ind}_P^G\sigma$ and let $\mathcal{H}'$ be the representation space for ρ', so $\mathcal{H}' = \{\beta \in C^\infty(G : V/V(P)) \mid \beta(pg) = \sigma(p)\beta(g) \quad (p \in P,\ g \in G)\}$.

Define $T : v \mapsto \beta_v$ from V to $\mathcal{H}'$ by setting $\beta_v(x) = (\pi(x)v)^-$ ($v \in V$, $x \in G$). The reader may check that T intertwines π and ρ'. Since $\Delta\beta_{v_0} = \bar{v}_0 \in V/V(P)$, we see that $\Delta | T(V)$ is surjective. Thus, we may set $\mathcal{H}_0 = TV \subset \mathcal{H}'$ and $U_0 = \Delta\mathcal{H}_0 = V/V(P)$ and use Theorem 2.3.1 to conclude that $V/V(P)$ is an admissible M-module.

§2.4. Dual Exponents, the Subrepresentation Theorem, and $\mathcal{E}'(G)$.

The main result of this section is essentially due to Jacquet in the case of $GL(n, \Omega)$; it depends upon Corollary 2.3.6. In terms of "dual exponents" we shall define the subset $\mathcal{E}'(G) \subset \mathcal{E}(G)$; the Bruhat theory (cf., §2.5) applies to $\mathcal{E}'(G)$.

For this section let π be an irreducible admissible representation of G acting in a vector space V. If $P = MN$ is a parabolic subgroup of G and σ is a representation of M, then we frequently regard σ as a representation of P.

Let (P, A) $(P = MN)$ be a p-pair of G and let $(\bar{P}, A)$ $(\bar{P} = M\bar{N})$ be the opposite p-pair. Suppose that there exists a nonzero functional $\varphi \in V'$ and a quasi-character $\chi \in \mathcal{X}(A)$ such that ${}^t\pi(\bar{n})\varphi = \varphi$ for all $\bar{n} \in \bar{N}$ and ${}^t\pi(a)\varphi = \delta_{\bar{P}}^{\frac{1}{2}}(a)\chi(a)\varphi$ for all $a \in A$. In this case we call χ a <u>dual exponent</u> of π with respect to (P, A). We write $\mathcal{Y}_\pi(P, A)$ for the set of all dual exponents of π with respect to (P, A). In the first two sections of Chapter 3, we shall define a notion of exponent with respect to (P, A) and prove that "exponents" and "dual exponents" are the same thing.

Since $G = \bar{P}K$ for some open compact subgroup K of G (§0.6), V and, hence, $V/V(\bar{N})$ are $\bar{P}$-modules of finite type. According to Corollary 2.3.6, the quotient module $V/V(\overline{N})$ is an admissible M-module. It follows from Lemma 1.5.4 and the fact that $V/V(\overline{N})$ is finitely generated that $\mathcal{Y}_\pi(P, A)$ is a finite set.

Let $\chi \in \mathcal{Y}_\pi(P, A)$. Let $V'_\chi = \{\varphi \in V' \mid {}^t\pi(\overline{n})\varphi = \varphi \quad (\overline{n} \in \overline{N})$ and $({}^t\pi(a) - \delta_{\overline{P}}^{\frac{1}{2}}(a)\chi(a))^{n(\varphi)}\varphi = 0 \quad (a \in A)\}$. The integer $d = d(\chi) = \max_{\varphi \in V'_\chi} n(\varphi)$ is called the multiplicity of χ in $\mathcal{Y}_\pi(P, A)$. Since $V/V(\overline{N})$ is an admissible M-module of finite type, it follows easily that $d(\chi) < \infty$ for each element $\chi \in \mathcal{Y}_\pi(P, A)$.

Clearly, we may write $(V/V(\overline{N}))' = \oplus V'_\chi$ and $V/V(\overline{N}) = \oplus \overline{V}_\chi$, both with direct sums over $\chi \in \mathcal{Y}_\pi(P, A)$ -- define $\overline{V}_\chi$ to be the quotient module whose dual is V'_χ.

A p-pair (P, A) of G is called π-minimal if $\mathcal{Y}_\pi(P, A) \neq \phi$ and $\mathcal{Y}_\pi(P', A') = \phi$ for every p-pair $(P', A') \lneqq (P, A)$. Since, in particular, $\mathcal{Y}_\pi(G, Z) \neq \phi$, there always exist π-minimal p-pairs.

An element $\chi \in \mathcal{Y}_\pi(P, A)$ ((P, A) any p-pair of G) is called a critical exponent of π or a π-critical exponent if: For every $(P', A') \lneqq (P, A)$ and $\chi' \in \mathcal{Y}_\pi(P', A')$ such that $\chi' | A = \chi$ we have $d(\chi') < d(\chi)$. A p-pair (P, A) of G is called a π-critical p-pair if $\mathcal{Y}_\pi(P, A)$ contains a critical exponent. Obviously, if (P, A) is π-minimal, then (P, A) is π-critical too. The converse is true (Corollary 5.3.2.3; cf. also Lemma 3.3.3 for the case in which π is unitary).

Theorem 2.4.1. Let (P, A) $(P = MN)$ be a p-pair of G and $\chi \in \mathcal{Y}_\pi(P, A)$. Then there exists an irreducible admissible representation σ of M such that:

(1) $\{\chi\} = \mathcal{Y}_\sigma(M, A)$;

(2) $\pi \subset \mathrm{Ind}_{\overline{P}}^G(\delta_{\overline{P}}^{\frac{1}{2}}\sigma)$.

Proof. We consider only the case $(P, A) \neq (G, Z)$. Let $(\overline{V}_\chi, \overline{\pi})$ denote the $\overline{P}$-module quotient of V which is dual to the subspace $V'_\chi \subset V'$. Every element

$\bar{v} \in \bar{V}_\chi$ satisfies the relations $\bar{\pi}(\bar{n})\bar{v} = \bar{v}$ $(\bar{n} \in \bar{N})$ and $(\bar{\pi}(a) - \delta_{\bar{P}}^{\frac{1}{2}}(a)\chi(a))^{d(\chi)}\bar{v} = 0$ $(a \in A)$. Clearly, since $V/V(\bar{N})$ is finitely generated, $\bar{V}_\chi$ is finitely generated. Therefore, by Lemma 1.5.3, $\bar{V}_\chi$ has irreducible $\bar{P}$ quotient modules. Let $(\bar{\bar{V}}, \sigma)$ be an irreducible $\bar{P}$-module quotient of $\bar{V}_\chi$. Obviously, (1) $\{\chi\} = \mathcal{Y}_\sigma(M, A)$. To prove (2) use Corollary 1.7.11 and the canonical $\bar{P}$-module mapping $T : V \to \bar{\bar{V}}$.

Corollary 2.4.2. Let (P, A) be a π-critical p-pair of G and $\chi \in \mathcal{Y}_\pi(P, A)$ a critical exponent. Then there exists an irreducible admissible J-supercuspidal representation σ of M such that:

(1) $\{\chi\} = \mathcal{Y}_\sigma(M, A)$,

and (2) $\pi \subset \mathrm{Ind}_{\bar{P}}^{G}(\delta_{\bar{P}}^{\frac{1}{2}}\sigma)$.

Proof. Let $d = d(\chi)$ be the multiplicity of χ. Let $\bar{V}_\chi$ be as in Lemma 2.4.1 relative to χ, also $\bar{\pi}$ as before. Set $\bar{V}_{j,\chi} = \{\bar{v} \in \bar{V}_\chi \mid (\bar{\pi}(a) - \delta_{\bar{P}}^{\frac{1}{2}}(a)\chi(a))^j\bar{v} = 0$ for all $a \in A\}$, $0 \leq j \leq d$. By hypothesis, $\bar{V}_\chi \neq \bar{V}_{d-1,\chi}$.

It is obviously enough to show that $\bar{V}_\chi/\bar{V}_{d-1,\chi}$ has J-supercuspidal quotients, then to argue as in Lemma 2.4.1. Clearly, for every $a \in A$, $(\bar{\pi}(a) - \delta_{\bar{P}}^{\frac{1}{2}}(a)\chi(a))^{d-1}$ induces an M-module morphism of $\bar{V}_\chi/\bar{V}_{d-1,\chi}$ into $\bar{V}_{1,\chi}$. Choose $a \in A$ such that the image W is not zero. It is enough to show that W is J-supercuspidal. Since W, being the image of a finitely generated M-module, is finitely generated, W has irreducible quotients; every quotient of W is also a quotient of $\bar{V}_\chi/\bar{V}_{d-1,\chi}$. By Lemma 2.2.1, if W is J-supercuspidal, then so is every quotient of W.

Let $(P, A) \succ (P', A')$ $(P' = M'N')$ and let $\chi' \in \mathcal{Y}_\pi(P', A')$ satisfy $\chi' | A = \chi$. Let $\bar{w} \in W$. Then $\bar{w} = (\bar{\pi}(a) - \delta_{\bar{P}}^{\frac{1}{2}}(a)\chi(a))^{d-1}\bar{v}$ for some $\bar{v} \in \bar{V}_\chi$. Therefore, the projection $\bar{\bar{w}}$ of $\bar{w}$ to $W(\bar{N}')$ is of the form $(\bar{\bar{\pi}}(a) - \delta_{\bar{P}'}^{\frac{1}{2}}(a)\chi'(a))^{d-1}\bar{\bar{v}}$, where $\bar{\bar{v}}$ is the projection of $\bar{v}$ on $\bar{V}_\chi(\bar{N}')$ and $\bar{\bar{\pi}}$ denotes the representation of M' on $\bar{V}_\chi(\bar{N}')$. Since χ is π-critical, $\bar{\bar{w}} = 0$. This implies the corollary.

In terms of dual exponents we define a distinguished subset $\mathcal{E}'(G)$ of the set $\mathcal{E}(G)$ of all irreducible unitary admissible representations of G. A unitary admissible class $C(\pi)$ will belong to $\mathcal{E}'(G)$ if, for every p-pair $(P, A) \neq (G, Z)$, the set $\mathcal{Y}_\pi(P, A) \cap \hat{A} = \phi$. Since π is J-supercuspidal if and only if $\mathcal{Y}_\pi(P, A) = \phi$ for every p-pair $(P, A) \neq (G, Z)$, it follows that ${}^0\mathcal{E}(G) \subset \mathcal{E}'(G)$. Theorem 4.4.4 implies that $\mathcal{E}_2(G) \subset \mathcal{E}'(G)$.

§2.5. Irreducibility and Intertwining Numbers for Certain Induced Representations.

Let (P_i, A_i) $(i = 1, 2)$ be p-pairs of G with $P_i = M_i N_i$ the corresponding Levi decompositions. Let σ_i be an irreducible admissible representation of M_i in a vector space V_i. Extend σ_i to P_i by setting $\sigma_i(mn) = \sigma_i(m)$ for all $m \in M_i$ and $n \in N_i$. Let $\delta_i = \delta_{P_i}$ be the modular function for P_i (§1.2). Form $\mathrm{Ind}_{P_i}^G(\delta_i^{\frac{1}{2}}\sigma_i) = \pi_i$, an admissible representation of G. In this section we shall apply the general results developed in §1.9 to the representations π_i. The main results of the section are Theorems 2.5.8 and 2.5.9.

Let $P_1 \times P_2$ operate on G by setting $y \cdot (p_1, p_2) = p_1^{-1} y p_2$ for $y \in G$ and $(p_1, p_2) \in P_1 \times P_2$. For any orbit $\mathcal{O}$ and $y \in \mathcal{O}$ let $P_1 \times P_2(y)$ denote the stabilizer of y in $P_1 \times P_2$. There is an Ω-isomorphism of algebraic varieties from $\underset{\sim}{P_1 \times P_2(y)} \backslash \underset{\sim}{P_1 \times P_2}$ onto $\underset{\sim}{\mathcal{O}}$, so the corresponding sets of Ω-points are homeomorphic in the p-adic topology.

We write d or $d(\mathcal{O})$ for the dimension of the algebraic variety $\underset{\sim}{\mathcal{O}}$. Let $\mathcal{O}(\nu)$ denote the union of all orbits of dimension no greater than $\nu \geq 0$; let $\mathcal{O}(-1)$ be the empty set. The set $\underset{\sim}{\mathcal{O}}(\nu)$ is a Zariski closed Ω-set, so the set $\mathcal{O}(\nu)$ of Ω-points is closed with respect to the p-adic topology.

Let A_0 be a maximal split torus of G, so $W(G/A_0) = W(A_0)$ is the Weyl group of G, a finite group. Each orbit of $P_1 \times P_2$ in G contains all the representatives for some nonempty subset of $W(A_0)$, so the number of orbits is finite-- not greater than $[W(A_0)]$.

Let $E = V_1 \otimes V_2$. Let $\mathcal{T}$ be the space of all E-distributions T^0 on G such that $T^0(\lambda(p_1)\rho(p_2)\alpha) = T^0(\delta_1(p_1)^{\frac{1}{2}}\delta_2(p_2)^{\frac{1}{2}}\sigma_1(p_1^{-1}) \otimes \sigma_2(p_2^{-1})\alpha)$ $((p_1, p_2) \in P_1 \times P_2)$ for all $\alpha \in C_c^{\infty}(G : E)$. If $\mathcal{O}$ is an orbit of dimension d, write $\mathcal{T}(\mathcal{O})$ for the vector space consisting of all $T^0 \in \mathcal{T}$ with support in $\mathcal{O} \cup \mathcal{O}(d-1)$ (a closed set in either the Zariski or p-adic topologies). Write $\mathcal{T}_\nu$ for the space consisting of those T^0 with support in $\mathcal{O}(\nu)$. Clearly, $\mathcal{T} = \sum_{\mathcal{O}} \mathcal{T}(\mathcal{O})$ and $\sum_{\mathcal{O}} \dim(\mathcal{T}(\mathcal{O})/\mathcal{T}_{d(0)-1} \geq \dim(\mathcal{T})$.

According to Theorem 1.9.4 there is an injection of $\mathcal{B}(\pi_1, \pi_2)$ into $\mathcal{T}$. According to Theorem 1.9.5 there is, for each orbit $\mathcal{O}$ and any $y \in \mathcal{O}$, an injection of $\mathcal{T}(\mathcal{O})/\mathcal{T}_{d(\mathcal{O})-1}$ into a subspace $E'(y) \subset E'$ which we shall redefine presently and study in detail. It will be seen that, when $A_1 = A_2^y$, there is an injection $E'(y) \hookrightarrow \mathcal{B}(\sigma_1, \sigma_2^y)$. Exploiting these composed injections, we shall prove Theorems 2.5.8 and 2.5.9.

Let (P_0, A_0) be a minimal p-pair of G. Without loss of generality assume that (P_1, A_1) and (P_2, A_2) are standard with respect to (P_0, A_0) and that $y \in W(A_0)$, so (P_2^y, A_2^y) is semi-standard. The theory of Borel and Tits ([2c]) implies that ${}^*P_1 = M_1 \cap P_2^y$ is a parabolic subgroup of M_1 and that ${}^*P_2 = M_2 \cap P_1^{y^{-1}}$ is a parabolic subgroup of M_2. For split components we may take, respectively, ${}^*A_1 = A_1 A_2^y$ and ${}^*A_2 = A_2 A_1^{y^{-1}}$. Write ${}^*P_i = {}^*M_i {}^*N_i$ for the Levi decomposition associated with the p-pair $({}^*P_i, {}^*A_i)$; it is clear that ${}^*M_1 = M_1 \cap M_2^y$,

${}^*N_1 = M_1 \cap N_2^y$, ${}^*M_2 = M_1^{y^{-1}} \cap M_2 = ({}^*M_1)^{y^{-1}}$ and ${}^*N_2 = M_2 \cap N_1^{y^{-1}}$. Note that ${}^*N_1 = \{1\}$ if and only if ${}^*A_1 = A_1$, if and only if $A_1 \supset A_2^y$. Similarly, ${}^*N_2 = \{1\}$ if and only if $A_2 \supset A_1^{y^{-1}}$, i.e., $A_2^y \supset A_1$. Thus, ${}^*N_1 = {}^*N_2 = \{1\}$ if and only if $A_1 = A_2^y$.

Let $E'(y)$ be the space of linear functionals φ on $E = V_1 \otimes V_2$ such that

$$\delta_1(p)^{\frac{1}{2}}\delta_2(p^{y^{-1}})^{\frac{1}{2}} < \varphi, \sigma_1(p)v_1 \otimes \sigma_2(p^{y^{-1}})v_2 > \; = \delta_{P_1 \cap P_2^y}(p) < \varphi, v_1 \otimes v_2 >$$

for all $p \in P_1 \cap P_2^y$ and $v_i \in V_i$ $(i = 1, 2)$. The reader can easily see that this is the same space as that defined in Theorem 1.9.5. Note that the above modular functions are products of absolute values of rational characters (§1.2.1), hence trivial on unipotent elements of $P_1 \cap P_2^y$. Furthermore, since $N_2 \subset \operatorname{Ker} \sigma_2$ and ${}^*n_1^{y^{-1}} \in N_2$, for any ${}^*n_1 \in {}^*N_1$ we have

$$< \varphi, \sigma_1({}^*n_1)v_1 \otimes \sigma_2({}^*n_1^{y^{-1}})v_2 > \; = \; < \varphi, \sigma_1({}^*n_1)v_1 \otimes v_2 >$$

$$= \; < \varphi, v_1 \otimes v_2 > \text{ for any } \varphi \in E'(y).$$

This implies that $\varphi = 0$ on $V_1({}^*P_1) \otimes V_2$. It follows that, if σ_1 is a J-supercuspidal representation and ${}^*N_1 \neq \{1\}$, then $\varphi = 0$. In any case φ vanishes identically on

$$V_1({}^*P_1) \otimes V_2 + V_1 \otimes V_2({}^*P_2) \subset E.$$

Summarizing, we have proved the first half of the following:

<u>Lemma</u> 2.5.1. If $\varphi \in E'(y)$, then φ vanishes on $V_1({}^*P_1) \otimes V_2 + V_1 \otimes V_2({}^*P_2)$.

Let $p = m \in {}^*M_1$, $v_1 \otimes v_2 \in E$, and $\varphi \in E'(y)$. Then

$$< \varphi, \sigma_1(m)v_1 \otimes \sigma_2(m^{y^{-1}})v_2 > \; = \delta_{*1}(m)^{\frac{1}{2}}\delta_{*2}(m^{y^{-1}})^{\frac{1}{2}} < \varphi, v_1 \otimes v_2 >$$

where δ_{*i} denotes the modular function for ${}^*P_i$ (i = 1 or 2).

To complete the proof of Lemma 2.5.1, we must prove the following:

Lemma 2.5.2. Let $m \in {}^*M_1 = M_1 \cap M_2^y$. Then $\delta_{P_1 \cap P_2^y}(m) = \delta_{*1}(m)^{\frac{1}{2}}\delta_{*2}(m^{y^{-1}})^{\frac{1}{2}}\delta_1(m)^{\frac{1}{2}}\delta_2(m^{y^{-1}})^{\frac{1}{2}}$, where δ_{*i} denotes the modular function for ${}^*P_i$ and δ_i the modular function for P_i (i = 1 or 2).

Proof. We use the fact (Borel-Tits [2c]) that there is a unique standard p-pair (P_i', A_i') such that $(P_i', A_i') \prec (P_i, A_i)$ (i = 1,2) and such that $A_i' = {}^*A_i$. It follows then that (P_i', A_i') has the Levi decomposition $P_i' = M_i'N_i'$, where $M_i' = {}^*M_i$ and $N_i' = {}^*N_iN_i$, also that $\delta_{P_i'}(m) = \delta_{*i}(m)\delta_i(m)$ for all $m \in M_i'$. To complete the proof of Lemma 2.5.2 we shall establish that, for $m \in M_1' = {}^*M_1 = M_1 \cap M_2^y$,

$$\delta_{P_1 \cap P_2^y}(m) = \delta_{P_1'}(m)^{\frac{1}{2}}\delta_{P_2'}(m^{y^{-1}})^{\frac{1}{2}}.$$

Although $P_1 \cap P_2^y$ is not in general a parabolic subgroup, we still have the semidirect product decomposition $P_1 \cap P_2^y = M_1 \cap M_2^y \cdot N_1' \cap (N_2')^y$, in which the first factor is reductive and the normal subgroup is unipotent.

It suffices to take $m = a \in A_1' = {}^*A_1 = A_1A_2^y$, since a rational character of ${}^*M_1$ is completely determined by its restriction to the split component ${}^*A_1$. Let us regard $\Sigma(P_0, A_0)$ as the set of positive roots in $\Sigma(G, A_0)$.

For $a \in A_1' = (A_2')^y$ we may write

$$\delta_{P_1 \cap P_2^y}(a) = \prod_{\alpha} |\alpha(a)|,$$

where α ranges over the root characters of A_1', with multiplicities, acting on the

Lie algebra of $N_1' \cap (N_2')^y$. On the other hand, $\delta_{P_1'}(a) = \prod_{\alpha' \in \Sigma_1'} |\alpha'(a)|$ and $\delta_{P_2'}(a^{y^{-1}}) = \prod_{\alpha' \in \Sigma_2'} |\alpha'(a^{y^{-1}})|$, where Σ_i' denotes the A_i'-roots, with multiplicities, occurring in the Lie algebra of N_i'. One sees that $\Sigma_1' \cap (\Sigma_2')^y$ is the set of roots of A_1' with respect to the Lie algebra of $N_1' \cap (N_2')^y$, also that, for $\alpha' \in \Sigma_2'$, we have $y\alpha' \in (\Sigma_2')^y \cap \Sigma_1'$ if and only if $y\alpha' > 0$. In this case, $\alpha'(a^{y^{-1}}) = y\alpha'(a) = \alpha(a)$ for some $\alpha \in \Sigma_1'$; otherwise $y\alpha'(a) = \alpha^{-1}(a)$ for some $\alpha \in \Sigma_1'$. That is, the roots $\alpha_1 \in \Sigma_1'$ and $y\alpha_2 \in (\Sigma_2')^y$ which occur both in the factorization of $\delta_{P_1'}$ and $\delta_{(P_2')^y}$ are exactly those which belong to $\Sigma_1' \cap (\Sigma_2')^y$. For the other roots $\alpha_2 \in \Sigma_2'$, we have $y\alpha_2 < 0$ and hence $\alpha_2(a^{y^{-1}}) = y\alpha_2(a) = \alpha(a^{-1})$ for some root $\alpha \in \Sigma_1'$ (since A_1' and A_2' are conjugate tori). This proves that $\delta_{P_1 \cap P_2^y}(m) = \delta_{P_1'}(m)^{\frac{1}{2}} \delta_{P_2'}(m^{y^{-1}})^{\frac{1}{2}}$ for $m \in {}^*M_1$ and completes the proof of Lemma 2.5.2.

As consequences of Lemma 2.5.1 we have the following corollaries. Let η_i denote the central character of σ_i, so $\sigma_i(ma) = \eta_i(a)\sigma_i(m)$ $(a \in A,\ m \in M_i)$, $i = 1$ or 2. See §2.2 and §2.4, respectively, for the notations ${}^0\mathcal{E}_{\mathbb{C}}$ and $\mathcal{E}'$.

<u>Corollary</u> 2.5.3. Assume that $A_2^y \not\subset A_1$. If $\sigma_1 \in \omega_1 \in {}^0\mathcal{E}_{\mathbb{C}}(M_1)$ or if $\sigma_1 \in \omega_1 \in \mathcal{E}'(M_1)$ and $\sigma_2 \in \omega_2 \in \mathcal{E}(M_2)$, then $E'(y) = (0)$.

<u>Proof</u>. We know that ${}^*N_1 = \{1\}$ if and only if $A_1 \supset A_2^y$. Therefore, under the above hypothesis, if $\sigma_1 \in \omega_1 \in {}^0\mathcal{E}_{\mathbb{C}}(M_1)$, then $V_1({}^*P_1) = V_1$, so $\varphi = 0$ by Lemma 2.5.1.

Now let $\sigma_1 \in \omega_1 \in \mathcal{E}'(M_1)$ and $\sigma_2 \in \omega_2 \in \mathcal{E}(M_2)$ and recall that ${}^*A_1 = A_1 A_2^y$. Using Lemma 2.5.1, we have, for $(v_i, a_i) \in V_i \times A_i$ $(i = 1, 2)$ and $\varphi \in E'(y)$,

$$< \varphi, \sigma_1(a_1a_2^y)v_1 \otimes v_2 > = \eta_1(a_1) < \varphi, \sigma_1(a_2^y)v_1 \otimes v_2 >$$
$$= \eta_1(a_1)\delta_{*1}(a_2^y)^{\frac{1}{2}}\delta_{*2}(a_2)^{\frac{1}{2}} < \varphi, v_1 \otimes \sigma_2(a_2^{-1})v_2 >$$
$$= \frac{\eta_1(a_1)}{\eta_2(a_2)}\delta_{*1}(a_2^y)^{\frac{1}{2}} < \varphi, v_1 \otimes v_2 >$$

(A_2 central in M_2 implies $\delta_{*2}(a_2) = 1$). Let $\varphi_1 \in (V_1/V_1({}^*N_1))'$ be defined by the relation $< \varphi_1, v_1 > = < \varphi, v_1 \otimes v_2 >$ with v_2 fixed. Then $< \varphi_1, \sigma_1(a_1a_2^y)v_1 > =$ $\delta_{*1}(a_2^y)^{\frac{1}{2}}\frac{\eta_1(a_1)}{\eta_2(a_2)} < \varphi_1, v_1 >$. If $\varphi_1 \neq 0$, it follows that there exists $\chi \in \mathcal{Y}_{\sigma_1}({}^*\bar{P}_1, {}^*A_1)$ (cf., §2.4 for this notation) such that $\chi(a_1a_2^y) = \eta_1(a_1)\eta_2(a_2^{-1})$ for all $a_1 \in A_1$ and $a_2 \in A_2$. But $A_2^y \not\subset A_1$, so $({}^*P_1, {}^*A_1) \neq (M_1, A_1)$. Since $\omega_1 \in \mathcal{E}'(M_1)$, χ is not unitary which contradicts the fact that both η_1 and η_2 must be unitary (they are central characters of irreducible unitary representations).

<u>Corollary</u> 2.5.4. If $A_1 \supset A_2^y$, then

$$< \varphi, \sigma_1(m)v_1 \otimes \sigma_2(m^{y^{-1}})v_2 > = \delta_{*2}(m^{y^{-1}}) < \varphi, v_1 \otimes v_2 >$$

for all $\varphi \in E'(y)$, $m \in M_1 \cap M_2^y$, and $v_1 \otimes v_2 \in E$.

<u>Proof.</u> In this case ${}^*N_1 = \{1\}$, so ${}^*P_1$ is reductive and $\delta_{*1} = 1$.

<u>Corollary</u> 2.5.5. Assume either that $\sigma_i \in \omega_i \in {}^0\mathcal{E}_{\mathbb{C}}(M_i)$ for $i = 1$ and 2 or that $\sigma_i \in \omega_i \in \mathcal{E}'(M_i)$ for $i = 1$ and 2. Then, $E'(y) \neq (0)$ implies $A_1 = A_2^y$.

<u>Proof.</u> Corollary 2.5.3.

<u>Corollary</u> 2.5.6. If $A_1 = A_2^y$, then $< \varphi, \sigma_1(m)v_1 \otimes \sigma_2(m^{y^{-1}})v_2 > = < \varphi, v_1 \otimes v_2 >$ for all $\varphi \in E'(y)$, $m \in M_1$, and $v_1 \otimes v_2 \in E$.

<u>Proof.</u> Corollary 2.5.4.

Lemma 2.5.7. Assume that either $\sigma_i \in \omega_i \in {}^0\mathcal{E}_{\mathbb{C}}(M_i)$ for $i = 1$ and 2 or that $\sigma_i \in \omega_i \in \mathcal{E}'(M_i)$ for $i = 1$ and 2. Let $\mathcal{O} \subset G$ be an orbit of dimension d. Then $\dim \mathcal{T}(\mathcal{O})/\mathcal{T}_{d-1} = 0$ unless there exists $y \in \mathcal{O}$ such that $A_1 = A_2^y$. If $A_1 = A_2^y$ for some $y \in \mathcal{O}$, then $\dim(\mathcal{T}(\mathcal{O})/\mathcal{T}_{d-1}) \leq I(\sigma_1, \sigma_2^y)$, where $\sigma_2^y(m^y) = \sigma_2(m)$ $(m \in M_2)$ (of course, $M_1 = M_2^y = yM_2y^{-1}$).

Proof. Theorem 1.9.5 implies the existence of an injection $\mathcal{T}(\mathcal{O})/\mathcal{T}_{d-1} \hookrightarrow E'(y)$ for any $y \in \mathcal{O}$. Corollary 2.5.5 implies that $E'(y) = 0$ unless $A_1 = A_2^y$ for some $y \in \mathcal{O} \cap W(A_0)$, which proves the first assertion of the lemma. Corollary 2.5.6 implies that, if $A_1 = A_2^y$ with $y \in \mathcal{O}$ and $\varphi \in E'(y)$, we may map φ to a form $B_\varphi \in \mathcal{B}(\sigma_1, \sigma_2^y)$. Set $B_\varphi(v_1, v_2) = \langle \varphi, v_1 \otimes v_2 \rangle$ and check that $B_\varphi(v_1, v_2) = B_\varphi(\sigma_1(m)v_1, \sigma_2(m^{y^{-1}})v_2)$ for any $v_1 \in V_i$ and $m \in M_1$. The mapping $\varphi \mapsto B_\varphi$ is clearly a linear injection, and this implies the final assertion of the lemma.

Recall that $W(A_1|A_2)$ denotes the set of mappings $s : A_2 \to A_1$ which extend to inner automorphisms of G.

Theorem 2.5.8. Assume that either $\sigma_i \in \omega_i \in {}^0\mathcal{E}_{\mathbb{C}}(M_i)$ for $i = 1$ and 2 or that $\sigma_i \in \omega_i \in \mathcal{E}'(M_i)$ for $i = 1$ and 2 (cf., §§2.2 and 2.4 for ${}^0\mathcal{E}_{\mathbb{C}}$ and $\mathcal{E}'$). If A_1 and A_2 are not conjugate, then $I(\pi_1, \pi_2) = 0$. Assume A_1 and A_2 are conjugate. For each $s \in W = W(A_1|A_2)$ choose a representative $y_s \in G$. Then

$$I(\pi_1, \pi_2) \leq \sum_{s \in W} I(\sigma_1, \sigma_2^{y_s}).$$

Proof. For each $s \in W(A_1|A_2)$ set $\mathcal{O}_s = P_1 y_s P_2$. Then, using Theorem 1.9.4 for the first inequality, Lemma 2.5.7 for the second, and Theorem 1.9.5 for the last, we have

$$I(\pi_1, \pi_2) \leq \dim \mathcal{V} \leq \sum_{s \in W} \dim(\mathcal{V}(\mathcal{O}_s)/\mathcal{V}_{d(\mathcal{O}_s)-1}) \leq \sum_{s \in W} I(\sigma_1, \sigma_2^{y_s}).$$

Let (P, A) be a p-pair with $P = MN$ its Levi decomposition. Let ω be a class of irreducible admissible representations of G. Let $\sigma \in \omega$ and let y_s in the normalizer of A represent $s \in W(G/A) = W(A)$. Set $\sigma^y(m) = \sigma(y^{-1}my)$. Then σ^y depends up to equivalence only on s. We write ω^s for the class of σ^y. Let $W(\omega)$ denote the subgroup of $W(A)$ consisting of those $s \in W(A)$ such that $\omega^s = \omega$. Then $W(\omega)$ is called the ramification group of ω and ω is said to be unramified if $[W(\omega)] = 1$.

Recall that ${}^0\mathcal{E}(M) \subset \mathcal{E}'(M)$.

Theorem 2.5.9. Let $\sigma \in \omega \in \mathcal{E}'(M)$ and set $\pi = \mathrm{Ind}_P^G(\delta_P^{\frac{1}{2}}\sigma)$. Then $I(\tilde{\pi}, \pi) \leq [W(\omega)]$. In particular, if ω is unramified, then π is irreducible.

Proof. That $I(\tilde{\pi}, \pi) \leq [W(\omega)]$ follows immediately from Theorem 2.5.8. Lemma 1.6.2(1) and the definition of $\tilde{\pi}$ imply that $I(\tilde{\pi}, \pi) \geq 1$. Thus, if $[W(\omega)] = 1$, then $I(\tilde{\pi}, \pi) = 1$, in which case Lemma 1.6.2(3) implies that π is irreducible.

§2.6. The Constant Term.

The purpose of this section is to introduce the constant term, to prove its existence and uniqueness.

Fix a minimal p-pair (P_0, A_0) and a sequence $\{K_j\}_{j=1}^{\infty}$ of compact open subgroups as provided by Theorem 2.1.1.

Let (P, A) be a standard p-pair of G with $P = MN$ its Levi decomposition. Let $\Sigma^0 = \Sigma^0(P, A)$ be the set of simple roots of P with respect to A. For $\alpha \in \Sigma^0$ let α denote the corresponding root character of A. Put $\gamma_P(a) =$

$\inf_{\alpha\in\Sigma^0} |\alpha(a)|$. For $t > 0$ define $A(P,t) = A^+(P,t) = A^+(t) = \{a \in A \mid \gamma_P(a) \geq t\}$. Note that $A^+(t) \subset A^+(t_0)$ if $t \geq t_0$. If $t \geq 1$ and $t_0 > 0$, then $A^+(t)A^+(t_0) \subset A^+(tt_0) \subset A^+(t_0)$. Thus, for any $t_0 \geq 1$, $A^+(t_0)$ is a semigroup (without unit). Sometimes we shall write A^+ for $A^+(1)$.

More generally, let (P', A') $(P' = M'N')$ be a standard p-pair such that $(P', A') \prec (P, A)$. Write Σ'^0 for the set of simple roots of (P', A'). Then $(P, A) = (P', A')_F$ for some $F \subset \Sigma'^0$. Given $t \geq 1$ let $A'(F,t)$ denote the set of all elements $a' \in A'$ such that $|\alpha(a')| \geq 1$ $(\alpha \in F)$ and $|\alpha(a')| \geq t$ $(\alpha \in \Sigma'^0 - F)$. Note that $A^+(t) = A'(F,t) \cap A$; in fact, if $a \in A$, then $\alpha(a) = 1$ for all $\alpha \in F$.

Let S be a subset of G such that $SA^+(t_0) \subset S$ for some $t_0 > 0$. Let U be a complex vector space. Let $f : S \to U$ be a function. We write $f \underset{P}{\sim} 0$ ("f is negligible along P.") if the following condition holds: For every compact subset ω of G there exists $t = t(\omega) \geq t_0$ such that $f(xa) = 0$ provided $x \in \omega \cap S$ and $a \in A^+(t)$. Given $f_i : S \to U$ $(i = 1, 2)$, we write $f_1 \underset{P}{\sim} f_2$ if $f_1 - f_2 \underset{P}{\sim} 0$. Sometimes, given a complex number L, we write $f(xa) \to L$, $a \underset{P}{\to} \infty$, if: For every $\varepsilon > 0$ there exists t_ε such that $|f(xa) - L| < \varepsilon$ for all $a \in A^+(t_\varepsilon)$. If $f \underset{P}{\sim} L$ with respect to the set $S = xA^+ \subset G$, then $f(xa) \to L$, $a \underset{P}{\to} \infty$. We also write $\phi = \lim_{a \underset{P}{\to} \infty} f$ exists on S if $f(xa) \to \phi(x)$, $a \underset{P}{\to} \infty$ for some function $\phi : S \to U$. In a given context the set S is usually implicitly defined.

Let U denote, as before, a complex vector space. For $f \in \mathcal{A}(G : U)$ define $f_{(P)} \in C^\infty(M : U)$ by setting $f_{(P)}(m) = \delta_P^{\frac{1}{2}}(m) f(m)$ $(m \in M)$.

<u>Theorem</u> 2.6.1. For any $f \in \mathcal{A}(G : U)$ there exists a unique element $f_P \in \mathcal{A}(M : U)$

such that $f_{(P)} \underset{P}{\approx} f_P$ on M.

More precisely, given a compact subset ω of M, we may choose $t = t(\omega, f) \geq 1$ such that $f_{(P)}(m) = f_P(m)$ for all $m \in \omega A'(F,t)\omega$.

The function f_P of Theorem 2.6.1 is called the <u>constant term of</u> f <u>along</u> P.

The following lemma implies the uniqueness of the constant term (assuming existence). The reader will find another proof of this lemma in [7c], Lemma 10, p. 32.

<u>Lemma</u> 2.6.2. Let U be a complex vector space. Let $\phi : A \to U$ be an A-finite function. Suppose there exists $t > 0$ such that $\phi(a) = 0$ for all $a \in A^+(t)$. Then ϕ is identically zero.

<u>Proof.</u> For $a \in A$ and fixed $a_0 \in A^+(t_0)$ $(t_0 > 1)$ define $\mu_a : \mathbb{Z} \to A$ by setting $\mu_a(n) = aa_0^n$. Define an operation of the group $\mathbb{Z}$ on $\phi \circ \mu_a$ by setting $m(\phi \circ \mu_a)(n) = (\phi \circ \mu_a)(n+m)$. Then $\phi \circ \mu_a$ is a $\mathbb{Z}$-finite function on $\mathbb{Z}$ which vanishes for n sufficiently large. Therefore, by Lemma 1.3.1, $\phi \circ \mu_a = 0$. This implies that $\phi(a) = 0$. Since a is arbitrary, we may conclude that $\phi = 0$.

<u>Corollary</u> 2.6.3. If $f \in \mathcal{A}(G : U)$, then there is at most one element $g \in \mathcal{A}(M : U)$ such that $f_{(P)} \underset{P}{\approx} g$.

We prove the following ostensibly more general statement.

<u>Corollary</u> 2.6.4. Let $g \in \mathcal{A}(M : U)$ and assume $g \underset{P}{\approx} 0$. Then $g = 0$.

<u>Proof.</u> Corollary 1.10.2 implies that g is A-finite. Lemma 2.6.2 implies that, if $g \underset{P}{\approx} 0$ and g is A-finite, then $g = 0$.

We proceed to establish the existence of f_P, to prove Theorem 2.6.1, via a series of lemmas.

Recall that, associated to a p-pair (P, A) of G, there is an "opposite" p-pair $(\overline{P}, A)$. We write $\overline{P} = M\overline{N}$ for its Levi decomposition. We have $\Sigma(G, A) = \Sigma(P, A) \cup \Sigma(\overline{P}, A)$, the first set being called the set of positive roots and the second the set of negative roots in $\Sigma(G, A)$.

For completeness we include the proof of the following lemma (cf., [7c], Lemma 11, p. 33).

<u>Lemma</u> 2.6.5. Let H be an open compact subgroup of A. Then A^+/H is a finitely generated semigroup.

<u>Proof.</u> Let $^0A = \bigcap_{\chi \in X(A)} \ker|\chi|$. Then, clearly, ^0A is the maximal compact subgroup of A. Moreover, ^0A is open, $^0A \subset A^+$, and $^0A/H$ *is finite, so it suffices* to prove our lemma for $A^+/{}^0A$.

Recall that $A/{}^0A$ is a free $\mathbb{Z}$-module of finite type. Therefore, letting $A_1 = \bigcap \ker|\alpha|$, $\alpha \in \Sigma^0(P, A)$, we know that $A_1/{}^0A$ is a finitely generated subgroup of $A^+/{}^0A$. It is sufficient to show that A^+/A_1 is a finitely generated semigroup.

Observe that to each simple root $\alpha \in \Sigma^0(P, A)$ there corresponds an element $a_\alpha \in A^+$ such that $|\alpha(a_\alpha)| > 1$ and $|\alpha'(a_\alpha)| = 1$, if $\alpha \neq \alpha' \in \Sigma^0(P, A)$. This follows from the correspondence between maximal p-pairs and elements of $\Sigma^0(P_0, A_0)$. Given $a \in A^+$, we may write $a = a_0 \prod a_\alpha^{n(a,\alpha)}$, where $n(a, \alpha) \geq 0$, the product is over all $\alpha \in \Sigma^0(P, A)$, and a_0 satisfies $1 \leq |\alpha(a_0)| < |\alpha(a_\alpha)|$ for all $\alpha \in \Sigma^0(P, A)$. Thus, A^+/H is generated by a finite set S which is a union of sets of representatives chosen in A^+ for the following sets:

(1) a finite set of generators for ${}^0A/H$;

(2) a finite set of generators for $A_1/{}^0A$;

(3) the set $\{a_\alpha \mid \alpha \in \Sigma^0(P,A)\}$; and

(4) a finite fundamental domain in A^+/A_1 for the set $\{a \in A \mid 1 \leq |\alpha(a)| < |\alpha(a_\alpha)|$ for all $\alpha \in \Sigma^0(P,A)\}$.

Fix $f \in \mathcal{A}(G : U)$ and let V denote the (admissible) G-module generated by f under ρ. By Corollary 2.3.6 the M-module $\overline{V} = V/V(\overline{N})$ is admissible. Let $v \mapsto \overline{v}$ denote the $\overline{P}$-module homomorphism of V onto $\overline{V}$.

Since $\overline{f}$ is A-finite, we may choose $\Phi = \{f = \phi_1, \ldots, \phi_r\} \subset V^r$ and a smooth representation χ of A by $r \times r$ complex matrices such that, for each $a \in A$, $\rho(a)\Phi - \Phi\chi(a) \in V^r(\overline{N}) = V(\overline{N})^r$. Thus, for any $a \in A$, there exists an open compact subgroup $\overline{N}(\Phi, a) \subset \overline{N}$ such that $\int_{\overline{N}'} [\Phi(x\overline{n}a) - \Phi(x\overline{n})\chi(a)]d\overline{n} = 0$ for all $x \in G$, provided $\overline{N}'$ is a subgroup of $\overline{N}$ which contains $\overline{N}(\Phi, a)$.

We may assume that $\Phi \in C(G/\!/K_j : U^r)$.

Lemma 2.6.6. Let $a \in A$ and $m_0 \in M$. If $\overline{N}_j^{m_0^{-1}} \supset \overline{N}(\Phi, a)$, then $\Phi(m_0 a) = \Phi(m_0)\chi(a)$.

Proof. Normalize the Haar measure on $\overline{N}_j$. Then:

$$\begin{aligned}
\Phi(m_0 a) - \Phi(m_0)\chi(a) &= \int_{\overline{N}_j} [\Phi(\overline{n} m_0 a) - \Phi(\overline{n} m_0)\chi(a)]d\overline{n} \\
&= \int_{\overline{N}_j} [\Phi(m_0 \overline{n}^{m_0^{-1}} a) - \Phi(m_0 \overline{n}^{m_0^{-1}})\chi(a)]d\overline{n} \\
&= \int_{\overline{N}_j^{m_0^{-1}}} [\Phi(m_0 \overline{n} a) - \Phi(m_0 \overline{n})\chi(a)]d\overline{n}\, \delta_P(m_0) \\
&= 0.
\end{aligned}$$

Corollary 2.6.7. Given any fixed compact subset ω of M and any element $a \in A$, we may choose $t = t(\omega, \Phi, a)$ such that $\Phi(ma) = \Phi(m)\chi(a)$ for all $m \in \omega A'(F,t)\omega$.

Proof. Given any finite subset $\{m_1, \ldots, m_r\}$ of M we may choose t such that, for all $m \in \bigcup_{1 \leq i, j \leq r} m_i A'(F,t) m_j$, the condition of Lemma 2.6.6 is fulfilled. Since ω is compact and $\Phi \in C(G/\!/K_j : U^r)$, a choice of t relative to some finite subset $\{m_1, \ldots, m_r\}$ of ω obviously suffices to prove the corollary.

Lemma 2.6.8. Let S be a finite set of generators for $A^+/A \cap M_j$ as provided by Lemma 2.6.5. Fix a compact subset ω of M. Set $t = \max_{s \in S} t(\omega, \Phi, s)$. Then $\Phi(ma) = \Phi(m)\chi(a)$ for all $m \in \omega A'(F,t)\omega$ and $a \in A^+$.

Proof. Let $s_0 = 1 \in A$. Set $a = s_1 \ldots s_\ell$ with $s_i \in S$ $(i = 1, \ldots, \ell)$. Then, for any $m \in \omega A'(F,t)\omega$, we have

$$\Phi(ma) - \Phi(m)\chi(a) = \sum_{i=1}^{\ell} [\Phi(mas_0^{-1} \ldots s_{i-1}^{-1})\chi(s_0 \ldots s_{i-1}) - \Phi(mas_1^{-1} \ldots s_i^{-1})\chi(s_1 \ldots s_i)].$$

By Corollary 2.6.7 each bracketed term is zero.

Corollary 2.6.9. $\lim_{a \xrightarrow[P]{} \infty} \Phi(ma)\chi^{-1}(a)$ exists for all $m \in M$. More precisely, if ω is a compact subset of M, then there exists $t = t(\Phi, \omega)$ such that $\Phi(ma)\chi^{-1}(a) = \Phi(m)$ for all $m \in \omega A'(F,t)\omega$ and $a \in A^+$.

Proof. This follows immediately from Lemma 2.6.8.

Set $\Theta(m) = (\theta_1^{(m)}, \ldots, \theta_r^{(m)}) = \lim_{a \xrightarrow[P]{} \infty} \Phi(ma)\chi^{-1}(a)$. Set $f_P(m) = \delta_P(m)^{\frac{1}{2}}\theta_1(m)$.

In view of Corollary 2.6.9 it is obvious that $f^{(P)} \underset{P}{\sim} f_P$; in fact, the stronger second statement of Theorem 2.6.1 is clear. To conclude the proof of Theorem 2.6.1

we need only the following lemma.

Lemma 2.6.10. $f_P \in \mathcal{A}(M : U)$.

Proof. Since $\delta_P^{\frac{1}{2}} \in C(M /\!/ K_M)$ for any compact subgroup K_M of M, it suffices, in order to check that $f_P \in C(M /\!/ M_\nu : U)$, to show that $\theta_1 \in C(M /\!/ M_j : U)$. This follows immediately from Corollary 2.6.9 combined with the fact that A centralizes M.

Notice that, for any $m \in M$ and $\bar{n} \in \bar{N}$, there exists $t > 0$ such that $\rho(\bar{n})\Phi(ma) = \Phi(ma\bar{n}) = \Phi(\bar{n}^{ma} ma) = \Phi(ma)$ provided $a \in A^+(t)$. It follows that $(\rho(\bar{n})f)_P = f_P$. This implies that the M-module homomorphism $f \mapsto \theta_1$ from V to $C^\infty(M : U)$ extends to a $\bar{P}$-module homomorphism which factors through $V/V(\bar{N})$. Thus, θ_1 generates an admissible M-module under ρ_M (cf., §2.3). It follows (cf., §1.10) that both θ_1 and f_P lie in $\mathcal{A}(M : U)$.

§2.7. Elementary Properties of the Constant Term.

Let $(P, A)(P = MN)$ and $(\bar{P}, A)(\bar{P} = M\bar{N})$ be opposite p-pairs of G. Let U be a complex vector space. Given $f \in \mathcal{A}(G : U)$, let $f_P \in \mathcal{A}(M : U)$ denote the constant term of f along P (cf., §2.6).

Lemma 2.7.1.

(1) $(\rho(\bar{n})f)_P = (\lambda(n)f)_P = f_P$ for all $n \in N$ and $\bar{n} \in \bar{N}$.

(2) $(\rho(m)f)_P = \delta_{\bar{P}}(m)^{\frac{1}{2}}\rho(m)f_P = \delta_P(m)^{-\frac{1}{2}}\rho(m)f_P$.

$(\lambda(m)f)_P = \delta_P(m)^{\frac{1}{2}}\lambda(m)f_P$.

Proof.

(1) We proved the first part of (1) in the proof of Lemma 2.6.10. The second part goes similarly.

(2) We shall check only that, for $m_0 \in M$, $(\rho(m_0)f)_P = \delta_P(m_0)^{-\frac{1}{2}}\rho(m_0)f_P$. The second part of (2) goes similarly.

Fixing a compact subset ω of M, we know that there exists $t > 0$ such that, for all $m \in \omega A^+(t)$, $(\rho(m_0)f)_P(m) = (\rho(m_0)f)_{(P)}(m) = \delta_P(m)^{\frac{1}{2}}f(mm_0) = \delta_P(m_0)^{-\frac{1}{2}}f_{(P)}(mm_0) = \delta_P(m_0)^{-\frac{1}{2}}\rho(m_0)f_P(m)$.

According to Lemma 2.6.2, two A-finite functions defined on A which coincide on $A^+(t)$ coincide on all of A. It follows that $(\rho(m_0)f)_P$ and $\delta_P(m_0)^{-\frac{1}{2}}\rho(m_0)f_P$ coincide on ω. Since ω was arbitrary, the two functions are equal on all of M.

Now recall the one-one correspondence between p-pairs (P', A') $(P' = M'N')$ of G such that $(P', A') \prec (P, A)$ and p-pairs $({}^*P, {}^*A)({}^*P = {}^*M\,{}^*N = M \cap M' \cdot M \cap N';\ {}^*A = A')$ of M. If $f \in \mathcal{A}(G : U)$, then there exist unique functions $f_P \in \mathcal{A}(M : U)$, $f_{P'} \in \mathcal{A}(M' : U)$, and $(f_P)_{{}^*P} \in \mathcal{A}({}^*M : U)$ such that $f \underset{P}{\approx} f_P$, $f \underset{P'}{\approx} f_{P'}$, and $f_P \underset{{}^*P}{\approx} (f_P)_{{}^*P}$. The following theorem gives an affirmative answer to a natural question.

Theorem 2.7.2. ("Transitivity of the Constant Term").

With the preceding notations: $(f_P)_{{}^*P} = f_{P'}$.

Proof. Fix a compact subset ω of M'. It follows from Theorem 2.6.1 that there exists $t > 0$ such that $\delta_{P'}(m'a')^{\frac{1}{2}}f(m'a') = f_{P'}(m'a')$ for all $m' \in \omega$ and $a' \in A'^+(t)$. Since $(P', A') \prec (P, A)$, there is a set of simple roots $F \subset \Sigma^0(P', A')$ such that $(P, A) = (P', A')_F$ (cf., §0.5). The set $A'^+(t) = A'(P', t) \subset A'(F, t)$

(cf., §2.6). Using Theorem 2.6.1 again, we have $\delta_P(m'a')^{\frac{1}{2}}f(m'a') = f_P(m'a')$ for all $m' \in \omega$ and $a' \in A'(F,t)$, t sufficiently large. From the fact that $N' = N \cdot {}^*N$, it follows that $\delta_{P'}(m') = \delta_P(m')\delta_{{}^*P}(m')$ for all $m' \in M'$. Therefore, $\delta_{{}^*P}(m'a')^{\frac{1}{2}}f_P(m'a') = \delta_{P'}(m'a')^{\frac{1}{2}}f(m'a')$ for all $m \in \omega$ and $a' \in A'(F,t)$. Clearly, $A'({}^*P,t) \supset A'(P',t)$, inasmuch as $\Sigma({}^*P,A') \subset \Sigma(P',A')$. Therefore, $\delta_{{}^*P}(m'a')^{\frac{1}{2}}f_P(m'a') = (f_P)_{{}^*P}(m'a')$ for all $m' \in \omega$ and $a' \in A'(P',t)$, t sufficiently large.

Now we use Lemma 2.6.2 and the fact that $(f_P)_{{}^*P} - f_{P'}$ is an A'-finite function which vanishes on $\omega A'(P',t)$. We conclude that $(f_P)_{{}^*P} = f_{P'}$ on $\omega A'$; since ω is an arbitrary compact subset of M', we have $(f_P)_{{}^*P} = f_{P'}$ on all of M'.

Now fix a maximal split torus A_0 of G and assume that A is an A_0-standard torus. Let K be an A_0-good maximal compact subgroup of G. Let (V,τ) be a smooth double representation of K (cf., §1.12). If $f \in \mathcal{A}(G,\tau)$, then it is obvious that $f_P \in \mathcal{A}(M,\tau_M)$, where $\tau_M = \tau | K \cap M$.

The following simple lemma asserts, essentially, that it is enough to know constant terms with respect to standard parabolic subgroups.

Recall that $(P^k, A^k) = (kPk^{-1}, kAk^{-1})$ ($k \in K$ or G).

<u>Lemma</u> 2.7.3. If $f \in \mathcal{A}(G,\tau)$, then $f_{P^k}(m^k) = \tau(k)f_P(m)\tau(k^{-1})$.

<u>Proof.</u> Since $\delta_{P^k}(m^k) = \delta_P(m)$, we have

$$f_{(P^k)}(m^k) = \delta_P(m)^{\frac{1}{2}} f(m^k)$$

$$= \tau(k) f_{(P)}(m) \tau(k^{-1}).$$

To complete the proof it is sufficient to observe that $m \in \omega A(P, t)$ $(t > 1)$ if and only if $m^k \in \omega^k A^k(P^k, t)$.

For $\alpha \in C_c^\infty(G : U)$ or $C_c^\infty(G, Z : U)$ we define $\alpha^P(m) = \delta_P(m)^{\frac{1}{2}} \int_N \alpha(mn) dn$. Then $\alpha^P \in C_c^\infty(M : U)$ or $C_c^\infty(M, Z : U)$, respectively. We omit the proof of the following easy lemma, which implies that $\alpha \mapsto \alpha^P$ is a P or M-module homomorphism.

Lemma 2.7.4. Let $\alpha \in C_c^\infty(G : U)$ or $C_c^\infty(G, Z : U)$. Then:

(1) $(\rho(n)\alpha)^P = \alpha^P$ $(n \in N)$.

(2) $(\rho(m)\alpha)^P = \delta_P(m)^{\frac{1}{2}} \rho(m) \alpha^P$ $(m \in M)$.

Now let (V, τ) be a double representation of K which satisfies associativity conditions and consider $\alpha \in C_c(G, \tau)$. The two following lemmas show, first, that $\alpha \mapsto \alpha^P$ is a ring homomorphism of $C_c(G, \tau)$ to $C_c(M, \tau_M)$ and, second, that the constant term map $f \mapsto f_P$ of $\mathcal{A}(G, \tau)$ to $\mathcal{A}(M, \tau_M)$ is a $C_c(G, \tau)$-module homomorphism.

Lemma 2.7.5. *If* $\alpha, \beta \in C_c(G, \tau)$, *then*

$$(\alpha * \beta)^P = \alpha^P \underset{M}{*} \beta^P.$$

Proof. $$(\alpha*\beta)^P(m) = \delta_P(m)^{\frac{1}{2}} \int_N \int_G \alpha(x) \beta(x^{-1}mn) dx dn$$

$$= \delta_P(m)^{\frac{1}{2}} \int_N \int_P \int_K \alpha(pk) \beta(k^{-1}p^{-1}mn) d_\ell p \, dk \, dn$$

(Assume $d_\ell p$ and dk are normalized measures!)

$$= \delta_P(m)^{\frac{1}{2}} \int_N \int_P \alpha(p) \beta(p^{-1}mn) d_\ell p \, dn$$

$$= \delta_P(m)^{\frac{1}{2}} \int_N \int_M \int_N \alpha(m_1 n_1)\beta(m_1^{-1} mn')dn_1 dm dn'$$

$$= \alpha^P \underset{M}{*} \beta^P(m).$$

Lemma 2.7.6. If $\alpha \in C_c(G,\tau)$ and $f \in \mathcal{A}(G,\tau)$, then

$$(\alpha * f)_P = \alpha^P \underset{M}{*} f_P \quad \text{and} \quad (f*\alpha)_P = f_P \underset{M}{*} \alpha^{\overline{P}}.$$

Proof. We shall check only the first relation, the proof of the second being essentially the same. Recall that Lemma 1.12.3 asserts that $\mathcal{A}(G,\tau)$ is a double module over $C_c(G,\tau)$. For simplicity normalize the Haar measure $dy = d_\ell p dk$ on G (cf., §1.2.2) such that $\int_K dk = \int_{P\cap K} d_\ell p = 1$.

Computing, we obtain

$$\alpha * f(x) = \int_G \alpha(y) f(y^{-1}x)dy$$

$$= \int_P \alpha(p) f(p^{-1}x) d_\ell p$$

$$= \int_{MN} \alpha(mn) f(n^{-1}m^{-1}x) dm dn.$$

For $m_0 \in M$, we have

$$(\alpha * f)_{(P)}(m_0) = \delta_P(m_0)^{\frac{1}{2}} \int_{MN} \alpha(mn) f(n^{-1}m^{-1}m_0) dm dn.$$

Taking $m_0 \in \omega A^+(t)$, where ω is a compact subset of M and t is sufficiently large, we may replace $(\alpha*f)_{(P)}(m_0)$ by $(\alpha*f)_P(m_0)$. Since $\operatorname{Supp} \alpha$ is compact, it follows from Theorem 2.6.1 and Lemma 2.7.1 that, for t sufficiently large,

$$(\alpha*f)_P(m_0) = \int_{MN} \alpha(mn)\delta_P(m)^{\frac{1}{2}} \cdot \delta_P(m_0 m^{-1})^{\frac{1}{2}} \lambda(n) f(m^{-1}m_0) dm dn$$

$$= \int_M \alpha^P(m) f_P(m^{-1}m_0) dm$$

$$= (\alpha^P \underset{M}{*} f_P)(m_0).$$

To conclude the proof it is sufficient to invoke Lemma 2.6.2 and the by now standard arguments concerning A-finite functions.

Now let $V = C^\infty(K \times K)$ and τ the double representation given as example 2) of §1.12. Recall the bijection $j : \mathcal{A}(G) \to \mathcal{A}(G, \tau)$, $j(f) = \underset{\sim}{f}$ with $\underset{\sim}{f}(k_1 : x : k_2) = f(k_1 x k_2)$ $(f \in \mathcal{A}(G);\ (k_1, k_2, x) \in K \times K \times G)$.

Lemma 2.7.7. Fix $f \in \mathcal{A}(G)$, k_1 and $k_2 \in K$. Put $\phi(x) = f(k_1 x k_2)$. Then $\phi_P(m) = (\underset{\sim}{f})_P(k_1 : m : k_2)$.

Proof. Clearly, $\phi \in \mathcal{A}(G)$ and $\phi(x) = \underset{\sim}{f}(k_1 : x : k_2)$ for all $x \in G$. Therefore, $\phi_{(P)}(m) = \underset{\sim}{f}_{(P)}(k_1 : m : k_2)$. Conclude that $\phi_P(m) = \underset{\sim}{f}_P(k_1 : m : k_2)$ $(m \in M)$ by an A-finiteness argument.

For $f \in \mathcal{A}(G)$ we have a mapping $\Phi_f : K \times K \to \mathcal{A}(G)$ defined by setting $\Phi_f(k_1, k_2)(x) = \underset{\sim}{f}(k_1 : x : k_2) = f(k_1 x k_2)$ $(x \in G)$. Since $f \in C(G/\!/K_0)$ for some open compact subgroup K_0 of G, the image of Φ_f is finite.

Corollary 2.7.8. Let $f \in \mathcal{A}(G)$ and set $jf = \underset{\sim}{f} \in \mathcal{A}(G, \tau)$. Then $\underset{\sim}{f}_P = 0$ if and only if $\phi_P = 0$ for each $\phi \in \operatorname{Im}\Phi_f$. Put another way, $\underset{\sim}{f}_P = 0$ for every p-pair (P, A) of G if and only if $\underset{\sim}{f}_P = 0$ for every standard p-pair.

Proof. The first statement is obvious. To prove the second apply Lemma 2.7.3.

Now recall Lemma 1.11.3 which asserts that, for π any irreducible admissible representation, the space $\mathcal{A}(\pi)$ is a simple double $C_c^\infty(G)$-module. Passing to $\mathcal{A}(\pi, \tau)$, we have a simple double $C_c(G, \tau)$-module (Lemma 1.12.2). We observe that Lemma 2.7.6 implies that $\underset{\sim}{f} \mapsto \underset{\sim}{f}_P$ may be regarded as a $C_c(G, \tau)$-module homomorphism. Therefore, either $f \mapsto \underset{\sim}{f}_P$ is injective on $\mathcal{A}(\pi)$ or $f \mapsto f_P$ sends $\mathcal{A}(\pi)$ identically to zero.

Let (V, τ) be any smooth unitary double representation of K which satisfies associativity conditions. For any $\alpha \in C(G, \tau)$ we set $\alpha^{\sim}(x) = \alpha(x^{-1})^{*}$ (cf. §1.1

<u>Lemma</u> 2.7.9. Let $\alpha \in C_c(G, \tau)$. Then $(\alpha^{\sim})^P = (\alpha^P)^{\sim}$.

<u>Proof.</u> Assume first that $V = C^{\infty}(K \times K)$ and that τ is as in §1.12, 2). Then:

$$(\alpha^{\sim})^P(k_1 : m : k_2) = \delta_P^{\frac{1}{2}}(m)\int_N \alpha^{\sim}(k_1 : mn : k_2)dn$$

$$= \delta_P^{\frac{1}{2}}(m)\int_N \bar{\alpha}(k_2^{-1} : n^{-1}m^{-1} : k_1^{-1})dn$$

$$= \delta_P^{\frac{1}{2}}(m^{-1})\int_N \bar{\alpha}(k_2^{-1} : m^{-1}n : k_1^{-1})dn$$

$$= (\bar{\alpha})^P(k_2^{-1} : m^{-1} : k_1^{-1}).$$

Since $\bar{\alpha}^P = (\alpha^P)^{-}$, we see that $(\alpha^{\sim})^P(k_1 : m : k_2) = (\alpha^P)^{\sim}(k_1 : m : k_2)$. To reduce the general case to the above, observe that, without loss of generality, we may assume (V, τ) is finite-dimensional and irreducible. In this case, (V, τ) may be imbedded as a subalgebra and subrepresentation of $C^{\infty}(K \times K)$ with the associativity conditions of $C^{\infty}(K \times K)$. Thus, the above calculation implies the general case of the lemma.

§2.8. <u>The Constant Term and Supercusp Forms.</u>

In this section we shall characterize the space of supercusp forms (cf., §2.2) in terms of the constant term. We shall also conclude the proof that, for admissible representations, supercuspidal equals J-supercuspidal.

Let A_0 be a maximal split torus of G and K an A_0-good maximal compact subgroup of G. Let (V, τ) be a smooth double representation of K in a vector space V. Write ${}^0\mathcal{A}(G, \tau) = \mathcal{A}(G, \tau) \cap ({}^0\mathcal{A}(G) \otimes V)$.

Fix $P_0 \in \mathcal{P}(A_0)$, $P_0 = M_0N_0$.

Theorem 2.8.1. Let $f \in \mathcal{A}(G, \tau)$. Then the following three conditions on f are equivalent:

(1) $f_P = 0$ for any standard p-pair $(P, A) \neq (G, Z)$.

(2) Supp f is compact mod Z.

(3) $f \in {}^0\mathcal{A}(G, \tau)$.

Proof. That (3) implies (2) implies (1) is immediate.

(1) ⇒ (2): Recall that $G = KA_0^+\omega K$, where ω is a finite subset of M_0. If (2) is false, then there exists $m \in \omega$ and a sequence $\{a_i\}_{i=1}^{\infty} \subset A_0^+$ such that:

1) $f(a_i m) \neq 0$; and

2) $\sup_{\alpha} |\alpha(a_i)| \to \infty$, $i \to \infty$, where "sup" is taken over $\alpha \in \Sigma^0(P_0, A_0)$. Let F be the set of all $\alpha \in \Sigma^0(P_0, A_0)$ such that $\sup_i |\alpha(a_i)| < \infty$. Then ${}^cF \neq \phi$, so $(P, A) = (P_0, A_0)_F$ is a proper p-pair of G. There exists $t > 0$ such that $a_i \in A_0(P, t)$ for i large. Therefore, $f_P(ma_i) = \delta_P(ma_i)^{\frac{1}{2}} f(ma_i) \neq 0$ for i large. We have shown that not (2) implies $f_P \neq 0$, i.e., not (2) implies not (1).

(2) ⇒ (3): Let $f \in \mathcal{A}(G, \tau)$ have compact support mod Z. Let $W = \{\rho(y)f\}$ be the space generated by the right translates of f. For $g \in W$ the $\bar{P}$-module mapping $\mu : g \mapsto g^{\bar{P}}$ of §2.7 is defined. Since $(\rho(\bar{n})g)^{\bar{P}} = g^{\bar{P}}$, $\ker \mu \supset W(\bar{N})$ (cf., §2.2 for notation). Since $W/W(\bar{N})$ is admissible (§2.3), it follows that $g^{\bar{P}} \in \mathcal{A}(M, \tau_M)$, $\tau_M = \tau | K \cap M$, which implies $g^{\bar{P}}$ is A-finite. But $g^{\bar{P}}$ is compactly supported on M/Z and $A \neq Z$, so $g^{\bar{P}} = 0$.

Corollary 2.8.2. Let $0 \neq f \in \mathcal{A}(G, \tau)$. Let (P, A) be a p-pair such that

1) $f_P \neq 0$, and

2) (P, A) is minimal with respect to 1).

Then f_P is a supercusp form.

Proof. This follows immediately from the transitivity of the constant term (Theorem 2.7.2) and condition (1) of Theorem 2.8.1.

Remark. Note that $f_G = f$, so there is always some p-pair (i.e., (G, Z)) for which $f_P \neq 0$ (provided $0 \neq f \in \mathcal{A}(G)$). It is a consequence of Theorem 2.8.1 that, if $f \in \mathcal{A}(G)$ and $f \notin {}^0\mathcal{A}(G)$, then $f_P \neq 0$ for some $(P, A) \neq (G, Z)$. Of course, f_P may be zero for all semi-standard p-pairs.

Corollary 2.8.3. An admissible representation π is supercuspidal if and only if it is J-supercuspidal.

Proof. Let π act in a vector space V. Assume that π is J-supercuspidal. Let (P, A) be a p-pair with $P = MN$ its Levi decomposition. By hypothesis, $V = V(\bar{N})$. For $v \in V$ and $\tilde{v} \in \tilde{V}$ let $f_{v,\tilde{v}}(x) = (\tilde{v}, \pi(x)v)$. Then $f_{v,\tilde{v}} \in \mathcal{A}(\pi)$. According to Theorem 2.8.1 it suffices to show that $(f_{v,\tilde{v}})_P = 0$. Since obviously $f_{\pi(y)v,\tilde{v}} = \rho(y)f_{v,\tilde{v}}$ and, by Lemma 2.7.1, $(\rho(\bar{n})f_{v,\tilde{v}})_P = (f_{v,\tilde{v}})_P$, we have $(f_{\pi(\bar{n})v,\tilde{v}})_P - (f_{v,\tilde{v}})_P = (f_{\pi(\bar{n})v-v,\tilde{v}})_P = 0$. Since $V(\bar{N}) = V$, $(f_{v,\tilde{v}})_P = 0$ for all $v, \tilde{v}$, so $\mathcal{A}(\pi) \subset {}^0\mathcal{A}(G)$.

The converse is Lemma 2.2.2.

In the following let π be an irreducible admissible representation of G. Refer to §2.4 for definitions and notation.

Corollary 2.8.4. Let (P, A) be a π-critical p-pair of G. If $\mathcal{Y}_\pi(P, A)$ contains a unitary critical exponent, then π is unitary.

<u>Proof.</u> The hypotheses combined with Corollary 2.4.2 and Corollary 2.8.3 imply that we may choose $\sigma \in \omega \in {}^0\mathcal{E}_{\mathbb{C}}(M)$ with central exponent $\chi_\omega \in \hat{A}$ such that $\pi \subset \mathrm{Ind}_{\overline{P}}^{G}(\delta_{\overline{P}}^{\frac{1}{2}}\sigma)$. It follows from Corollary 1.11.8 that $\omega \in {}^0\mathcal{E}(M)$ and from Corollary 1.7.9 that $\mathrm{Ind}_{\overline{P}}^{G}(\delta_{\overline{P}}^{\frac{1}{2}}\sigma)$ is unitary. Therefore, π is unitary too.

Chapter 3. Exponents and the Maass-Selberg Relations.

The purpose of this chapter is to prepare what we need and to prove the Maass-Selberg relations (Theorem 3.5.3) in the case where the inducing representation is supercuspidal. A more general version of this theorem (Theorem 4.6.3), in which the inducing representation is only required to be square integrable, is proved in the next chapter.

§3.1. Exponents.

Let G be, as usual, the set of Ω-points of a reductive Ω-group and let Z be the split component of G. For U a complex vector space consider $f \in \mathcal{A}(G : U)$. According to Corollary 1.10.2, f is Z-finite, so the subspace $W_f \subset \mathcal{A}(G : U)$ spanned by the Z-translates of f may be written as $\oplus W_{f,\chi}$, a sum over finitely many $\chi \in \mathfrak{X}(Z)$. Write f_χ for the "χ-projection" of f to $W_{f,\chi}$, so $f = \Sigma f_\chi$. In other notation we have $\mathcal{A}(G : U) = \oplus \mathcal{A}(G : U)_\chi$ (cf., §1.10), χ ranging over $\mathfrak{X}(Z)$.

More generally, let (P, A) be a p-pair with the Levi decomposition $P = MN$. Then $f_P \in \mathcal{A}(M : U)$ and we may write $f_P = \Sigma f_{P,\chi}$ for finitely many $\chi \in \mathfrak{X}(A)$. If π is an admissible irreducible representation of G and if $f_{P,\chi} \neq 0$ for some $f \in \mathcal{A}(\pi)$ and $\chi \in \mathfrak{X}(A)$, we call χ an exponent of π with respect to (P, A). Write $\mathfrak{X}_\pi(P, A)$ for the set of all exponents of π with respect to (P, A). We want to show that $\mathfrak{X}_\pi(P, A)$ is a finite set and that there is a positive integer d such that $(\chi(a)-\rho(a))^d f_{P,\chi} = 0$ for all $a \in A$ and $f \in \mathcal{A}(\pi)$. The least such integer $d = d(\chi)$ is called the multiplicity of χ with respect to π. In the next section we shall show that dual exponent = exponent and that the two notions of multiplicity agree.

Let $V = C^\infty(K \times K)$ and let τ be the double representation of K in V which was defined in §1.12. Consider the mapping $f \mapsto j(f) = \underset{\sim}{f}$ of $C^\infty(G)$ to $C^\infty(G, \tau)$ and recall that j defines bijections of $C_c^\infty(G)$ with $C_c(G, \tau)$ and $\mathcal{A}(\pi)$ with $\mathcal{A}(\pi, \tau)$. Since convolution is defined on the image of j in such a way that $j(f_1 * f_2) = j(f_1) * j(f_2)$ whenever $f_1 * f_2$ is defined, we may carry over the simple $C_c^\infty(G)$-module structure on $\mathcal{A}(\pi)$ and regard $\mathcal{A}(\pi, \tau)$ as a simple module either over $C_c^\infty(G)$ or $C_c(G, \tau)$. Given $\chi \in \mathfrak{X}(Z)$, we write $\mathcal{A}(G, \tau)_\chi$ for $\mathcal{A}(G : V)_\chi \cap \mathcal{A}(G, \tau)$ and $\mathcal{A}(G, \tau, \chi)$ for $\mathcal{A}(G : V, \chi) \cap \mathcal{A}(G, \tau)$.

<u>Theorem</u> 3.1.1. Let π be an irreducible admissible representation of G. Let (P, A) be a p-pair, $P = MN$, and let $\chi \in \mathfrak{X}(A)$. Let $\mu : \mathcal{A}(\pi) \to \mathcal{A}(M, \tau_M)_\chi$ be the composition of the mappings $f \mapsto \underset{\sim}{f} = j(f) \mapsto \underset{\sim}{f}_{P,\chi}$. Then μ is nonzero if and only if $\chi \in \mathfrak{X}_\pi(P, A)$, in which case μ is injective. If $\chi \in \mathfrak{X}_\pi(P, A)$, there is an integer $d = d(\chi) > 0$ such that, for all $a \in A$ and nonzero $f \in \mathcal{A}(\pi)$, the relations $(\rho(a) - \chi(a))^d \underset{\sim}{f}_{P,\chi} = 0$ and $(\rho(a) - \chi(a))^{d-1} \underset{\sim}{f}_{P,\chi} \neq 0$ both hold.

The integer $d(\chi)$ is the <u>multiplicity</u> of χ with respect to π and (P, A) and χ is said to be <u>simple</u> if $d(\chi) = 1$.

<u>Proof.</u> For $f \in \mathcal{A}(\pi)$ and $\alpha, \beta \in C_c^\infty(G)$ we have $\mu(\alpha * f * \beta) = \underset{\sim}{\alpha}^{(P)} \underset{M}{*} \mu(f) \underset{M}{*} \underset{\sim}{\beta}^{(\overline{P})}$, as follows from Lemma 2.6.6 and remarks in §1.12. (One must also note that convolution on M commutes with translation by A.) One sees that μ may be regarded as a double $C_c^\infty(G)$-module homomorphism. Inasmuch as $\mathcal{A}(\pi)$ is a simple double-module over $C_c^\infty(G)$ (§1.11), $\operatorname{Ker}\mu = (0)$ or $\mathcal{A}(\pi)$. If μ is injective

$\underset{\sim}{f}_{P,\chi} \neq 0$ for every nonzero $f \in \mathcal{A}(\pi)$; since $\underset{\sim}{f}_{P,\chi}(k_1^{-1} : m : k_2) = (\lambda(k_1)\rho(k_2)f)_{P,\chi}$ (Lemma 2.7.7), this implies that $\chi \in \mathfrak{X}_\pi(P, A)$. Conversely, $f_{P,\chi}(m) \neq 0$ implies $\underset{\sim}{f}_{P,\chi}(1 : m : 1) \neq 0$.

To see that $d(\chi)$ is independent of nonzero $f \in \mathcal{A}(\pi)$, observe that $(\rho(a)-\chi(a))^d \mu(\alpha * f * \beta) = \underset{\sim}{\alpha}^{(P)} \underset{M}{*} (\rho(a)-\chi(a))^d \underset{\sim}{f}_{P,\chi} * \underset{\sim}{\beta}^{(\bar{P})}$. Since $\mathcal{A}(\pi)$ is a simple module, it follows that $(\rho(a)-\chi(a))^d$ annihilates all nonzero $\mu(f)$ or none. This completes the proof of Theorem 3.1.1.

<u>Remarks</u>. (1) Let π be an irreducible admissible representation of G. The set $\mathfrak{X}_\pi(G, Z)$ consists of a single element, called the <u>central exponent of</u> π. Obviously the central exponent is simple.

(2) It may happen that, for $f \in \mathcal{A}(\pi)$ and $\chi \in \mathfrak{X}_\pi(P, A)$, $f_{P,\chi} = 0$. Since $\underset{\sim}{f}_{P,\chi} \neq 0$, it follows that $(\lambda(k_1)\rho(k_2)f)_{P,\chi} \neq 0$ for some $(k_1, k_2) \in K \times K$.

<u>Lemma</u> 3.1.2. Let $(P, A) \succ (P', A')$ be p-pairs. Let $\chi' \in \mathfrak{X}_\pi(P', A')$ and set $\chi = \chi' | A$. Then $\chi \in \mathfrak{X}_\pi(P, A)$.

<u>Proof.</u> Let $0 \neq f \in \mathcal{A}(\pi)$. We shall show that $\underset{\sim}{f}_{P,\chi} \neq 0$. Let $P = MN$ be the Levi decomposition of (P, A). Let $({}^*P, A')$ be the p-pair $(P' \cap M, A')$ of M. By Theorem 2.7.2, $\underset{\sim}{f}_{P'} = (\underset{\sim}{f}_P)_{{}^*P}$. Therefore, if

$$\underset{\sim}{f}_P = \sum_{\eta \in \mathfrak{X}_\pi(P,A)} \underset{\sim}{f}_{P,\eta},$$

then

$$\underset{\sim}{f}_{P'} = \Sigma (\underset{\sim}{f}_{P,\eta})_{{}^*P}.$$

Since

$$(\rho(a)\underset{\sim}{f}_{P,\eta})_{*_P} = \rho(a)(\underset{\sim}{f}_{P,\eta})_{*_P}$$

for any $a \in A$, we have

$$(\rho(a)-\eta(a))^d(\underset{\sim}{f}_{P,\eta})_{*_P} = 0 \quad (a \in A).$$

Therefore, $(\underset{\sim}{f}_{P,\eta})_{*_P} = \Sigma(\underset{\sim}{f}_{P,\eta})_{*_{P,\chi'}}$, where $\chi' \in \mathfrak{X}_\pi(P', A')$ and $\chi'|A = \eta$.

This implies that, if $\chi' \in \mathfrak{X}_\pi(P', A')$, then $\chi'|A \in \mathfrak{X}_\pi(P, A)$.

In the following, let A_0 be a maximal split torus of G and K an A_0-good maximal compact subgroup of G.

Lemma 3.1.3. Let (P, A) be a p-pair and $y \in K$. Let π be an irreducible admissible representation of G. Then $\mathfrak{X}_\pi(P^y, A^y) = \{\chi \in \mathfrak{X}(A^y) | \chi = \chi_1^y, \chi_1 \in \mathfrak{X}_\pi(P, A)\} = (\mathfrak{X}_\pi(P, A))^y$ (notation).

Proof. Let $f \in \mathcal{A}(\pi)$. Then $\underset{\sim}{f}_{P^y}(m^y) = \tau(y)\underset{\sim}{f}_P(m)\tau(y^{-1})$ (Lemma 2.7.3)

$$= \tau(y) \sum_{\chi_1 \in \mathfrak{X}_\pi(P,A)} \underset{\sim}{f}_{P,\chi_1}(m)\tau(y^{-1}).$$

Set

$$\underset{\sim}{f}_{P^y,\chi_1^y}(m^y) = \tau(y)\underset{\sim}{f}_{P,\chi_1}(m)\tau(y^{-1}).$$

Then

$$\underset{\sim}{f}_{P^y} = \sum_{\chi_1 \in \mathfrak{X}_\pi(P,A)} \underset{\sim}{f}_{P^y,\chi_1^y}.$$

If $(\rho(a)-\chi_1(a))^\ell \underset{\sim}{f}_{P,\chi_1} = 0$, then

$$(\rho(a^y)-\chi_1^y(a^y))^\ell \underset{\sim}{f}_{P^y,\chi_1^y}(m^y) = \tau(y) \cdot (\rho(a)-\chi_1(a))^\ell \underset{\sim}{f}_{P,\chi_1}(m)\tau(y^{-1}) = 0,$$

so $\chi_1^y \in \mathfrak{X}_\pi(P^y, A^y)$ whenever $\chi_1 \in \mathfrak{X}_\pi(P, A)$.

For any $f \in \mathcal{A}(G)$ we write $\mathfrak{X}_f(P, A) = \{\chi \in \mathfrak{X}(A) | \underset{\sim}{f}_{P,\chi} \neq 0\}$.

§3.2. Dual Exponents and Class Exponents.

In this section we shall introduce a notion of class exponent and study its properties. We shall also prove that the terms dual exponent (cf., §2.4) and exponent (cf., §3.1) have the same meaning.

Given the content of this section, we may say that a class $C(\pi) \in \mathcal{E}(G)$ belongs to $\mathcal{E}'(G)$ (cf. §2.4) if and only if, for every p-pair $(P, A) \neq (G, Z)$, we have $\mathfrak{X}_\pi(P, A) \cap \hat{A} = \phi$. In §4.4 we shall prove that $\mathcal{E}_2(G) \subset \mathcal{E}'(G)$.

Let A_0 be a maximal split torus of G and K an A_0-good maximal compact subgroup of G. Let (P, A) $(P = MN)$ be a semi-standard p-pair and let $\omega \in \mathcal{E}_{\mathbb{C}}(M)$. Given $f \in \mathcal{A}(G)$, let $W_{f,P}$ denote the M-module generated by $\{(\lambda(k_1)\rho(k_2)f)_P \mid (k_1, k_2) \in K \times K\}$ under ρ. Write $\rho_{f,P}$ for the resulting representation of M on $W_{f,P} \subset \mathcal{A}(M)$. If there exists $\sigma \in \omega$ such that $\sigma \prec \rho_{f,P}$, we call ω a <u>class exponent for</u> f with respect to (P, A). Write $\mathfrak{X}_f(P, A)$ for the set of class exponents for f relative to (P, A).

<u>Lemma</u> 3.2.1. Let $f \in \mathcal{A}(G)$ and $\chi \in \mathfrak{X}(A)$. If $\chi \in \mathfrak{X}_f(P, A)$, then $\chi = \chi_\omega$, the central exponent of some $\omega \in \mathfrak{X}_f(P, A)$.

<u>Proof.</u> If $\chi \in \mathfrak{X}_f(P, A)$, then there exists $(k_1, k_2) \in K \times K$ such that $(\lambda(k_1)\rho(k_2)f)_{P,\chi} \neq 0$. Consider the M-module W generated by $(\lambda(k_1)\rho(k_2)f)_{P,\chi}$. The class ω of any irreducible quotient module of M belongs to $\mathfrak{X}_f(P, A)$. Any such ω obviously has χ as its central exponent.

Let π be an irreducible admissible representation of G. Let (P, A) $(P = MN)$ be a p-pair and $\omega \in \mathcal{E}_{\mathbb{C}}(M)$. We say that ω is a <u>class exponent</u> for π if there exists $f \in \mathcal{A}(\pi)$ such that $\omega \in \mathfrak{X}_f(P, A)$. We write $\mathfrak{X}_\pi(P, A)$ for the set of class exponents for π with respect to (P, A).

Let $(\overline{P}, A)(\overline{P} = M\overline{N})$ denote the p-pair opposite to (P, A). Recall that $\mathcal{Y}_\pi(P, A)$ denotes the set of dual exponents of π with respect to (P, A) (cf. §2.4).

<u>Lemma</u> 3.2.2. Let π be an irreducible admissible representation of G in a vector space V. Then:

(1) $\mathfrak{X}_\pi(P, A) \subset \mathcal{Y}_\pi(P, A)$.

(2) If $\omega \in \mathfrak{X}_\pi(P, A)$, then there is an irreducible subquotient of $V/V(\overline{N})$ which is of class $\delta_P^{-\frac{1}{2}}\omega$.

<u>Proof.</u> It follows from Lemma 3.2.1 that (2) implies (1). To prove (2) note that, given any nonzero $\tilde{v} \in \tilde{V}$, we may define $i_{\tilde{v}} : V \hookrightarrow \mathcal{A}(\pi)$ by setting $i_{\tilde{v}}(v) = f_{\tilde{v}, v}$, where $f_{\tilde{v}, v}(x) = \langle \tilde{v}, \pi(x)v \rangle$ $(x \in G,\ v \in V)$. Then $i_{\tilde{v}}(\pi(g)v) = f_{\tilde{v}, \pi(g)v} = \rho(g)f_{\tilde{v}, v} = \rho(g)i_{\tilde{v}}(v)$, so $i_{\tilde{v}}$ intertwines π with ρ. Clearly, $i_{\tilde{v}}$ is injective for any nonzero $\tilde{v}$. Composing $i_{\tilde{v}}$ with the mapping $f \mapsto \delta_P^{-\frac{1}{2}} f_P$, we obtain an M-module mapping $V/V(\overline{N}) \to \mathcal{A}(M)$ (Lemma 2.7.1). Since any $f \in \mathcal{A}(G)$ is of the form $\sum_{i=1}^{r} f_{\tilde{v}_i, v_i}$, (2) follows easily.

Now fix a minimal p-pair (P_0, A_0) and assume that $(P, A) \succ (P_0, A_0)$. Let $\{K_j\}_{j=1}^{\infty}$ be a fundamental sequence of open compact subgroups of G as in Theorem 2.1.1. Let K_0 be any open compact subgroup of G. If $f \in C(G/K_0)$ is right Hecke-finite (cf. §1.10), then, setting $f_j(x) = \int_{K_j} f(kx)dk$, we obviously define an element $f_j \in \mathcal{A}(G)$.

In the following let χ be a smooth representation of A in a finite-dimensional vector space U.

Lemma 3.2.3. Let $f \in C(G/K_0 : U)$ be right Hecke-finite and assume that $f(\bar{n}ax) = \delta_{\bar{P}}(a)^{\frac{1}{2}}\chi(a)f(x)$ $(\bar{n} \in \bar{N},\ a \in A,\ x \in G)$. Then $(f_j)_P(m_0) = \delta_P(m_0)^{\frac{1}{2}}\int_{M_j} f(mm_0)dm$ $(m_0 \in M)$.

Proof. Given $a \in A(P, t)$ with t sufficiently large, we have

$$\begin{aligned}(f_j)_P(m_0a) &= \delta_P(m_0a)^{\frac{1}{2}}\int_{K_j} f(km_0a)dk \\ &= \delta_P(m_0a)^{\frac{1}{2}}\int_{\bar{N}_j\times M_j\times N_j} f(\bar{n}mnm_0a)d\bar{n}dmdn \\ &= \delta_P(m_0a)^{\frac{1}{2}}\int_{M_j} f(mm_0a)dm,\end{aligned}$$

assuming t so large that $m_0^{-1}a^{-1}N_jm_0a \subset K_0$. It follows that, for a as above,

$$(f_j)_P(m_0a) = \delta_P(m_0)^{\frac{1}{2}}\chi(a)\int_{M_j} f(mm_0)dm.$$

To conclude we use the standard argument:
$F(a) = (f_j)_P(m_0a) - \delta_P^{\frac{1}{2}}(m_0)\chi(a)\int_{M_j} f(mm_0)dm$ is an A-finite function which vanishes for all $a \in A(P,t)$, so $F(1) = 0$ (Lemma 2.6.2).

Remark. If f satisfies the hypotheses of Lemma 3.2.3, then $f_P(ma) = \chi(a)f_P(m)$ $(a \in A,\ m \in M)$.

Theorem 3.2.4. Let π be an irreducible admissible representation of G in a vector space V. Let $\omega \in \mathcal{E}_{\mathbb{C}}(M)$. If there is an irreducible subquotient of $V/V(\bar{N})$ which is of class $\delta_{\bar{P}}^{\frac{1}{2}}\omega$, then $\omega \in \mathfrak{X}_\pi(P, A)$. More precisely, if $h \in \mathcal{A}(M)$ is a matrix-coefficient for $V/V(\bar{N})$, then there is an element $f^1 \in \mathcal{A}(\pi)$ such that $f_P^1 = \delta_P^{\frac{1}{2}}h$.

Proof. Let $h \in \mathcal{A}(M)$ be a matrix coefficient for the admissible M-module $V/V(\bar{N})$. We may choose an M-smooth element $\varphi \in (V/V(\bar{N}))' \subset V'$ and an element $v \in V$ such that $<\varphi, \pi(m)v> = h(m)$. As φ is A-finite, we may choose a base $\varphi = \varphi_1, \ldots, \varphi_r$ for the subspace of $(V/V(\bar{N}))^\sim$ which is spanned by the A-translates of φ. Let $f(x) = (<\varphi_1, \pi(x)v>, \ldots, <\varphi_r, \pi(x)v>) = (f^1(x), \ldots, f^r(x))$ $(x \in G)$. Then, letting U denote $\mathbb{C}^r$, it is clear that we may define a smooth representation $\chi : A \to GL(r, \mathbb{C})$ such that all the hypotheses of Lemma 3.2.3 are satisfied. Indeed, since φ is M-smooth, $\delta_P^{\frac{1}{2}}(m)f(m) = (f_j)_P(m)$ for K_j sufficiently small. It is clear that $f_j^1 \in \mathcal{A}(\pi)$ and $(f_j^1)_P = \delta_P^{\frac{1}{2}}h$.

If $\delta_{\bar{P}}^{\frac{1}{2}}\omega$ occurs in $V/V(\bar{N})$, then we may choose $h \in \mathcal{A}(\delta_{\bar{P}}^{\frac{1}{2}}\omega)$. Thus, $f_P^1 \in \mathcal{A}(\omega)$. To conclude that $\omega \in \mathfrak{X}_\pi(P, A)$ use Lemma 1.11.4.

Corollary 3.2.5. Let π be as in Theorem 3.2.4. Then:

(1) $\mathcal{Y}_\pi(P, A) = \mathfrak{X}_\pi(P, A)$.

(2) Let $\omega \in \mathcal{E}_{\mathbb{C}}(M)$. Then $\omega \in \mathfrak{X}_\pi(P, A)$ if and only if, for any $h \in \mathcal{A}(\omega)$, there exists $f \in \mathcal{A}(\pi)$ such that $f_P = h$.

(3) Let $\chi \in \mathfrak{X}_\pi(P, A)$. Then the multiplicity of χ as an element of $\mathfrak{X}_\pi(P, A)$ equals the multiplicity of χ as an element of $\mathcal{Y}_\pi(P, A)$.

Proof. Parts (1) and (2) follow immediately from Lemma 3.2.2 and Theorem 3.2.4. To prove (3) pick $\varphi \in \dot{V}'_\chi$ (cf. §2.4) and $v \in V$ such that $<({}^t\pi(a)-\chi(a)\delta_{\bar{P}}^{\frac{1}{2}}(a))^{d(\chi)-1}\varphi, v> \neq 0$. Let $h(m) = \delta_P^{\frac{1}{2}}(m)<\varphi, \pi(m)v>$. By (2) there exists $f \in \mathcal{A}(\pi)$ such that $f_P = h$. We have $(\delta_P^{\frac{1}{2}}(a)\rho(a)-\chi(a))^{d(\chi)-1}f_P(1) = \delta_P^{\frac{1}{2}}(a)<({}^t\pi(a)-\chi(a)\delta_{\bar{P}}^{\frac{1}{2}}(a))^{d(\chi)-1}\varphi, v> \neq 0$, so multiplicity of χ in $\mathfrak{X}_\pi(P, A)$ is no less than the multiplicity of χ in $\mathcal{Y}_\pi(P, A)$. The reverse inequality follows easily from properties of the homomorphism $V/V(\bar{N}) \to \mathcal{A}(\delta_{\bar{P}}^{\frac{1}{2}}\omega)$.

Let $\chi \in \mathfrak{X}(A)$ and $\sigma \in \omega \in \mathcal{E}_{\mathbb{C}}(M)$. Set $\chi^*(a) = \bar{\chi}(a^{-1})$ $(a \in A)$ and recall the definition of $\sigma^* \in \omega^* \in \mathcal{E}_{\mathbb{C}}(M)$ (cf. §1.4).

Corollary 3.2.6. Let π be an irreducible admissible unitary representation of G and $(P, A)(P = MN)$ a p-pair of G. Then $\mathfrak{X}_\pi(\bar{P}, A) = (\mathfrak{X}_\pi(P, A))^*$ and $\mathcal{X}_\pi(\bar{P}, A) = (\mathcal{X}_\pi(P, A))^*$.

Proof. Let $\omega \in \mathcal{X}_\pi(P, A)$. Then, by Corollary 3.2.5, there exists $f \in \mathcal{A}(\pi)$ such that $f_P \in \mathcal{A}(\omega)$. Since π is unitary $f^\sim \in \mathcal{A}(\pi)$ $(f^\sim(x) = \bar{f}(x^{-1}))$. There exists $t > 0$ such that, for $m \in \omega A(P, t)$, $\bar{f}_P(m) = \delta_P^{\frac{1}{2}}(m)\bar{f}(m) = \delta_{\bar{P}}^{\frac{1}{2}}(m^{-1})f^\sim(m^{-1}) = f^\sim_{\bar{P}}(m^{-1})$. It follows that $f^\sim_{\bar{P}} \in \mathcal{A}(\omega^*)$, in fact, that $\omega \in \mathcal{X}_\pi(P, A)$ if and only if $\omega^* \in \mathcal{X}_\pi(\bar{P}, A)$. Clearly, $\chi_{\omega^*} = \chi_\omega^*$, so $\mathfrak{X}_\pi(\bar{P}, A) = (\mathfrak{X}_\pi(P, A))^*$, too.

Recall the mapping $j : \mathcal{A}(\pi) \to \mathcal{A}(\pi, \tau)$ where (V, τ) is $C^\infty(K \times K)$ with the usual double representation defined on it. Note that $j(f)^\sim = j(f^\sim)$, that $j(f)^\sim(k_1 : x : k_2) = j(f)^-(k_2^{-1} : x^{-1} : k_1^{-1})$ or $j(f)^\sim(x) = j(f)(x^{-1})^*$.

Corollary 3.2.7. Let π, $(P, A)(P = MN)$ be as in Corollary 3.2.6. Let $f \in \mathcal{A}(\pi)$. Then, for $m \in M$, $(f_P)^-(m) = (f^\sim)_{\bar{P}}(m^{-1})$ and $(\underset{\sim}{f}_P)^*(m) = (\underset{\sim}{f}^\sim)_{\bar{P}}(m^{-1})$; i.e., $(\underset{\sim}{f}_{P,\chi})^*(m) = (\underset{\sim}{f}^\sim)_{\bar{P},\chi^*}(m^{-1})$ for any $\chi \in \mathfrak{X}_\pi(P, A)$; if $\underset{\sim}{f}_{P,\chi} \in \mathcal{A}(\omega, \tau_M)$, then $(\underset{\sim}{f}^\sim)_{\bar{P},\chi^*} \in \mathcal{A}(\omega^*, \tau_M)$.

Proof. The first statement was actually verified during the proof of Corollary 3.2.6. The others follow formally.

As a final result along the lines of Corollaries 3.2.6 and 3.2.7, we prove:

Corollary 3.2.8. Let π be an irreducible admissible representation of G; let $\tilde{\pi}$ be its contragredient. Let (P, A) (P = MN) be a p-pair, $(\overline{P}, A)$ $(\overline{P} = M\overline{N})$ its opposite. Then $\mathfrak{X}_\pi(P, A) = \mathfrak{X}_{\tilde{\pi}}(\overline{P}, A)^{-1}$ and $\mathfrak{X}_\pi(P, A) = \mathfrak{X}_{\tilde{\pi}}(\overline{P}, A)^{\sim}$. For any $\chi \in \mathfrak{X}_\pi(P, A)$ and $\chi^{-1} \in \mathfrak{X}_{\tilde{\pi}}(\overline{P}, A)$, the corresponding multiplicities satisfy $d(\chi, \pi; P, A) = d(\chi^{-1}, \tilde{\pi}; \overline{P}, A)$.

Proof. For any $f \in \mathcal{A}(\pi)$, set $f'(x) = f(x^{-1})$. Then $f' \in \mathcal{A}(\tilde{\pi})$. For any $m \in M$ and $a \in A(P, t)$, $t >> 1$, we have $f_P(ma) = \delta_P(ma)^{\frac{1}{2}} f(ma) = \delta_{\overline{P}}(m^{-1}a^{-1})^{\frac{1}{2}} f'(m^{-1}a^{-1}) =$ $= f'_{\overline{P}}(m^{-1}a^{-1})$. Thus $f_P(m) = f'_{\overline{P}}(m^{-1})$ for all $m \in M$. By Corollary 3.2.5(2), there exists $f \in \mathcal{A}(\pi)$ such that $f_P \in \mathcal{A}(\omega)$ for any $\omega \in \mathfrak{X}_\pi(P, A)$; if $f_P \in \mathcal{A}(\omega)$, then $f'_{\overline{P}} \in \mathcal{A}(\tilde{\omega})$. Finally,

$$\prod_{\chi \in \mathfrak{X}_\pi(P, A)} (\rho(a) - \chi(a))^{d(\chi, \pi; P, A)} f_P = \prod_{\chi \in \mathfrak{X}_\pi(P, A)} (\rho(a^{-1}) - \chi^{-1}(a^{-1}))^{d(\chi, \pi; P, A)} f'_{\overline{P}}$$

$$= 0.$$

The corollary follows.

§3.3. On Exponents and Induced Representations.

Let (P, A) (P = MN) be a p-pair and let $(\overline{P}, A)$ $(\overline{P} = M\overline{N})$ be its opposite. If σ is a representation of M, we frequently regard σ as a representation of P by setting $\sigma(mn) = \sigma(m)$ $(m \in M,\ n \in N)$.

In the proof of the following theorem we make use of ideas which are developed in greater detail in §5.4.

Theorem 3.3.1. Let π be an irreducible admissible representation of G in a vector space V. Let $\sigma \in \omega \in {}^0\mathcal{E}_{\mathbb{C}}(M)$. Then the following are equivalent:

(1) $\omega \in \mathfrak{X}_\pi(P, A)$.

(2) There is a direct summand of $V/V(\overline{N})$ which has a finite composition series (as an M-module) and each composition factor is of class $\delta_{\overline{P}}^{\frac{1}{2}}\omega$.

(3) $\pi \subset \mathrm{Ind}_{\overline{P}}^{G}(\delta_{\overline{P}}^{\frac{1}{2}}\sigma)$.

(4) π occurs as a quotient representation of $\mathrm{Ind}_P^G(\delta_P^{\frac{1}{2}}\sigma)$.

Proof. Certainly, (2) implies (1), and, by Corollary 1.7.11, (2) implies (3) and (3) implies (1). To complete the proof that (1), (2), and (3) are equivalent we shall show that (1) implies (2).

Consider the space $\delta_{\overline{P}}^{-\frac{1}{2}}V/V(\overline{N})$. It is an M-module direct sum $\oplus \overline{V}_\chi$ over $\chi \in \mathfrak{X}_\pi(P, A)$. It suffices to show that if (1) is true, then there is a non-zero sub-module $\overline{V}_\omega \subset \overline{V}_\chi$ with the following properties (let $\overline{\pi}$ denote the representation of M in any of the sub-modules of $\overline{V}_\chi$):

(1) $\overline{V}_\omega$ is a direct summand of $\overline{V}_\chi$.

(2) Every irreducible sub-quotient of $\overline{V}_\omega$ is of class ω.

(3) ω does not occur as a sub-quotient of $\overline{V}_\chi/\overline{V}_\omega$.

(4) Let $0 \leq \ell \leq d \leq d(\chi, \pi)$, where d is the least integer such that $(\chi(a) - \overline{\pi}(a))^d \overline{V}_\omega = 0$ for all $a \in A$. Set $\overline{V}_\ell = \{v \in \overline{V}_\omega \mid (\chi(a) - \overline{\pi}(a))^\ell v = 0\}$. Then $\overline{V}_\ell/\overline{V}_{\ell-1}$ is semi-simple, $1 \leq \ell \leq d$.

Let (W, μ) be an admissible M-module and assume that $\mu(a)w = \chi(a)$ for all $a \in A$ and $w \in W$. Observe that $\mathcal{A}(\omega)$ is a semi-simple algebra with respect to the product $\varphi * \psi(m) = \int_{M/A} \varphi(x)\psi(x^{-1}m)dx^*$ $(\varphi, \psi \in \mathcal{A}(\omega))$. We may set

$\mu(\varphi)w = \int_{M/A}\varphi(x^{-1})\mu(x)dx^* \cdot w$ $(w \in W,\ \varphi \in \mathcal{A}(\omega))$ and regard (W, μ) as a module over $\mathcal{A}(\omega)$. It follows easily from the theory of modules over semi-simple rings that W is the direct sum of a sub-module annihilated by $\mathcal{A}(\omega)$ and of (finitely many, by admissibility) simple $\mathcal{A}(\omega)$-modules. Using Lemma 1.11.5, we conclude easily that each simple $\mathcal{A}(\omega)$-module is an M-module of class ω. Let d be the least integer such that $(\bar{\pi}(a)-\chi(a))^d\bar{V}_\chi$ has no sub-quotients of class ω for all $a \in A$. Let $\bar{V}_{\chi,\ell}$ be the subspace of $\bar{V}_\chi$ annihilated by $(\bar{\pi}(a)-\chi(a))^\ell$ for all $a \in A$. As we have observed above, $\bar{V}_{\chi,\ell}/\bar{V}_{\chi,\ell-1}$, $1 \le \ell \le d$, is a direct sum of a finite direct sum of M-modules of class ω and of a sub-module with no components of class ω. Using elementary linear algebra, one may easily define $\bar{V}_\omega$ having all the required properties.

To see finally that (3) and (4) are equivalent, use Corollary 3.2.8, which asserts that $\omega \in \mathfrak{X}_\pi(P, A)$ if and only if $\tilde{\omega} \in \mathfrak{X}_{\tilde{\pi}}(\bar{P}, A)$. Thus, $\pi \subset \mathrm{Ind}_{\bar{P}}^G(\delta_{\bar{P}}^{\frac{1}{2}}\sigma)$ if and only if $\tilde{\pi} \subset \mathrm{Ind}_P^G(\delta_P^{\frac{1}{2}}\tilde{\sigma})$. However, $\pi \subset \mathrm{Ind}_{\bar{P}}^G(\delta_{\bar{P}}^{\frac{1}{2}}\sigma)$ if and only if $\tilde{\pi}$ is a quotient of $\mathrm{Ind}_{\bar{P}}^G(\delta_{\bar{P}}^{\frac{1}{2}}\tilde{\sigma})$, by Corollary 1.7.8; similarly, $\tilde{\pi} \subset \mathrm{Ind}_P^G(\delta_P^{\frac{1}{2}}\tilde{\sigma})$ if and only if π is a quotient of $\mathrm{Ind}_P^G(\delta_P^{\frac{1}{2}}\sigma)$.

Now let A be a standard torus of G with $Z_G(A) = M$. Let $\sigma \in \omega \in {}^0\mathcal{E}(M)$ and assume that ω is unramified. Then, for any $P \in \mathcal{P}(A)$, the induced representation $\pi = \mathrm{Ind}_P^G(\delta_P^{\frac{1}{2}}\sigma)$ is an irreducible unitary representation of G (Theorem 2.5.9). Write $C(P, \omega)$ or $C(\pi)$ for the class of π. Note that, for any $s \in W(G/A)$, ω^s is unramified also. The following theorem will justify the additional notations: $C(\pi) = C(M, \omega) = C_M^G(\omega) = C(\omega) = C(\omega^s)$ $(s \in W(G/A))$.

<u>Theorem</u> 3.3.2. Let $P_1, P_2 \in \mathcal{P}(A)$ and $s_1, s_2 \in W(G/A)$. Then $C(P_1, \omega^{s_1}) = C(P_2, \omega^{s_2})$.

Proof. The equality $C(P_1, \omega) = C(P_1^s, \omega^s)$ $(s \in W(G/A))$ being purely formal (Lemma 1.7.3), it will be sufficient to show that $C(P_1, \omega) = C(P_2, \omega)$.

Assume that $\dim A/Z = 1$. In this case $\mathcal{P}(A) = \{P, \overline{P}\}$, a set consisting of two opposite parabolic subgroups. Since $C(P, \omega)$ is irreducible, Theorem 3.3.1 implies that $C(P, \omega) = C(\overline{P}, \omega)$.

Now assume $\dim A/Z \geq 2$. Let $\mathfrak{a} = \mathrm{Hom}(X(A), \mathbb{R})$ denote the "real Lie algebra" of A. Let $\mathfrak{a}'$ be the set of all $H \in \mathfrak{a}$ such that $\alpha(H) \neq 0$ for every root α. Then $\mathfrak{a}'$ breaks up into chambers which correspond one-one to elements of $\mathcal{P}(A)$. It suffices to consider P_1 and P_2 corresponding to adjacent chambers. In this case there exists a semi-standard p-pair $(P', A') \succ (P_i, A)$ $(i = 1, 2)$ with $\dim A' = \dim A - 1$. There is a single simple root $\alpha \in \Sigma^0(P_1, A)$ such that $-\alpha \in \Sigma^0(P_2, A)$. We may write $(P', A') = (P_i, A)_{\{\pm\alpha\}}$, i.e., $A' = (A \cap \ker \alpha)^0$, $M' = Z_G(A')$, and $P' = M' \cdot N_1 = M' \cdot N_2$ $(P_i = Z_G(A) N_i = M N_i$ $(i = 1, 2))$.

We shall show that $\mathrm{Ind}_{P_1}^G (\delta_{P_1}^{\frac{1}{2}} \sigma) \sim \mathrm{Ind}_{P_2}^G (\delta_{P_2}^{\frac{1}{2}} \sigma)$. Set ${}^*P_i = M' \cap P_i$. Then $({}^*P_i, A)$ is a maximal p-pair in M', i.e., $\dim A/A' = 1$. This is the preceding situation. Let δ_{*i} be the modular function of ${}^*P_i$. We know that the class of $\sigma_i' = \mathrm{Ind}_{*P_i}^{M'} (\delta_{*i}^{\frac{1}{2}} \sigma)$ is independent of $i = 1, 2$. Clearly, this implies that $\mathrm{Ind}_{P_1}^{P'} (\delta_{*1}^{\frac{1}{2}} \sigma) \sim \mathrm{Ind}_{P_2}^{P'} (\delta_{*2}^{\frac{1}{2}} \sigma)$. Since $\mathrm{Ind}_{P_i}^{P'} (\delta_{P'}^{\frac{1}{2}} \delta_{*i}^{\frac{1}{2}} \sigma) \sim \delta_{P'}^{\frac{1}{2}} \mathrm{Ind}_{P_i}^{P'} (\delta_{*i}^{\frac{1}{2}} \sigma)$ (Lemma 1.7.1) and $\delta_{P_i} = \delta_{P'} \delta_{*i}$, we may conclude that $\mathrm{Ind}_{P_1}^{P'} (\delta_{P_1}^{\frac{1}{2}} \sigma) \sim \mathrm{Ind}_{P_2}^{P'} (\delta_{P_2}^{\frac{1}{2}} \sigma)$. The required result now follows by transitivity of induction (Lemma 1.7.2).

Lemma 3.3.3. Let π be an irreducible admissible unitary representation of G. Assume that there exists a p-pair $(P, A) (P = MN)$ such that $\mathfrak{X}_\pi(P, A)$ contains

$\omega \in {}^0\mathcal{E}(M)$. Let $\chi_\omega = \chi$ be the central exponent of ω. Then, for any p-pair $(P_1, A_1)(P_1 = M_1N_1)$:

(1) $\mathcal{X}_\pi(P_1, A_1) \subset \{\chi_1 \in \hat{A}_1 | \chi_1 = \chi \circ s$ for some $s \in W(A|A_1)\}$. In particular, if $W(A|A_1) = \phi$, then $\mathcal{X}_\pi(P_1, A_1) = \phi$.

(2) If $A_1 \sim A$, then $\mathcal{X}_\pi(P_1, A_1) \subset \{\omega_1 \in {}^0\mathcal{E}(M_1) | \omega_1 = \omega^{s^{-1}}$ for some $s \in W(A|A_1)\}$.

(3) If (P_1, A_1) is a π-critical p-pair, then (P_1, A_1) is π-minimal.

Proof. Assume (P_1, A_1) is minimal with the property: There exists $\chi_1 \in \mathcal{X}_\pi(P_1, A_1)$ such that $\chi_1 \neq \chi^{s^{-1}}$ for any $s \in W(A|A_1)$. Then χ_1 is obviously a π-critical exponent (cf. §2.4), so Corollary 2.4.2 implies the existence of $\sigma_1 \in \omega_1 \in {}^0\mathcal{E}_{\mathbb{C}}(M_1)$ such that: (1) $\chi_{\omega_1} = \chi_1$ and (2) $\pi \subset \operatorname{ind}_{\bar{P}_1}^G(\delta_{\bar{P}_1}^{\frac{1}{2}} \sigma_1) = \pi_1$. However, we know that, for $\sigma \in \omega \in {}^0\mathcal{E}(M)$, $\pi \subset \operatorname{Ind}_{\bar{P}}^G(\delta_{\bar{P}}^{\frac{1}{2}}\sigma) = \pi_0$, a unitary representation of G. It follows that $I(\tilde{\pi}_0, \pi_1) \geq 1$, so Theorem 2.5.8 implies $A = A_1^y$, for some $y \in K$, and $\sigma_1 = \sigma^{y^{-1}} \in {}^0\mathcal{E}(M_1)$ $(M_1^y = M)$. Thus, $\chi_1 = \chi^{y^{-1}} = \chi \circ s, s \in W(A|A_1)$, contradicting the assumption above. All the conclusions of Lemma 3.3.3 are now obviously true.

Corollary 3.3.4. Let $(P, A)(P = MN)$ be a p-pair of G. Let $\omega \in {}^0\mathcal{E}(M)$ and assume that ω is unramified. Let $\pi \in C_M^G(\omega)$. Then:

(1) $\mathcal{X}_\pi(P, A) = \{\omega^s | s \in W(G/A)\}$.

Let (P_1, A_1) be a p-pair of G. Then:

(2) $\mathcal{X}_\pi(P_1, A_1) = \{\chi \circ s | s \in W(A|A_1)\}$.

In particular, if $W(A|A_1) = \phi$, then $\mathcal{X}_\pi(P_1, A_1) = \phi$.

Proof. Statement (1) follows immediately from Theorems 3.3.1 and 3.3.2 and Lemma 3.3.3. Statement (2) follows from (1), Lemma 3.3.3, and Lemma 3.1.2.

§3.4. Simple Classes and Negligibility.

Let A be a special torus of G and let M be its centralizer. Then M is the Levi factor associated to A and any $P \in \mathcal{P}(A)$. Let $\omega \in {}^0\mathcal{E}(M)$ and assume that ω is unramified. Then, if $\sigma \in \omega$, we know that $\mathrm{Ind}_P^G(\delta_P^{\frac{1}{2}}\sigma) = \pi$ is an irreducible unitary representation of G whose class $C_M^G(\omega)$ is independent of the choice of $P \in \mathcal{P}(A)$ (Theorem 3.3.2).

A class $\omega \in {}^0\mathcal{E}(M)$ will be called a simple class if:

(1) ω is unramified in G, and

(2) for every $P \in \mathcal{P}(A)$ and every $\eta \in \mathcal{X}_\pi(P, A)$ $(C(\pi) = C_M^G(\omega))$ the exponent η is simple (cf. §3.1).

In Chapter 5, §§5.3 and 5.4, we shall show that, in fact, (1) implies (2), i.e., simple = unramified. The proof will involve both the Eisenstein integral and the analytic structure on ${}^0\mathcal{E}_{\mathbb{C}}(M)$. The first step in the proof involves showing that there exists a subset of ${}^0\mathcal{E}_{\mathbb{C}}(M)$, which is dense in the complex analytic topology and consists of simple classes. This is sufficient, when combined with the Maass-Selberg relations, to yield the functional equations for the Eisenstein integral and the general result simple = unramified.

The Maass-Selberg relations themselves require no complex structure on ${}^0\mathcal{E}_{\mathbb{C}}(M)$ either for their formulation or proof. We prove the Maass-Selberg relations in this chapter without knowing that there exist simple classes; that there exist situations in which the Maass-Selberg relations are valid we shall establish in Chapter 5.

Our first goal in this section is Theorems 3.4.4 and 3.4.5. The preceding lemmas are used in the proof.

Lemma 3.4.1. If $\omega \in {}^0\mathcal{E}(M)$ is simple and (P_1, A_1) is any p-pair, then every $\eta \in \mathfrak{X}_\pi(P_1, A_1)$ is simple.

Proof. If not, there exists a p-pair $(P_1, A_1)(P_1 = M_1N_1)$ which is minimal with the property: There exists $\eta \in \mathfrak{X}_\pi(P_1, A_1)$ such that the multiplicity $d(\eta) > 1$. The exponent η is critical, so Corollary 2.4.2 implies the existence of $\sigma_1 \in \omega_1 \in {}^0\mathcal{E}_{\mathbb{C}}(M_1)$ such that: (1) $\chi_{\omega_1} = \eta$ and (2) $\pi \subset \mathrm{Ind}_{\underline{P}_1}^G(\delta_{\underline{P}_1}^{\frac{1}{2}}\sigma_1)$. Taking $\tilde{\pi}_1 = \mathrm{Ind}_{\underline{P}_1}^G(\delta_{\underline{P}_1}^{\frac{1}{2}}\tilde{\sigma}_1)$, where $\tilde{\sigma}_1$ is the contragredient of σ_1, we have $I(\tilde{\pi}_1, \pi) \geq 1$. It follows from Theorem 2.5.8 that A_1 is conjugate to A. Without loss of generality assume $A_1 = A$ and $P_1 \in \mathcal{P}(A)$. The lemma then follows from the definition of simple class.

Recall that, for any $\chi \in \mathfrak{X}(Z)$, $\mathcal{A}(G, \chi) = \{\psi \in \mathcal{A}(G) \mid \psi(xz) = \chi(z)\psi(x)\}$ and that ${}^0C_c^\infty(G) \subset C_c^\infty(G)$ consists of all those functions f such that: For any parabolic subgroup $P = MN$ of G, there exists a compact open subgroup $N(f)$ of N for which $\int_{N(f)} f(xn)dn = 0$ for all $x \in G$.

Lemma 3.4.2. Let $f \in {}^0C_c^\infty(G)$ and $\psi \in \mathcal{A}(G, \chi)$. Then $f*\psi \in {}^0\mathcal{A}(G, \chi)$.

Proof. It follows from Lemma 1.10.9 that $f*\psi \in \mathcal{A}(G)$; in fact, it is clear that $f*\psi \in \mathcal{A}(G, \chi)$. It follows from Lemma 2.7.6 that

$$(\underset{\sim}{f}*\underset{\sim}{\psi})_P = \underset{\sim}{f}^P \underset{M}{*} \underset{\sim}{\psi}_P = 0,$$

so $f*\psi \in {}^0\mathcal{A}(G, \chi)$.

Now let $\chi \in \hat{Z}$.

Lemma 3.4.3. Let V be a ρ-stable and admissible subspace of $\mathcal{A}(G, \chi)$. Suppose there exist functions $f_0 \in V$ and $\phi_0 \in {}^0C_c^\infty(G)$ such that

$$\int_G \bar{\phi}_0(x) f_0(x) dx \neq 0.$$

Then there exists a ρ-stable subspace $V_0 \subsetneq V$ such that the representation $\bar{\rho}$ of G on V/V_0 is irreducible and supercuspidal.

Proof. It follows from Lemma 3.4.2 and the hypothesis above that $\tilde{\phi}_0 * f_0 = \phi_1$ is a nonzero element of ${}^0\mathcal{A}(G, \chi)$ $(\tilde{\phi}_0(x) = \phi_0(x^{-1})^-)$. In ${}^0\mathcal{A}(G, \chi)$ we have the scalar product $\|f\|^2 = \int_{G/Z} |f(x)|^2 dx^*$ $(f \in {}^0\mathcal{A}(G, \chi))$. The representation ρ of G in the space ${}^0\mathcal{A}(G, \chi)$ is quasi-admissible and unitary, so the subspace W of ${}^0\mathcal{A}(G, \chi)$ generated by ϕ_1 under ρ is both admissible and unitary. It follows from Lemma 1.6.5 that W is the direct sum of finitely many irreducible admissible unitary submodules: $W = W_1 \oplus \ldots \oplus W_r$; for every $i = 1, \ldots, r$, W_i is a supercuspidal G-module, because there is a G-module projection $W \to W_i$ which maps ϕ_1 on a generator ϕ_{1i} of W_i. To complete the proof of Lemma 3.4.3 compose the obvious G-module homomorphisms $V \xrightarrow{\alpha} W \xrightarrow{\beta} W_i$, for any i, and let $V_0 = \ker(\beta \circ \alpha)$.

Let $f \in \mathcal{A}(G)$ and let $(P, A)(P = MN)$ be a p-pair of G. We shall call f_P negligible and write $f_P \sim 0$ if $\int_M \bar{\phi}(m) f_P(m) dm = 0$ for every $\phi \in {}^0C_c^\infty(M)$.

Let $\omega \in {}^0\mathcal{E}(M)$ be a simple class and let $C(\omega)$ denote the corresponding induced class $C_M^G(\omega) \in \mathcal{E}(G)$. Let (P_1, A_1) be any p-pair of G.

Theorem 3.4.4. If $f \in \mathcal{A}(C(\omega))$, then $f_{P_1} \sim 0$ unless A_1 is conjugate to A in G.

Proof. Assume A_1 is not conjugate to A. We must show that $\int_{M_1} \bar{\phi}(m) f_{P_1}(m) dm = 0$ for all $\phi \in {}^0C_c^\infty(M_1)$ $(P_1 = M_1 N_1)$. We know that $f_{P_1} = \sum_{\eta \in \mathfrak{X}_\pi(P_1, A_1)} f_{P_1, \eta}$, so it suffices to show that $\int_{M_1} \bar{\phi}(m) f_{P_1, \eta}(m) dm = 0$

for all $\eta \in \mathcal{X}_{\pi}(P_1, A_1)$. Since ω is simple, η is simple (Lemma 3.4.1). Writing $\chi = \chi_{\omega}$ for the central exponent of ω, we know that $\eta = \chi \circ s$ for some $s \in W(A|A_1)$ (Lemma 3.3.3), so η is unitary.

Let V be the representation space of $\pi = \mathrm{Ind}_P^G(\delta_P^{\frac{1}{2}}\sigma)$ $(\sigma \in \omega)$. If the theorem is false, there is $(v_0, \tilde{v}_0) \in V \times \tilde{V}$ such that the theorem is false for $f = f_{v_0, \tilde{v}_0}$, where $f_{v_0, \tilde{v}_0}(x) = \langle \tilde{v}_0, \pi(x)v_0 \rangle$ $(x \in G)$. Write $f_v(x) = \langle \tilde{v}_0, \pi(x)v \rangle$ $(v \in V)$. Set $\mu(v) = (f_v)_{P_1, \eta}$. Then $\mu : V \to \mathcal{A}(M_1, \eta)$ is a $\bar{P}_1$-module homomorphism. Let $\mathcal{B} = \mu(V)$. Then $\mathcal{B}$ is an admissible M_1-module of finite type under ρ_1. Applying Lemma 3.4.3, we conclude: There exists an irreducible supercuspidal representation σ_1 of M_1 on a quotient space of $\mathcal{B}$. It follows from Corollary 1.7.11 that $\pi \subset \mathrm{Ind}_{\bar{P}_1}^G(\delta_{\bar{P}_1}^{\frac{1}{2}}\sigma_1)$, whence from Theorem 2.5.8 that A_1 is conjugate to A. Contradiction.

Now fix a maximal split torus A_0 in G. Let K be an A_0-good maximal compact subgroup of G. Let τ be a unitary smooth double representation of K in a vector space V. Write $(\ ,\ )_V$ for the scalar product in V, $\|\ \|_V$ for the norm. Let (P, A) be a semi-standard p-pair of G with the Levi decomposition P = MN. If $f \in \mathcal{A}(G, \tau)$, then $f_P \in \mathcal{A}(M, \tau_M)$. In this context we define $f_P \sim 0$ to mean $\int_M (\phi, f_P(m))_V dm = 0$ for all $\phi \in {}^0C_c(M, \tau_M)$.

<u>Theorem</u> 3.4.5. If $f \in \mathcal{A}(G, \tau)$ and $f_P \sim 0$ for all standard p-pairs (P, A), then $f = 0$.

<u>Proof.</u> We argue by induction on the semi-simple rank of G. We need consider only the case $f_P = 0$ for all $(P, A) \neq (G, Z)$. This implies $f \in {}^0\mathcal{A}(G, \tau)$ (Theorem 2.8.1). Let $\chi_1, \ldots, \chi_\ell$ be generators of X(G). For $T \geq 1$ define

$G_T = \{x \in G \mid T^{-1} \leq |\chi_i(x)| \leq T \ (i = 1, \ldots, \ell)\}$. Then $G_T \cap Z$ is compact. Let Φ_T be the characteristic function of G_T. Set $f_T = f\Phi_T$. Then $f_T \in {}^0C_c(G, \tau)$, since $KG_TK = G_T$ and $NG_TN = G_T$ for any unipotent radical N in G. If $f_G = f \sim 0$, then $0 = \int_G (f(x), f_T(x))_V dx = \int_{G_T} \|f(x)\|_V^2 dx$ for all $T \geq 1$. It follows that $f = 0$.

<u>Lemma</u> 3.4.6. Fix $\omega \in {}^0\mathcal{E}(M)$ and assume that ω is simple. Fix $f \in \mathcal{A}(G, \tau, C(\omega))$. Then, if (P', A') is a semi-standard p-pair, $f_{P'} \sim 0$ unless A' is conjugate to A. If $P \in \mathcal{P}(A)$, then $f_P \in \sum_{s \in W} \mathcal{A}(M, \tau_M, \omega^s)$.

<u>Proof.</u> The first part is Theorem 3.4.4 restated for $f \in \mathcal{A}(G, \tau, C(\omega))$; the present case is an immediate consequence of the earlier (i.e., $f \in \mathcal{A}(G) \otimes V$, so $f_P \neq 0$ would imply that the same holds for some scalar-valued function).

Let us prove the second part. Let χ be the central exponent of $\sigma \in \omega$. Set $\pi = \mathrm{Ind}_P^G(\delta_P^{\frac{1}{2}}\sigma)$, so $\pi \in C(\omega)$. Let $f \in \mathcal{A}(\pi)$. For $\eta \in \mathfrak{X}_\pi(P, A)$ define $W_\eta = \{s \in W(A) \mid \eta = \chi \circ s\}$. It suffices to show that $f_{P,\eta} \in \sum'_{s \in W_\eta} \mathcal{A}(M, \tau_M, \omega^s)$, since $f_P = \sum_{\eta \in \mathfrak{X}_\pi(P, A)} f_{P,\eta}$. Certainly (P, A) is π-minimal and this implies that $f_{P,\eta} \in {}^0\mathcal{A}(M, \tau_M, \eta)$ (η is unitary, by Lemma 3.3.3).

Let $\mathcal{H}$ be the representation space of π. Fix $0 \neq \tilde{h}_0 \in \tilde{\mathcal{H}}$ and define $f_h(x) = \langle \tilde{h}_0, \pi(x)h \rangle$ ($h \in \mathcal{H}$). Define $\mu(h) = (f_h)_{P,\eta} \in {}^0\mathcal{A}(M, \eta)$. It is enough to show that $\mu(\mathcal{H}) \subset \sum_{s \in W_\eta} \mathcal{A}(\omega^s)$.

Since (1) ${}^0\mathcal{A}(M, \eta)$ has a pre-hilbert structure, and (2) $\mu(\mathcal{H})$ is an M-module of finite type, Lemma 1.6.5 implies that $\mu(\mathcal{H}) = U_1 + \ldots + U_r$, a finite orthogonal sum of irreducible submodules. It suffices to show that

$\epsilon \sum_{s \in W_\eta} \mathcal{A}(\omega^s)$. Let E_i be the orthogonal projection of U on U_i. Then

$(E_i \circ \mu)$ is a $\bar{P}$-homomorphism of $\mathcal{H}$ to $\delta_{\bar{P}}^{\frac{1}{2}} U_i$. Let $\sigma_i \in \omega_i \in {}^0\mathcal{E}(M)$ be

alized in U_i. It follows from Corollary 1.7.11 that $\pi \subset \mathrm{Ind}_{\bar{P}}^{G}(\delta_{\bar{P}}^{\frac{1}{2}}\sigma_i)$, whence from

heorem 2.5.8 that $\omega_i = \omega^s$ for some $s \in W(G/A)$. It is, however, clear that

$_i = \eta$, so $s \in W_\eta$.

We introduce the following notation. Given $\pi \in \omega \in \mathcal{E}(G)$

:f. §1.5), we write $\mathcal{A}(\pi) = \mathcal{A}(G,\omega) = \mathcal{A}(\omega)$; let $\mathcal{A}(G,\tau,\omega) = \mathcal{A}(\pi,\tau)$. If $\Omega \subset \mathcal{E}(G)$

rite $\mathcal{A}(G,\Omega) = \sum_{\omega\in\Omega} \mathcal{A}(G,\omega)$ and $\mathcal{A}(G,\tau,\Omega) = \sum_{\omega\in\Omega} \mathcal{A}(G,\tau,\omega)$; both sums are direct.

Fix a Haar Measure dx^* on G/Z. Then, if $\omega \in {}^0\mathcal{E}(G)$, $\mathcal{A}(\omega,\tau)$ has

natural pre-hilbert space structure: $\|f\|^2 = \int_{G/Z} |f(x^*)|^2 dx^*$, and either right

: left translation is a unitary representation of G on $\mathcal{A}(\omega)$. If $\Omega \subset {}^0\mathcal{E}(G)$, we

ıt a direct sum of pre-hilbert spaces structure on $\mathcal{A}(G,\Omega)$.

As before, let A be a special torus of G and M its centralizer.

et $W = W(G/A)$ be the Weyl group of A. We shall call a subset $\Omega \subset {}^0\mathcal{E}(M)$

imple if every $\omega \in \Omega$ is simple and the mapping $(s,\omega) \mapsto \omega^s$ of $W \times \Omega \to {}^0\mathcal{E}(M)$

injective. This implies that $\Omega^{s_1} \cap \Omega^{s_2} = \phi$ unless $s_1 = s_2 \in W(G/A)$.

Let $\ell \geq \dim Z$. Let J be a finite set and suppose that for every $j \in J$

e have fixed a standard torus A_j such that:

) $\dim A_i = \ell$ for all $i \in J$; and

) A_i and A_j are <u>not</u> conjugate in G if $i \neq j$. Set $M_j = Z(A_j)$ and

$_j = W(G/A_j) = W(A_j)$. Let $\Omega_j \subset {}^0\mathcal{E}(M_j)$, $\Omega = \coprod_{j\in J} \Omega_j$. Write $\mathcal{A}(G,\tau,\Omega)$ for the

et of all $f \in \mathcal{A}(G,\tau)$ which satisfy the following conditions:

1) If (P', A') is a semi-standard p-pair, then $f_{P'} \sim 0$ unless A' is conjugate to some A_j (in which case we may assume $A' = A_j$).

2) If $P \in \mathscr{P}(A_j)$, then $f_P \in \sum_{s \in W_j} \mathscr{A}(M_j, \tau_{M_j}, \Omega_j^s)$. We say Ω is simple if Ω_j is simple for every j.

If Ω is simple, then $\omega \mapsto C(\omega) = C_{M_j}^G(\omega)$ defines an injective mapping $\Omega_j \to \mathcal{E}(G)$, also of $\Omega \to \mathcal{E}(G)$. We set $C(\Omega_j) = C_{M_j}^G(\Omega_j) = \bigcup_{\omega \in \Omega_j} C(\omega)$ and $C(\Omega) = \coprod_{j \in J} C(\Omega_j)$.

Lemma 3.4.7. For any Ω the space $\mathscr{A}(G, \tau, \Omega)$ is a double module over $C_c(G, \tau)$.

Proof. Let $\alpha \in C_c(G, \tau)$ and $f \in \mathscr{A}(G, \tau, \Omega)$. We shall check only that $g = \alpha * f \in \mathscr{A}(G, \tau, \Omega)$; the argument on the other side goes similarly. Let (P, A_j) $(P = MN)$ be a semi-standard p-pair. Then $g_P = \alpha^P \underset{M}{*} f_P$ (Lemma 2.7.6). If $f_P \in \sum_{s \in W_j} \mathscr{A}(M, \tau_M, \Omega_j^s)$, then, by Lemma 1.10.9, the same is true for g_P (i.e., 2) holds for g_P).

We must prove: If $f_P \sim 0$, then $g_P \sim 0$, too. For this observe that, for any $\phi \in C_c(M, \tau_M)$, $\int_M (\phi(m), \alpha^P * f_P(m))_V dm = \int_M ((\alpha^P)^\sim * \phi(m), f_P(m))_V dm$. Since ${}^0C_c(M, \tau_M)$ is an ideal in $C_c(M, \tau_M)$ and since $(\alpha^P)^\sim = (\alpha^\sim)^P$ (Lemma 2.7.9), it follows that $f_P \sim 0$ implies that the right-hand integral is zero for all $\phi \in {}^0C_c(M, \tau_M)$. Therefore, $g_P \sim 0$ too.

Theorem 3.4.8. Let Ω, as defined above, be simple. Then $\mathscr{A}(G, \tau, \Omega) = \mathscr{A}(G, \tau, C(\Omega))$.

Proof. To show that $\mathcal{A}(G, \tau, C(\Omega)) \subset \mathcal{A}(G, \tau, \Omega)$ it suffices to show that, for any $\omega \in \Omega_j \subset \Omega$, $\mathcal{A}(G, \tau, C(\omega)) \subset \mathcal{A}(G, \tau, \Omega)$. This is implied by Lemma 3.4.6. In the following we shall show that $\mathcal{A}(G, \tau, \Omega) \subset \mathcal{A}(G, \tau, C(\Omega))$.

First we observe that it is sufficient to consider the case $V = C^\infty(K \times K)$ and the usual double representation on V. To see this let $j : C^\infty(G) \to C(G, \tau)$ be the map which sends $f \mapsto \underset{\sim}{f} = jf$, where $\underset{\sim}{f}(x) = f(k_1 x k_2)$, and set $\mathcal{A}(G, \Omega) = j^{-1}(\mathcal{A}(G, \Omega, \tau))$. Suppose that our theorem is true in the special case $(V = C^\infty(K \times K), \tau)$. Then $\mathcal{A}(G, \Omega) = j^{-1}(\mathcal{A}(G, \tau, \Omega)) \subset j^{-1}(\mathcal{A}(G, \tau, C(\Omega))) = \mathcal{A}(C(\Omega))$. If (V, τ) is arbitrary, then $\mathcal{A}(G, \tau, \Omega) = (\mathcal{A}(G, \Omega) \otimes V) \cap C(G, \tau) \subset (\mathcal{A}(C(\Omega)) \otimes V) \cap C(G, \tau) = \mathcal{A}(G, \tau, C(\Omega))$. In other words, it follows that the theorem is true for a general (V, τ).

Now let $(V, \tau) = (C^\infty(K \times K), \tau)$. To show that $\mathcal{A}(G, \tau, \Omega) \subset \mathcal{A}(G, \tau, C(\Omega))$ we shall show that every ρ-stable subspace of the left side lies in the right. For this purpose we introduce a scalar product on $\mathcal{A}(G, \tau, \Omega)$.

Let $f \in \mathcal{A}(G, \Omega)$. For $P \in \mathcal{P}(A_j)$ and $s \in W(A_j)$ write $\underset{\sim}{f}_{P,s} = \underset{\sim}{f}_{P,\omega^s}$, where $\underset{\sim}{f}_{P,\omega^s}$ is the component of $\underset{\sim}{f}_P \in \mathcal{A}(M, \tau_M, \omega^s)$, $\omega \in \Omega_j \subset \Omega$. Now define $\|f\|^2 = \sum_j \sum_{P \in \mathcal{P}(A_j)} \sum_{s \in W(A_j)} \|\underset{\sim}{f}_{P,s}\|^2$ We define $\|\underset{\sim}{f}_{P,s}\|^2 =$ $\int_{Z(A_j)/A_j} \|\underset{\sim}{f}_{P,s}(m^*)\|_V^2 dm^*$. Clearly, $\|f\| = 0$ implies that, for all j, $\underset{\sim}{f}_{P,s} = 0$ for all $P \in \mathcal{P}(A_j)$ and $s \in W(A_j)$, so $\|f\| = 0$ implies $\underset{\sim}{f}_P = 0$ for $P \in \mathcal{P}(A_j)$ and $s \in W(A_j)$. Therefore, $\| f \| = 0$ implies $\underset{\sim}{f}_P \sim 0$ for all semi-standard (P, A), so $f = 0$ (Theorem 3.4.5).

We want to show that $\mathcal{A}(G,\Omega)$ is a unitary double G-module (ρ and λ) with respect to the above norm. We have already observed that $\mathcal{A}(G,\tau,\Omega)$ is a double module over $C_c(G,\tau)$ (Lemma 3.4.7). This clearly implies that $\mathcal{A}(G,\Omega)$ is a double G-module.

Let (,) denote the scalar product in $\mathcal{A}(G,\Omega)$. The following lemma shows that the representation ρ is unitary. The proof for λ goes similarly and is omitted.

Lemma 3.4.9. Let $f, g \in \mathcal{A}(G,\Omega)$ and $\alpha \in C_c^\infty(G)$. Then $(f*\alpha, g) = (f, g*\tilde{\alpha})$, where $\tilde{\alpha}(x) = \bar{\alpha}(x^{-1})$.

Proof. Note that, for any p-pair (P,A) with $P = MN$, $j(f*\alpha)_P = \underset{\sim}{f}_P \underset{M}{*} \underset{\sim}{\alpha}^{\bar{P}}$ (Lemma 2.7.6), so $j(f*\alpha)_{P,s} = \underset{\sim}{f}_{P,s} * \underset{\sim}{\alpha}^{\bar{P}}$. It follows from Lemma 2.7.9 that $(\underset{\sim}{\alpha}^{\bar{P}})^{\sim} = (\underset{\sim}{\tilde{\alpha}})^{\bar{P}}$; recall also that $j(\tilde{\alpha}) = \underset{\sim}{\tilde{\alpha}}$. Thus, $(f*\alpha, g) = \sum_j \sum_{P \in \mathcal{P}(A_j)} \sum_{s \in W_j} (j(f*\alpha)_{P,s}, \underset{\sim}{g}_{P,s})$ and
$$
\begin{aligned}
(j(f*\alpha)_{P,s}, \underset{\sim}{g}_{P,s}) &= (\underset{\sim}{f}_{P,s} \underset{M}{*} \underset{\sim}{\alpha}^{(\bar{P})}, \underset{\sim}{g}_{P,s}) \\
&= (\underset{\sim}{f}_{P,s}, \underset{\sim}{g}_{P,s} \underset{M}{*} (\underset{\sim}{\alpha}^{(\bar{P})})^{\sim}) \\
&= (\underset{\sim}{f}_{P,s}, \underset{\sim}{g}_{P,s} \underset{M}{*} (\underset{\sim}{\tilde{\alpha}})^{\bar{P}}) \\
&= (\underset{\sim}{f}_{P,s}, j(g*\tilde{\alpha})_{P,s}),
\end{aligned}
$$
so $(f*\alpha, g) = (f, g*\tilde{\alpha})$.

Now fix $f \in \mathcal{A}(G,\Omega)$ and let U be the space generated by f under ρ. We want to show that $U \subset \mathcal{A}(G, C(\Omega))$. Since U is admissible, cyclic, and unitary, it follows from Lemma 1.6.5 that U is the direct sum of finitely many irreducible subspaces. It suffices to show that any irreducible component lies in $\mathcal{A}(G, C(\Omega))$.

Without loss of generality assume that U is irreducible and nonzero. Then $\|f\| \neq 0$, so there is an A_j, a parabolic subgroup $P \in \mathcal{P}(A_j)$, and an $s \in W_j$ such that $f_{P,s} \not\equiv 0$. Let ϕ be the representation in the space U (i.e., $\rho|U$). Then $U \subset \mathcal{A}(\phi)$ (Lemma 1.11.5 combined with the definition: $\mathcal{A}(G,\Omega) = \Sigma\mathcal{A}(G,\omega)$). It is sufficient to prove that $\phi \in C(\Omega_j) \subset C(\Omega)$.

It follows from our hypotheses that $f_P \in \sum_{t \in W_j} \mathcal{A}(M, \tau_M, \Omega_j^t)$, also that this sum is direct. Let $E_{s,\omega^s} : \sum_{t \in W_j} \mathcal{A}(M, \tau_M, \Omega_j^t) \to \mathcal{A}(M, \tau_M, \omega^s)$, where we assume that $\omega \in \Omega_j$ is chosen such that $E_{s,\omega^s} \neq 0$. Define $E_{s,\omega^s} : \sum_{t \in W_j} \mathcal{A}(\Omega_j^t) \to \mathcal{A}(\omega^s)$. If $E_{s,\omega^s}(f_{P,s}) \neq 0$, then $E_{s,\omega^s} f_{P,s}(k_1 : m : k_2) \neq 0$ for some $(k_1, k_2) \in K \times K$, so $E_{s,\omega^s}(\lambda(k_1^{-1})\rho(k_2)f)_P \neq 0$. This means that $\omega^s \in \mathfrak{X}_\phi(P, A_j)$. Therefore, it follows from Theorem 3.3.1, since $U \subset \mathcal{A}(\phi)$, that $\phi \in C_M^G(\omega^s) = C_M^G(\omega) \in C(\Omega_j)$. This completes the proof of Theorem 3.4.8.

<u>Corollary</u> 3.4.10. If $\Omega \subset {}^0\mathcal{E}(M)$ is simple with respect to G, then $\mathcal{A}(G,\tau,\Omega) = \sum_{\omega \in \Omega} \mathcal{A}(G,\tau,\omega) = \sum_{\omega \in \Omega} \mathcal{A}(G,\tau,C(\omega))$.

§3.5. The Maass-Selberg Relations.

We begin by formulating and proving some easy consequences of the theory developed in the preceding section. We then apply these results in the proof of the Maass-Selberg relations.

Fix a maximal split torus A_0 of G and an A_0-good maximal compact subgroup K of G. Let (V, τ) be a smooth double representation of K. Let (P, A) and (P', A') be semi-standard p-pairs of G with the respective Levi decompositions $P = MN$ and $P' = M'N'$. Let $\gamma \in W(A|A')$ have representative $y_\gamma \in G$. We may always assume that y_γ represents an element of $W(G/A_0)$. Let $A_\gamma = A^{y_\gamma^{-1}}$ and $M_\gamma = M^{y_\gamma^{-1}}$. Then $A' \subset A_\gamma$ and $M' \supset M_\gamma$. Let $\omega \in {}^0\mathcal{E}(M)$. Then $\omega^{y_\gamma^{-1}} \in {}^0\mathcal{E}(M_\gamma)$. If ω is simple with respect to G (cf. §3.4), then $\omega^{y_\gamma^{-1}}$ is simple both with respect to G and M'. Note that the standard tori of M' are the standard tori of G which contain A'; however, tori which are conjugate in G may not be conjugate in M'. Let $\{M_\gamma, A_\gamma\}_{\gamma \in \Gamma}$ be a complete set of conjugacy class representatives with respect to M' for $(M, A)^G \cap M'$. Then $W(A|A') = \bigcup_{\gamma \in \Gamma} y_\gamma W(G/A_\gamma)/W(M'/A_\gamma)$.

<u>Theorem</u> 3.5.1. Let $\omega \in {}^0\mathcal{E}(M)$ be a simple class. Let (P', A') be any semi-standard p-pair. Then, for any $f \in \mathcal{A}(G, \tau, \omega)$, $f_{P'} \in \sum_{\gamma \in W(A|A')} \mathcal{A}(M', \tau_{M'}, c_{M_\gamma}^{M'}(\omega^{y_\gamma^{-1}}))$, the sum being direct. In particular, if $W(A|A') = \phi$, then $f_{P'} = 0$.

<u>Proof.</u> Let $f \in \mathcal{A}(G, \tau, \omega)$. If $f_{P'} \neq 0$, then $\mathcal{X}_\pi(P', A') \neq \phi$, so $W(A|A') \neq \phi$ (Corollary 3.3.4). Assume $W(A|A') \neq \phi$. Let $({}^*P, {}^*A)$ be any semi-standard p-pair of (M', A') and let $(P, {}^*A)$ be the semi-standard p-pair of G such that ${}^*P = P \cap M'$. If ${}^*A$ is not conjugate in M' to A_γ for some $\gamma \in \Gamma$, then ${}^*A$ is not conjugate to A in G. Therefore, $(f_{P'})_{{}^*P} = f_P \sim 0$. Assume ${}^*A = A_\gamma$. Then $(f_{P'})_{{}^*P} = f_P \in \sum_{s \in W(G/A_\gamma)} \mathcal{A}(M_\gamma, \tau_{M_\gamma}, \omega^{y_s^{-1} y_\gamma^{-1}})$. Therefore, by Theorem 3.4.8,

$$f_{P'} \in \sum_{\gamma \in \Gamma} \sum_{s \in W(G/A_\gamma)/W(M'/A_\gamma)} \mathcal{A}(M', \tau_{M'}, C^{M'}_{M_\gamma}(\omega^{y_s^{-1} y_\gamma^{-1}}))$$

$$= \sum_{\gamma \in W(A|A')} \mathcal{A}(M', \tau_{M'}, C^{M'}_{M_\gamma}(\omega^{y_\gamma^{-1}})).$$

To see that the sum is direct it is sufficient to observe that $C^{M'}_{M_\gamma}(\omega^{y_\gamma^{-1}}) = C^{M'}_{M_{\gamma'}}(\omega^{y_{\gamma'}^{-1}})$ if and only if $y_\gamma = y_{\gamma'} y_t$ with $t \in W(M'/A_\gamma)$. This follows from Theorem 2.5.8.

Now let $A' \subset A$, so that $M' \supset A$. Set $W = W(G/A)$ and $W' = W(M'/A)$. Let $y_s \in G$ be a representative for $s \in W$. Let $\omega \in {}^0\mathcal{E}(M)$ be simple with respect to G. If $\gamma = s \in W(G/A)$, then $M_\gamma = M$ and $C^{M'}_M(\omega^{y_s^{-1}})$ depends only on the coset $sW(M'/A)$. Let $f_{P',s}$ denote the component of $f_{P'}$ in $\mathcal{A}(M', \tau_{M'}, C^{M'}_M(\omega^s))$. Let E_s be the projection of $\sum_{t \in W} \mathcal{A}(M, \tau_M, \omega^t)$ on $\mathcal{A}(M, \tau_M, \omega^s)$. Write $f_{P,s} = E_s(f_P)$.

<u>Corollary</u> 3.5.2. Let $({}^*P, A)$ be the semi-standard p-pair $(M' \cap P, A)$ of M'. If $f \in \mathcal{A}(G, \tau, \omega)$, then

$$f_{P,s} = E_s((f_{P',s})_{*_P}).$$

<u>Proof.</u> Of course, we have $(f_{P'})_{*_P} = f_P \in \sum_{t \in W} \mathcal{A}(M, \tau_M, \omega^t)$, so the above makes sense. Let $f_{P',\gamma^{-1}}$ be the component of $f_{P'} \in \mathcal{A}(M', \tau_{M'}, C^{M'}_{M_\gamma}(\omega^{y_\gamma^{-1}}))$, so $f_{P'} = \Sigma f_{P',\gamma^{-1}}$. By Theorem 3.4.8 $\mathcal{A}(M', \tau_{M'}, C^{M'}_{M_\gamma}(\omega^{y_\gamma^{-1}})) = \mathcal{A}(M', \tau_{M'}, \omega^{y_\gamma^{-1}})$. Therefore, unless $M_\gamma = M$, we have $(f_{P',\gamma^{-1}})_{*_P} = 0$. It follows that

$(f_{P'})_{*_P} = \sum_{t \in W(G/A)/W(M'/A)} (f_{P',t})_{*_P}$. Since

$(f_{P',t})_{*_P} \in \sum_{u \in W(M'/A)} \mathcal{A}(M, \tau_M, \omega^{ut})$, it follows that $E_s((f_{P',t})_{*_P}) = 0$ unless $t \in sW(M'/A)$, in which case $E_s((f_{P',s})_{*_P}) = E_s((f_{P'})_{*_P}) = E_s(f_P) = f_{P,s}$.

In the following theorem A denotes a standard torus and M its centralizer. We fix a Haar measure dm^* on M/A, so that there is a fixed norm on ${}^0\mathcal{A}(M)$:

$$\|\varphi\|^2 = \int_{M/A} |\varphi(m)|^2 dm^* \quad (\varphi \in {}^0\mathcal{A}(M)).$$

The pair (V, τ) denotes, as before, an arbitrary smooth unitary double representation of K which satisfies associativity conditions.

<u>Theorem</u> 3.5.3 (Maass-Selberg relations). *Assume that* $\omega \in {}^0\mathcal{E}(M)$ *is simple with respect to* G. *For* $f \in \mathcal{A}(G, \tau, \omega)$ *and* $P \in \mathcal{P}(A)$ *write* $f_{P,s}$ *for the component of* f_P *in* $\mathcal{A}(M, \tau_M, \omega^s)$. *Then*

$$\|f_{P_1, s_1}\| = \|f_{P_2, s_2}\|$$

for all P_1 *and* $P_2 \in \mathcal{P}(A)$ *and* s_1 *and* $s_2 \in W = W(G/A)$.

<u>Proof.</u> We first show that it is sufficient to consider the case $s_1 = s_2 = 1$. According to Lemma 2.7.3, $f_{P^y}(m^y) = \tau(y)f_P(m)\tau(y^{-1})$ $(m \in M,\ y \in K)$. Let W operate on $\mathcal{A}(M, \tau_M)$ by setting $s\phi(m) = \tau(y)\phi(y^{-1}my)\tau(y^{-1})$ for $m \in M$ and $y \in K$ any representative for $s \in W$. If $\phi \in \mathcal{A}(M, \tau_M, \omega)$, then one can check easily that $s\phi \in \mathcal{A}(M, \tau_M, \omega^s)$. It follows that $f_{P^s} = sf_P = \sum_{t \in W} sf_{P,t} = \sum_{t \in W} f_{P^s, st}$ (i.e., $s(\omega^t) = \omega^{st}$), so $\|t^{-1}f_{P,t}\| = \|f_{P^{t^{-1}},1}\|$. On the other hand, it is obvious that $\|t^{-1}f_{P,t}\| = \|f_{P,t}\|$, so $\|f_{P,t}\| = \|f_{P^{t^{-1}},1}\|$ -- it is sufficient to consider $s_1 = s_2 = 1$.

Recall that $\mathfrak{a}$ denotes $\mathrm{Hom}(X(A), \mathbb{R}) = \mathrm{Hom}(X(M), \mathbb{R})$, the "real Lie algebra of A", and $\mathfrak{a}'$ denotes the regular elements in $\mathfrak{a}$, i.e., those elements $x = \Sigma H(m) \otimes r_m$ such that $\alpha(x) = \Sigma r_m < \alpha, H(m)> \neq 0$ for all roots $\alpha \in \mathfrak{a}^* = X(M) \otimes \mathbb{R}$ $(|\alpha(m)| = q^{<\alpha, H(m)>})$. The connected components of $\mathfrak{a}'$ correspond one-one to the elements of $\mathcal{P}(A)$. It suffices to prove that $\|f_{P_1,1}\| = \|f_{P_2,1}\|$ for any P_1 and $P_2 \in \mathcal{P}(A)$ which share a wall. We shall prove this by induction on the parabolic rank of P, i.e., on $\dim \underset{\sim}{A}/\underset{\sim}{Z}$.

Before proceeding to the main part of the proof we prove the following lemma.

Lemma 3.5.4. It suffices to establish the Maass-Selberg relations (Theorem 3.5.3) in the case $V = C^{\infty}(K \times K)$ and τ the usual double representation on V.

Proof. Let (V^0, τ^0) be an arbitrary smooth double unitary representation of K. Let $f \in \mathcal{A}(G, \tau^0, \omega)$. Then we may write f in the form $f(g) = \sum_{i=1}^{r} f_i(g)v_i$, where $v_1, \ldots, v_r$ are orthonormal elements of V^0 and $f_i(g) \in \mathcal{A}(G, \omega)$. Assuming that $\int_{M/A} \|\varphi_{P,s}(m^*)\|_V^2 dm^*$ is independent of $P \in \mathcal{P}(A)$ and $s \in W(A)$ for $\varphi \in \mathcal{A}(G, \omega)$, we want to show that $\int_{M/A} \|f_{P,s}(m^*)\|_{V^0}^2 dm^*$ is also independent of P and s.

Notice that $f_P(m) = \sum_{i=1}^{r} f_{i,P}(m)v_i$ and $f_{P,s}(m) = \sum_{i=1}^{r} f_{i\,P,s}(m)v_i$. Moreover, $\|f_{P,s}(m)\|_{V^0}^2 = \sum_{i=1}^{r} |f_{i\,P,s}(m)|^2$. For $(k_1, k_2) \in K \times K$ we have

$(\lambda(k_1^{-1})\rho(k_2)f)(g) = \sum_{i=1}^{r} f_i(k_1 g k_2)v_i = \sum_{i=1}^{r} f_i(g)\tau^0(k_1)v_i\tau^0(k_2)$, so

$\|(\lambda(k_1^{-1})\rho(k_2)f)_{P,s}(m)\|_{V^0}^2 = \|f_{P,s}(m)\|_{V^0}^2.$

Let $f_i \in C(G/\!\!/K_0)$ $(i = 1, \ldots, r)$, where K_0 is an open subgroup of K. Then

$$\|\underset{\sim}{f}_{i\,P,s}(m)\|_V^2 = \int_{K \times K} |\underset{\sim}{f}_{i\,P,s}(k_1 : m : k_2)|^2 dk_1 dk_2$$

$$= \text{meas}(K_0 \times K_0) \sum_{(k_1, k_2) \in K \times K/K_0 \times K_0} |(\lambda(k_1^{-1})\rho(k_2)f_i)_{P,s}(m)|^2.$$

Therefore,

$$\sum_{i=1}^{r} \|\underset{\sim}{f}_{i\,P,s}(m)\|_V^2 = \text{meas}(K \times K)\, \|f_{P,s}(m)\|_{V^0}^2.$$

By hypothesis $\int_{M/A} \sum_{i=1}^{r} \|\underset{\sim}{f}_{i\,P,s}(m^*)\|_V^2 dm^*$ is independent of P and s. It follows that the same is true for $\int_{M/A} \|f_{P,s}(m^*)\|_{V^0}^2 dm^*$.

For the remainder of the proof of Theorem 3.5.3 we may assume that $(V, \tau) = C^\infty(K \times K)$ with τ as in §1.12, example 2.

<u>Lemma</u> 3.5.5. The Maass-Selberg relations hold for p-pairs of p-rank one.

<u>Proof.</u> This is the case $\dim A/Z = 1$ and $\mathcal{P}(A) = \{P, \bar{P}\}$. It suffices to show that $\|f_{P,1}\| = \|f_{\bar{P},1}\|$. Recall that $\mathcal{A}(G, \omega) = j^{-1}(\mathcal{A}(G, \tau, \omega)) = j^{-1}(\mathcal{A}(G, \tau, C(\omega))) = \mathcal{A}(C(\omega))$. For $f \in \mathcal{A}(C(\omega))$ we have $f^\sim \in \mathcal{A}(C(\omega))$, $f^\sim(x) = \bar{f}(x^{-1})$, and $j(f^\sim) = \underset{\sim}{f}^\sim = \underset{\sim}{f}(x^{-1})^*$, where $v^*(k_1 : k_2) = \bar{v}(k_2^{-1} : k_1^{-1})$ $(v \in V = C^\infty(K \times K))$. It follows from Corollary 3.2.7 and the fact that $\omega = \omega^*$ that $(\underset{\sim}{f}^\sim)_{P,1}(m) = (\underset{\sim}{f})_{\bar{P},1}(m^{-1})^*$. This implies that $\|(\underset{\sim}{f}^\sim)_{P,1}\| = \|\underset{\sim}{f}_{\bar{P},1}\|$. We want this equality to hold without the tilde.

Let $\mathcal{H}$ denote the representation space of $\pi \in C(\omega)$, so $\mathcal{H}$ is a pre-hilbert space. Let $\mathcal{T} = \text{End}^0(\mathcal{H})$. In §1.11 we associated to π a natural admissible representation π_2 of $G \times G$ on $\mathcal{T}$ (i.e., $\pi_2(x, y)T = \pi(x)T\pi(y^{-1})$ for $(x, y) \in G \times G$ and $T \in \mathcal{T}$). The representation π_2 is obviously unitary with respect to the norm $\|T\|^2 = \mathrm{tr}(T^*T)$; π_2 is irreducible, because π is irreducible (Lemma 1.11.3). To prove the Maass-Selberg relations we introduce, for each $P \in \mathcal{P}(A)$, a new norm $\|\ \|_P$ on $\mathcal{T}$. We show that π_2 is unitary with respect to $\|\ \|_P$ for each P, which implies that the norms differ by constant factors. We then show that $\|\ \|_P = \|\ \|_{\overline{P}}$, and deduce from the equality of the norms that $\|\underset{\sim}{f}_{P,1}\| = \|\underset{\sim}{f}_{\overline{P},1}\|$.

To define $\|\ \|_P$ recall that, to $T \in \mathcal{T}$, there corresponds a function $f_T \in \mathcal{A}(\pi)$ such that $f_T(x) = \mathrm{tr}(\pi(x)T)$ (cf. §1.11). Set $\|T\|_P^2 = \|j(f_T)_{P,1}\|^2$. Since $\omega \in \mathcal{E}_\pi(P, A)$ (Corollary 3.3.4(1)), $j(f)_{P,1} \neq 0$ for some $f \in \mathcal{A}(\pi)$. Since $T \mapsto f_T$ is a bijective mapping of $\mathcal{T}$ onto $\mathcal{A}(\pi)$ (Lemma 1.11.1), $\|T\|_P = \|j(f_T)_{P,1}\| \neq 0$ for some $T \in \mathcal{T}$ -- $\|\ \|_P$ is not identically zero as a norm on $\mathcal{T}$.

Write $J(T) = j(f_T)_{P,1}$. Let $(\ ,\ )_P$ denote the scalar product associated to the norm $\|\ \|_P$.

Lemma 3.5.6. π_2 is unitary with respect to $\|\ \|_P$; $\|T\|_P = 0$ if and only if $0 = T \in \mathcal{T}$.

Proof. Let $\alpha, \beta \in C_c^\infty(G)$ and $S, T \in \mathcal{T}$. Write $\alpha'(x) = \alpha(x^{-1})$ and $\beta'(x) = \beta(x^{-1})$. To show that π_2 is unitary with respect to $\|\ \|_P$ it suffices to check that $(\pi(\alpha)S\pi(\beta), T)_P = (S, \pi(\alpha^\sim)T\pi(\beta^\sim))_P$. Since $f_{\pi(\alpha)S\pi(\beta)} = \beta' * f_S * \alpha'$ (Lemma 1.11.2) and $J(\pi(\alpha)S\pi(\beta)) = \underset{\sim}{\beta}'^{P} \underset{M}{*} J(S) * \underset{\sim}{\alpha}'^{\overline{P}}$ (Lemma 2.7.6; also note that $\Sigma\mathcal{A}(\omega, \tau_M, \omega^s)$ is a direct sum of M-modules), we have $(\pi(\alpha)S\pi(\beta), T)_P = (\underset{\sim}{\beta}'^{P} \underset{M}{*} J(S) * \underset{\sim}{\alpha}'^{\overline{P}}, J(T)) = (J(S), (\underset{\sim}{\beta}'^{P})^\sim * J(T) * (\underset{\sim}{\alpha}'^{\overline{P}})^\sim) = (J(S), (\underset{\sim}{\beta}'^\sim)^P * J(T) * (\underset{\sim}{\alpha}'^\sim)^{\overline{P}}) = (S, \pi(\alpha^\sim)T\pi(\beta^\sim))_P$. That $\|T\|_P = 0$ if and only if $T = 0$ follows from the fact that π_2 is irreducible on $\mathcal{T}$, the fact that $\|T\| = 0$ if and only if $T = 0$, and Lemma 1.6.2(2).

We now know that $\|T\|_P = c(P)\|T\|$ ($c(P)$ a constant). We want to show that $c(P) = c(\bar{P})$. For this observe that $f_{\tilde{T}}(x) = \bar{f}_T(x^{-1}) = \overline{tr}(\pi(x^{-1})T) = tr(T^* \pi^*(x^{-1})) = tr(T^*\pi(x)) = f_{T^*}(x)$. We have already observed that $\|(jf)_{\bar{P},1}\| = \|j(\tilde{f})_{P,1}\|$ for any $f \in \mathcal{A}(\pi)$, so $\|T\|_P = \|T^*\|_{\bar{P}}$. However, it is obvious that $\|T\| = \|T^*\|$, so $\|T\|_P = \|T\|_{\bar{P}}$. This proves Lemma 3.5.5.

Now assume $\dim A/Z \geq 2$. It is sufficient to prove that the Maass-Selberg relations hold for adjacent P_1 and $P_2 \in \mathcal{P}(A)$. Let $\alpha \in \Sigma^0(P_1, A)$ and $-\alpha \in \Sigma^0(P_2, A)$. Let $(P', A') = (P_i, A)_{\{\alpha\}}$ $(i = 1, 2)$, so $A' \subset A$ and $\dim A' = \dim A - 1$. We have $(P', A') \succ (P_i, A)$, $i = 1, 2$. Let $P' = M'N'$ be the Levi decomposition of (P', A') and set ${}^*P_i = P_i \cap M'$. Then $({}^*P_i, A)$ is a semi-standard p-pair of M' with $\dim A/A' = 1$. Thus, $({}^*P_2, A)$ and $({}^*P_1, A)$ are opposite maximal p-pairs of M'.

Fix $f \in \mathcal{A}(G, \tau, \omega)$ and set $g = f_{P',1} \in \mathcal{A}(M', \tau_{M'}, C_M^{M'}(\omega)) = \mathcal{A}(M', \tau_{M'}, \omega)$. Since ω is simple in G, it is also simple in M', so Lemma 3.5.5 implies that $\|g_{{}^*P_1,1}\| = \|g_{{}^*P_2,1}\|$. Now we apply Corollary 3.5.2 and conclude that $f_{P_i,1} = g_{{}^*P_i,1}$ $(i = 1, 2)$. Thus $\|f_{P_1,1}\| = \|f_{P_2,1}\|$; the proof of Theorem 3.5.3 is completed.

<u>Corollary</u> 3.5.7. The mapping $f \mapsto f_{P,s}$ from $\mathcal{A}(G, \tau, \omega)$ to $\mathcal{A}(M, \tau_M, \omega^s)$ is *injective for all* $P \in \mathcal{P}(A)$ *and* $s \in W(G/A)$.

Chapter 4. The Schwartz Spaces.

The Schwartz spaces, $\mathcal{C}(G)$ and $\mathcal{C}_*(G)$, will be introduced and studied in this chapter. We shall see that $\mathcal{C}(G)[\mathcal{C}_*(G)]$ is a complete locally convex Hausdorff space. The space $\mathcal{C}(G)$ is an algebra under convolution on G; for every $\chi \in \hat{Z}$ there is a space $\mathcal{C}_*(G,\chi) \subset \mathcal{C}_*(G)$ such that $\mathcal{C}_*(G,\chi)$ is an algebra under convolution on G/Z. There are inclusion mappings, continuous in the relevant topologies: $C_c^\infty(G) \hookrightarrow \mathcal{C}(G) \hookrightarrow L^2(G)[C_c^\infty(G,\chi) \hookrightarrow \mathcal{C}_*(G,\chi) \hookrightarrow L^2(G,\chi)]$. The K-finite matrix coefficients of the discrete series of G lie in $\mathcal{C}_*(G)$, wave packets in $\mathcal{C}(G)$ (cf. §5.5.1). Indeed, it is a nontrivial problem to construct wave packets which lie in $C_c^\infty(G)$ or in $L^1(G)$. For M the centralizer of a standard torus, there are morphisms $\mathcal{C}(G) \to \mathcal{C}(M)$ and $\mathcal{C}_*(G) \to \mathcal{C}_*(M)$ which have importance in various situations.

We shall define a notion of "tempered" distribution, i.e., continuous linear functional on $\mathcal{C}(G)$. It is known in the case of real groups and conjectured for p-adic groups that those and only those irreducible admissible unitary representations whose characters are tempered distributions "occur" in the Plancherel formula for G (cf. [7f], however).

The first three sections of this chapter are largely technical. They may be skipped at a first reading, used for reference as needed. §4.1 collects miscellaneous facts and definitions. §4.2 proves the basic facts about the spherical function Ξ. §4.3 proves various inequalities involving the Ξ-function; these are needed in order that we may give complete proofs of many later results.

§§4.4-4.6 contain the heart of the chapter. §4.4 introduces the Schwartz spaces and discusses their connection with the discrete series of G.

§4.5 introduces the "weak constant term" and proves the analogues of various earlier results, preparing the way for the general Maass-Selberg relations. §4.6 provides the connecting link between the discussion in Chapter 3 leading to the Maass-Selberg relations and a more general theory.

The unifying feature of the last two sections lies in their connection with the problem of representing discrete series characters by functions. These two sections are inessential in that we do not later apply the results. Based on the existence of the Steinberg character for a p-adic group, §4.7 shows that an integrable function represents the restriction of any discrete series character to the elliptic elements of G. Howe's result (§4.8) implies that the character of any irreducible admissible representation of G (char $\Omega = 0$), regarded as a distribution on the open set G' of regular elements of G, is represented by a locally constant function (Corollary 4.8.2).

§4.1. Some Preliminaries.

This section introduces some definitions and proves a few technical lemmas for later use. The reader will probably want to scan this section at a first reading and to refer to it as necessary.

Fix a minimal p-pair (P_0, A_0) $(P_0 = M_0N_0)$ of G and let K be an A_0-good maximal compact subgroup of G. Recall the real Lie algebra $\mathfrak{a}_0 = \mathrm{Hom}(X(A_0), \mathbb{R})$ of A_0. We have the mapping $H_0 : M_0 \to \mathrm{Hom}(X(A_0), \mathbb{Z})$ defined such that, for $\chi \in X(M_0) \subset X(A_0)$, we have $|\chi(m)| = q^{\langle \chi, H_0(m) \rangle}$ $(m \in M_0)$. We know that H_0 maps M_0 to a lattice in $\mathfrak{a}_0$. We write ${}^0M_0 = \ker H_0$ and

$M_0^+ = \{m \in M_0 \mid \langle\alpha, H_0(m_0)\rangle \geq 0$ for all $\alpha \in \Sigma(P_0, A_0)\}$. Then $G = \coprod_{m \in M_0^+/{}^0M_0} KmK$.

It follows that, for $f \in C_c(G)$, we have (assuming normalized measures)

$$\int_G f(x)dx = \sum_{m \in M_0^+/{}^0M_0} \int_{KmK} f(x)dx$$

$$= \sum_{m \in M_0^+/{}^0M_0} \mu(KmK) \int_{K \times K} f(k_1 m k_2)dk_1 dk_2.$$

Let S be a set and f_1 and f_2 functions from S to the positive real numbers. If there exists a constant $c > 0$ such that $f_1(s) \leq cf_2(s)$ for all $s \in S$, we write $f_1 \prec f_2$ on S. If S is understood, we write simply $f_1 \prec f_2$. If both $f_1 \prec f_2$ and $f_2 \prec f_1$, we write $f_1 \asymp f_2$.

Recall that the modular function δ_{P_0} of P_0 is defined such that $d_r p = \delta_{P_0}(p) d_\ell p$, where $d_\ell p$ and $d_r p$ are, respectively, normalized left and right Haar measures on P_0.

Lemma 4.1.1. $\mu(KmK) \asymp \delta_{P_0}(m)$ $(m \in M_0^+)$.

Proof. Let $K_1 = \bar{N}_1 M_1 N_1$ be an open compact subgroup of G as provided by Theorem 2.1.1. Since $[K:K_1]^{-2}\mu(KmK) \leq \mu(K_1 m K_1) \leq \mu(KmK)$, it is sufficient to show that $\mu(K_1 m K_1) \asymp \delta_{P_0}(m)$ $(m \in M_0^+)$. Since $K_1 m K_1 = \coprod_{k \in K_1/K_1 \cap K_1^m} kmK_1$, we have $\mu(K_1 m K_1) = \mu(K_1)[K_1 : K_1 \cap K_1^m] = \mu(K_1)^2 \mu(K_1 \cap K_1^m)^{-1}$. Since $M_0^+/{}^0M_0 \xleftrightarrow{1-1} \omega A_0^+/{}^0A_0$, where ω is a finite subset of M_0 and ${}^0A_0 = A_0 \cap {}^0M_0$, it is sufficient to show that $\mu(K_1 a K_1) \asymp \delta_{P_0}(a)$ $(a \in A_0^+)$. For this we note that $K_1^a = \bar{N}_1^a M_1 N_1^a$, so $K_1 \cap K_1^a = \bar{N}_1^a M_1 N_1$ for $a \in A_0^+$. Therefore, $\mu(K_1)/\mu(K_1 \cap K_1^a) = \mu(\bar{N}_1)/\mu(\bar{N}_1^a) = \delta_{P_0}(a)$ $(a \in A_0^+)$. This proves our lemma.

Given $x \in M(n,\Omega)$, we set $|x| = \max_{1 \leq i,j \leq n} |x_{ij}|$. If $x \in GL(n,\Omega)$, we define $\|x\| = \max(|x|, |x^{-1}|)$. We also define $\sigma(x) = \log_q \|x\|$. If $G = \underset{\sim}{G}(\Omega)$, where $\underset{\sim}{G}$ is an Ω-group, then we may regard G as a subgroup of $GL(n,\Omega)$. Thus, both $\| \ \|$ and σ are defined for points of G. If K is any compact subgroup of G, then we may even choose an embedding $G \hookrightarrow GL(n,\Omega)$ such that $K \subset GL(n,\mathcal{O})$ ($\mathcal{O}$ the integers of Ω). This implies that $\|k_1 x k_2\| = \|x\|$ for all $k_1, k_2 \in K$ and $x \in G$. We shall usually assume that $\| \ \|$ and σ are defined by such an embedding. It is easy to see that, up to $\asymp$, $1+\sigma$ does not depend upon the embedding; the topology induced by $\| \ \|$ does not depend upon the embedding.

Letting Z denote the split component of G, we also define $\sigma_*(x) = \inf_{z \in Z} \sigma(xz)$ $(x \in G)$. It is clear that $\sigma(xy) \leq \sigma(x)+\sigma(y)$ and that $\sigma_*(xy) \leq \sigma_*(x)+\sigma_*(y)$ $(x, y \in G)$.

Lemma 4.1.2. Let T be a split torus of dimension ℓ. Then

$$\sum_{t \in T/{}^0T} (1+\sigma(t))^{-(\ell+\epsilon)} < \infty.$$

Proof. The group $X(T)$ of rational characters of T is a free abelian group of rank ℓ. Let $\chi_1, \ldots, \chi_\ell$ be a set of generators of $X(T)$. The mapping $t \mapsto (\chi_1(t), \ldots, \chi_\ell(t))$ defines an isomorphism of T with a group of diagonal matrices. It follows that there exist $t_1, \ldots, t_\ell$ in T such that $|\chi_i(t_j)| = q^{\delta_{ij}}$ $(1 \leq i, j \leq \ell)$; we may regard $t_1, \ldots, t_\ell$ as generators of $T/{}^0T$. For any $t \in T$ set $|\chi_i(t)| = q^{m_i(t)}$. Clearly, $1 + \sigma(t) \asymp 1 + \max_{1 \leq i \leq \ell} |m_i(t)|$. One sees that $(1+\max_i |m_i(t)|)^{-(\ell+\epsilon)} \leq \prod_{i=1}^{\ell} (1+|m_i(t)|)^{-(1+\frac{\epsilon}{\ell})}$, also that

$\left(\sum_{m \in \mathbb{Z}} (1+|m|)^{-(1+\frac{\epsilon}{\ell})} \right)^{\ell} = \sum_{t \in T/{}^0T} \prod_{i=1}^{\ell} (1+|m_i(t)|)^{-(1+\frac{\epsilon}{\ell})}$. The left side converges, so

$\sum_{t \in T/{}^0T} (1+\max_i |m_i(t)|)^{-(\ell+\epsilon)} < \infty$, which proves the lemma.

Lemma 4.1.3. Let T be a split torus and Z a subtorus of T. Put $\sigma_*(t) = \inf_{z \in Z} \sigma(tz)$. Then $\sum_{t \in T/{}^0TZ} (1+\sigma_*(t))^{-r} < \infty$ for any $r > \dim T - \dim Z$.

Proof. There exists a subtorus $\underset{\sim}{T}_1$ of $\underset{\sim}{T}$ such that $\underset{\sim}{T}_1$ is defined (and, hence, split) over Ω and such that $T = T_1 Z$ and $T_1 \cap Z$ is finite. Therefore, $T/{}^0T \cdot Z \cong T_1/{}^0T_1$. It is clear that $1+\sigma(t_1 z) \asymp 1+\sigma(t_1)+\sigma(z)$ for any $t_1 \in T_1$ and $z \in Z$, so $1+\sigma_*(t_1) \asymp 1+\sigma(t_1)$. By the preceding lemma $\sum_{t_1 \in T/{}^0T_1} (1+\sigma(t_1))^{-r} < \infty$ for $r > \dim T_1$ and this implies the present lemma.

Lemma 4.1.4. Let A be the set of Ω-points of a split torus of G. Let φ be a continuous finite-dimensional complex-valued representation of A by unipotent linear transformations. Choose $r \geq 1$ such that $(\varphi(a)-1)^r = 0$ for all $a \in A$. Then there exist positive constants c_1 and c_2 such that $c_1 \leq |\varphi(a)| \leq c_2 (1+\sigma(a))^r$ for all $a \in A$.

Proof. Since A is an abelian group, we may assume that φ is a representation of A by upper triangular matrices. Since ${}^0A$ is compact, ${}^0A \subset \ker \varphi$. Thus, φ is essentially a representation of $A/{}^0A \cong \mathbb{Z}^{\ell}$ for some positive integer ℓ. The lemma has an obvious formulation for the case of a single unipotent matrix and its powers (cf. Lemma 1.4.6); its generalization to the present case is easy.

We omit the proofs of the following statements (cf. §1.4).

Corollary 4.1.5. Let ϕ be a continuous finite-dimensional complex-valued representation of A. Write $\mathfrak{X}_\phi(A) = \{\chi \in \mathfrak{X}(A) | \chi \prec \phi\}$. Let $a \in A$. Then $|\phi(a^n)| \to 0$, $n \to \infty$ if and only if $\chi(a^n) \to 0$, $n \to \infty$ for all $\chi \in \mathfrak{X}_\phi(A)$.

Corollary 4.1.6. Let ψ be an A-finite function on A. Fix a_0, $a \in A$. If $\mathfrak{X}_\psi(A) \subset \hat{A}$, and if $\psi(a_0 a^n) \to 0$, $n \to \infty$, then $\psi(a_0) = 0$.

Corollary 4.1.7. Let $f \in \mathcal{A}(G)$. Let (P, A) $(P = MN)$ be a p-pair of G. Fix $a \in A$ and $m \in M$. The following statements are equivalent.

(1) $\underset{\sim}{f}_P(ma^n) \to 0$, $n \to \infty$.

(2) $\underset{\sim}{f}_{P,\chi}(ma^n) \to 0$, $n \to \infty$ for all $\chi \in \mathfrak{X}_f(P, A)$.

(3) $\chi(a^n) \to 0$, $n \to \infty$ for all $\chi \in \mathfrak{X}_f(P, A)$.

The following statements are also equivalent.

(1)' $\underset{\sim}{f}_P(ma) \to 0$, $a \underset{P}{\to} \infty$.

(2)' $\underset{\sim}{f}_{P,\chi}(ma) \to 0$, $a \underset{P}{\to} \infty$ for all $\chi \in \mathfrak{X}_f(P, A)$.

(3)' $\chi(a) \to 0$, $a \underset{P}{\to} \infty$, for all $\chi \in \mathfrak{X}_f(P, A)$.

If (1), (2), and (3) hold for all $a \in A^+(P, t)$, $t > 1$, then (1)', (2)', and (3)' hold. If (1)', (2)', (3)' hold, then (1), (2), and (3) hold for all $a \in A^+(P, t)$, $t > 1$.

§4.2. The Spherical Function Ξ.

Let (P_0, A_0) $(P_0 = M_0 N_0)$ be a minimal p-pair of G and K an A_0-good maximal compact subgroup of G. Since M_0/A_0 is compact, the trivial representation $\sigma = 1$ is a supercuspidal representation of M_0.

Let $\pi_0 = \mathrm{Ind}_{P_0}^{G}(\delta_{P_0}^{\frac{1}{2}})$. The representation space $\mathcal{H}$ of π_0 may be identified with $C^\infty(K \cap P_0 \backslash K)$. There is a natural pre-hilbert space structure on $C^\infty(K \cap P_0 \backslash K)$ derived from $L^2(K)$ and π_0 is unitary with respect to this metric. Write $(\ ,\ )$ for the scalar product in $\mathcal{H}$. Assume that the Haar measure on K is normalized. Then the function $h_0 \equiv 1$ on K is a normalized K-invariant element of $\mathcal{H}$. It is obvious that 1_K occurs in $\pi_0|K$ exactly once. Set

$$\Xi(x) = (h_0, \pi_0(x)h_0)$$

$$= \int_K h_0(k)h_0(kx)dk$$

$$= \int_K h_0(kx)dk.$$

It is clear that $\Xi(x)$ is a positive real-valued function which satisfies $\Xi(kxk') = \Xi(x)$ for all k and $k' \in K$ and $x \in G$. One also sees that $\Xi(x) = \Xi(x^{-1})$ and that Ξ satisfies the functional equation $\int_K \Xi(xky)dk = \Xi(x)\,\Xi(y)$ $(x, y \in G)$.

Since $G = P_0K$, we may write $kx = nmk'$ $(n \in N_0, m \in M_0$, and $k' \in K)$, so $h_0(kx) = \delta_{P_0}^{\frac{1}{2}}(m)$. On the other hand, $x^{-1}k^{-1} = k'^{-1}m^{-1}n^{-1}$, so $\delta_{P_0}(m^{-1}) = \delta_{P_0}(x^{-1}k^{-1})$, which implies that $h_0(kx) = \delta_{P_0}(x^{-1}k^{-1})^{-\frac{1}{2}}$. (Recall that, by definition, $\delta_{P_0}(kp) = \delta_{P_0}(p)$ $(k \in K,\ p \in P_0)$.) Thus we may write

$$\Xi(x) = \int_K \delta_{P_0}(x^{-1}k^{-1})^{-\frac{1}{2}}dk$$

$$= \int_K \delta_{P_0}(xk)^{-\frac{1}{2}}dk.$$

Obviously the central exponent of "1" is also the trivial representation "1" of A_0. It follows from Lemma 3.3.3 that, if (P, A) is a p-pair of G, then $\mathcal{X}_{\pi_0}(P, A) = \{1\}$. Let w denote the order of the Weyl group $W(G/A_0)$ of G. Then it follows from Corollary 5.5.3.7 that $(\rho(\alpha)-\hat{\alpha}(1))^w\, \Xi_P = 0$ for any $\alpha \in C_c^\infty(A)$, (P, A) a semi-standard p-pair. Equivalently, $(\rho(a)-1)^w\, \Xi_P = 0$ for all $a \in A$. It follows from the same corollary that w is the least number for which these relations hold. In any case, it follows from the A-finiteness of Ξ_P for $P \in \mathcal{P}(A)$ and from the fact that $\mathfrak{X}_\Xi(P, A) = \{1\}$ (cf. §3.1) that there is an integer $r_0 > 0$ for which the next theorem is true. This is all we shall need for applications.

Theorem 4.2.1. Let r_0 be a sufficiently large positive integer (e.g., r_0 = mult. of "1" in $\mathfrak{X}_\Xi(P_0, A_0)$). Then there exist positive constants c_1 and c_2 such that $c_1 \leq \delta_{P_0}(m)^{\frac{1}{2}}\, \Xi(m) \leq c_2(1+\sigma(m))^{r_0}$ for all $m \in M_0^+$.

Proof. Since $M_0^+/{}^0M_0 = \omega A_0^+$, where ω is a finite subset of M_0, and since ${}^0M_0$ is compact (so we may assume that ${}^0M_0 \subset K$), it suffices to fix $m \in M_0$ and to prove that there exist c_1 and c_2 such that $c_1 \leq \delta_{P_0}(ma)^{\frac{1}{2}}\, \Xi(ma) \leq c_2(1+\sigma(ma))^{r_0}$ $(a \in A_0^+)$.

Pick r_0 at least as large as the multiplicity of the exponent "1" $\in \mathfrak{X}_\Xi(P_0, A_0)$. There is a compact subset U of A_0^+ such that, for $a \notin U$,

$\delta_{P_0}(ma)^{\frac{1}{2}}\Xi(ma) = \Xi_P(ma)\delta_{{}^*P_0}^{\frac{1}{2}}(ma)$ $(P = MN;\ {}^*P_0 = P_0 \cap M)$ for some $(P, A) \succ (P_0, A_0)$ (Theorem 2.6.1). It is clearly sufficient to prove that for any standard p-pair (P, A) there exist constants c_1 and c_2 such that $0 < c_1 \leq \Xi_P(ma) \leq c_2(1+\sigma(ma))^{r_0}$. Since $\mathfrak{X}_\Xi(P, A) \subset \mathfrak{X}_{\pi_0}(P, A) \subset \{1\}$ and the multiplicity of "1" with respect to (P, A) is no greater than r_0, we may conclude that the space of A-translates of Ξ_P is of dimension no greater than r_0. On this space we have a unipotent representation of A. Our theorem is thus an immediate consequence of Lemma 4.1.4.

Let (P, A) $(P = MN)$ and $(\bar{P}, A)$ $(\bar{P} = M\bar{N})$ be opposite p-pairs of G. Then, as representations for $P\backslash G$, one may take either $K \cap P\backslash K$ or, except for closed nowhere dense set, $\bar{N}$. It follows from §1.2.2 that the K-invariant measure on $P\backslash G$ is given by $\delta_P(\bar{n})^{-1}d\bar{n}$ on $\bar{N}$, where we assume that

$$\int_{\bar{N}} \delta_P(\bar{n})^{-1} d\bar{n} = 1.$$

<u>Lemma</u> 4.2.2. $\delta_{P_0}^{\frac{1}{2}}(m_0)\ \Xi(m_0) = \int_{\bar{N}_0} \delta_{P_0}^{-\frac{1}{2}}(\bar{n}_0^{\,m_0})\delta_{P_0}^{-\frac{1}{2}}(\bar{n}_0)d\bar{n}_0.$

<u>Proof.</u> $\Xi(m_0) = \int_K \delta_{P_0}^{-\frac{1}{2}}(m_0 k)dk$

$$= \int_{\bar{N}_0} \delta_{P_0}^{-\frac{1}{2}}(m_0\kappa(\bar{n}_0))\delta_{P_0}^{-1}(\bar{n}_0)d\bar{n}_0,$$

where $\bar{n}_0 \in \kappa(\bar{n}_0)\mu(\bar{n}_0)N_0$ $(\kappa(\bar{n}_0) \in K,\ \mu(\bar{n}_0) \in M_0)$. Thus

$$\Xi(m_0) = \int_{\bar{N}_0} \delta_{P_0}^{-\frac{1}{2}}(m_0\bar{n}_0\mu^{-1}(\bar{n}_0))\delta_{P_0}^{-1}(\bar{n}_0)d\bar{n}_0 \quad \text{or}$$

$$\delta_{P_0}^{\frac{1}{2}}(m_0)\ \Xi(m_0) = \int_{\bar{N}_0} \delta_{P_0}^{-\frac{1}{2}}(\bar{n}_0^{\,m_0})\delta_{P_0}^{-\frac{1}{2}}(\bar{n}_0)d\bar{n}_0.$$

<u>Lemma</u> 4.2.3. Let ω be a compact subset of G. Then $\sup_{y_1, y_2 \in \omega} \Xi(y_1 x y_2) \asymp \Xi(x)$ for all $x \in G$.

<u>Proof.</u> It suffices to show that there is a constant c such that $\Xi(y_1 x y_2) \leq c\ \Xi(x)$ for any y_1, y_2, and all $x \in G$. Using the fact that $\Xi(x) = \Xi(x^{-1})$ one sees that it is even sufficient to show that there exists c such that $\Xi(yx) \leq c\ \Xi(x)$ for any fixed y and all $x \in G$.

We have $\Xi(yx) = \int_K q^{-\langle \rho, H_0(yxk)\rangle} dk$, $(\rho = \rho_{P_0})$. Since there exists an element $k_1(xk) \in K$ such that $k_1(xk)^{-1}xk \in P_0$, we may write $H_0(yxk) = H_0(yk_1(xk)) + H_0(xk)$. It follows that

$$\Xi(yx) = \int_K q^{-\langle \rho, H_0(yk_1(xk)) + H_0(xk)\rangle} dk$$

$$\leq \max_{k_1 \in K} q^{-\langle \rho, H_0(yk_1)\rangle} \int_K q^{-\langle \rho, H_0(xk)\rangle} dk$$

$$\leq c\ \Xi(x).$$

Let $(P, A) \succ (P_0, A_0)$ $(P = MN)$. We let Ξ_M denote the function associated to M, just as Ξ was defined for G, i.e., $M = {}^*P_0 K_M$, etc. We extend Ξ_M to a function on G by setting $\Xi_M(kmn) = \Xi_M(m)$ $(k \in K,\ m \in M,$ and $n \in N)$.

Assume Haar measures on K and $K_M = K \cap M$ are normalized.

Lemma 4.2.4. $\Xi(x) = \int_K \Xi_M(xk)\delta_P(xk)^{-\frac{1}{2}} dk$.

Proof. Let $f \in C_c(G/\!/K)$. Then

$$\int_G f(x)\ \Xi(x)dx = \int_G f(x)\delta_{P_0}(x)^{-\frac{1}{2}}dx$$

$$= \int_{K \times M_0 \times N_0} f(m_0 n_0)\delta_{P_0}(m_0)^{+\frac{1}{2}} dk\,dm_0\,dn_0$$

$$= \int_{M_0} f^{P_0}(m_0)dm_0.$$

On the other hand, setting $g = f^P$ and ${}^*P_0 = P_0 \cap M$, we have

$$\int_G f(x)\,\Xi_M(x)\delta_P(x)^{-\frac{1}{2}}dx = \int_{MN} f(mn)\,\Xi_M(m)\delta_P(m)^{+\frac{1}{2}}dmdn$$

$$= \int_M f^P(m)\,\Xi_M(m)dm$$

$$= \int_{M_0} g^{*P_0}(m_0)dm_0$$

$$= \int_{M_0} f^{P_0}(m_0)dm_0,$$

since $g^{*P_0} = f^{P_0}$. Since $\int_K \Xi_M(xk)\delta_P(xk)^{-\frac{1}{2}}dk$ defines a function in $C(G/\!/K)$, the lemma is obviously true.

Lemma 4.2.5. There exists $r > 0$ such that

(1) $\int_G \frac{\Xi(x)^2}{(1+\sigma(x))^r}dx < \infty$ and (2) $\int_{G/Z} \frac{\Xi(x^*)^2}{(1+\sigma_*(x^*))^r}dx^* < \infty.$

Proof. We know that $\Xi(x) = \Xi(k_1 x k_2)$ for all $x \in G$, k_1 and $k_2 \in K$. We may assume that $\sigma(k_1 x k_2) = \sigma(x)$. It follows that

$$\int_G \frac{\Xi(x)^2}{(1+\sigma(x))^r}dx = \sum_{m\in M_0^+/{}^0M_0} \mu(KmK)\frac{\Xi(m)^2}{(1+\sigma(m))^r}$$

$$\asymp \sum \mu(KmK)\delta_{P_0}(m)^{-1}\frac{(\delta_{P_0}(m)^{\frac{1}{2}}\,\Xi(m))^2}{(1+\sigma(m))^r}$$

$$\leq c \sum_{m\in M_0^+/{}^0M_0} \frac{(1+\sigma(m))^{2r_0}}{(1+\sigma(m))^r} \quad \text{(Theorem 4.2.1 and Lemma 4.1.1)}$$

$$\leq c' \sum_{a\in A_0} (1+\sigma(a))^{2r_0-r}.$$

If $r-2r_0 > \dim A_0$, then, by Lemma 4.1.2, this sum converges. This proves (1). The proof of (2) goes similarly. It involves Lemma 4.1.3.

§4.3. Inequalities.

In this section we shall use the theory of "rational representations" of reductive algebraic groups in order to prove certain useful inequalities. We begin with a brief summary of the facts we shall need from this theory. For more details the reader may consult [2b] and [2c]. (In [2c] results are formulated only for the case $\operatorname{char}\Omega = 0$; the general case is, however, not essentially different. It is easy to generalize the discussion in [2b], p. A36-37, in order, for example, to prove the existence of a unique irreducible strongly rational representation with highest weight corresponding to an element of $X(M_0)$; cf. below.)

Let $\underset{\sim}{G}$ be a reductive Ω-group. Let (P_0, A_0) $(P_0 = M_0N_0)$ be a minimal p-pair of G. Let $\underset{\sim}{V}$ be a finite-dimensional Ω-vector space. A morphism of algebraic groups $\pi : \underset{\sim}{G} \to \underset{\sim}{GL}(\underset{\sim}{V})$ is called a strongly rational representation of $\underset{\sim}{G}$ (or G) if it is defined over Ω and if there is a line in $\underset{\sim}{V}[V]$ stabilized by $\pi|\underset{\sim}{P}_0[\pi|P_0]$. It is known that irreducible strongly rational representations of G correspond, up to equivalence, one-one with elements of $X(M_0)/W(A_0)$, hence with a certain sublattice cone of the lattice cone of dominant weights with respect to $\Sigma(P_0, A_0)$.

Let λ_α denote the fundamental weight such that $\langle\lambda_\alpha, \alpha'\rangle = \delta_{\alpha,\alpha'}$ $(\alpha, \alpha' \in \Sigma^0(P_0, A_0))$. Let $\lambda = \Sigma m_\alpha\lambda_\alpha$ be a dominant weight which corresponds to an element $\chi_\lambda \in X(M_0)$. Let $F \subset \Sigma^0(P_0, A_0)$ be defined such that $m_\alpha = 0$ if and only if $\alpha \notin F$. Let $\pi_\lambda : G \to GL(V)$ be the strongly rational representation of G associated to λ and write v_λ for a corresponding highest weight vector in V. Then the stabilizer in G of the line spanned by v_λ is the parabolic subgroup associated to the p-pair $(P, A) = (P_0, A_0)_F$. For $a \in A_0$ we have $\pi(a)v_\lambda = \chi_\lambda(a)v_\lambda$, where $|\chi_\lambda(a)| = q^{<\lambda, H(a)>}$.

Let π be a rational representation of G whose highest weight is λ. We may pick an Ω-base for V with respect to which $\pi|A_0$ is diagonalized. Let $\lambda = \lambda_1, \ldots, \lambda_r$ be all the weights of π with respect to A_0. Then, for $i \geq 1$, we may write $\lambda_i = \lambda - \Sigma m_{\alpha, i}\alpha$ where $\alpha \in \Sigma(P_0, A_0)$ and $m_{\alpha, i} \geq 0$.

Fix an Ω-basis of V : $\{v_1, \ldots, v_t\}$. We define a p-adic Banach space structure on V by setting $|v| = \max_{1 \leq i \leq t} |c_i|$, where $v = \sum_{i=1}^{t} c_i v_i$ $(c_i \in \Omega)$. For any $T \in \mathrm{End}_\Omega(V)$ set $|T| = \max_{i,j} |T_{ij}|$, where (T_{ij}) represents T relative to the given basis.

For any p-pair $(P, A) \succ (P_0, A_0)$ the modular function $\delta_P = |2\rho|$, where $2\rho \in X(M)$ corresponds to the weight $2\rho_P \in \mathfrak{a}_0^*$. Thus, for any standard p-pair (P, A), the weight $2\rho_P = \Sigma\alpha$ $(\alpha \in \Sigma(P, A))$ is always the highest weight of a rational representation of G.

Let (P, A) $(P = MN)$ be a standard p-pair and $\mathfrak{a}$ the real Lie algebra of A. Recall that $+\mathfrak{a}$ denotes the cone consisting of all linear combinations of elements of $\Sigma(P, A)$ with nonnegative coefficients. Let $(\overline{P}, A)$ $(\overline{P} = M\overline{N})$ be the p-pair opposite to (P, A).

Lemma 4.3.1. There exists an element $H_0 \in \mathfrak{a}$ such that $H_P(\overline{n}) \in +\mathfrak{a} + H_0$ for all $\overline{n} \in \overline{N}$.

Proof. Let $\{\alpha_1, \ldots, \alpha_s\} = \Sigma^0(P, A)$ (simple roots) and let $\{\lambda_1, \ldots, \lambda_s\}$ be the dual set of fundamental weights, so that $\lambda_i(\alpha_j) = \delta_{ij}$ $(1 \leq i, j \leq s)$. It is sufficient to show that, for any $i = 1, \ldots, s$, there is a constant $C(i)$ such that $\langle\lambda_i, H_P(\bar{n})\rangle \geq C(i)$ for all $\bar{n} \in \bar{N}$; one can then choose $-H_0 \in +\mathfrak{a}$ such that $\langle\lambda_i, H_P(\bar{n})-H_0\rangle \geq 0$ for all $i = 1, \ldots, s$, so $H_P(\bar{n}) \in +\mathfrak{a}+H_0$.

In order to prove the existence of $C(i)$ as above, we consider a strongly rational representation $\pi = \pi_i$ of G in an Ω-vector space V which has highest weight $m_i\lambda_i = \lambda$, a multiple of the fundamental highest weight λ_i (fixed $i \in (1, \ldots, s)$). We let $v_1 = v_\lambda$, a highest weight vector, and choose a base $v_1, \ldots, v_t$ with respect to which $\pi|A_0$ is diagonalized. As before, we give V the structure of a p-adic Banach space.

Recall that, if $x \in G$, we may write $x = \kappa(x)\mu(x)n(x)$ $(\kappa(x) \in K$, $\mu(x) \in M$, and $n(x) \in N)$. It follows that $|\pi(\bar{n})v_1| = |\pi(\kappa(\bar{n})\mu(\bar{n}))v_1| \leq C|\pi(\mu(\bar{n}))v_1| = C_q^{m_i\langle\lambda_i, H_P(\bar{n})\rangle}$ for any $\bar{n} \in \bar{N}$ and some $C \geq 1$, since K is compact. On the other hand, $\pi(\bar{n})v_1 = v_1 + \sum_{i=2}^{r} T_{1i}(\bar{n})v_i$, so $|\pi(\bar{n})v_1| \geq 1$. Therefore, setting $C^{-1/m_i} = C_i$, we have $q^{\langle\lambda_i, H_P(\bar{n})\rangle} \geq C_i$, which proves the lemma.

Remark: J. Tits has proved the following:

Theorem. Let π be a rational representation of G in a vector space V. Let K be an A_0-good maximal compact subgroup of G and (P, A) a semi-standard p-pair of G. Then there exists a basis $v_1, \ldots, v_n$ of V which consists of eigenvectors for A and with respect to which $\pi(K) \subset GL(n, \mathcal{O})$ ($\mathcal{O}$ the integers of Ω).

This theorem implies that the constant C in Lemma 4.3.1 may be taken to be

$C = 1$. In other words $H_P(\overline{N}) \subset +\mathfrak{a}$.

Tits has also suggested the following direct proof that $H_P(\overline{N}) \subset +\mathfrak{a}$, based on 4.4.4, Proposition, Part (i), p. 80 of [4c]:

First, it is enough to consider the case $(P, A) = (P_0, A_0)$ and $\overline{N} = \overline{N}_0$. To see this let $(P_0, A_0) \prec (P, A)(P = MN)$. We have $x = km_0n_0 = km_0n_1n = kmn$, where $m_0 \in M_0 \subset M$, $n_1 \in N_0 \cap M$, and $n \in N$, $n_0 \in N_0$, $k \in K$. It follows that $H_P(x) = H_P(m_0)$. If $H_{P_0}(m_0) \in +\mathfrak{a}_0$, then $H_P(m_0) \in +\mathfrak{a}$, since fewer conditions need be fulfilled.

The proposition cited says essentially that, if $Km_1K \cap Km_2N_0 \neq \phi$ with $m_1 \in M_0^+$, then $m_1m_2^{-1} \in +M_0$ (the dual chamber). Let $\overline{n} \in \overline{N}_0$ and assume that $\overline{n} \in Km_0N_0$ $(m_0 \in M_0)$. We want to show that $m_0 \in +M_0$. We know that, if $x \in M_0^+(t)$, $t >> 1$, then $x\overline{n}x^{-1} \in K$. Therefore, $x^{-1}Kx \cap Km_0N_0 \neq \phi$, or $Kx^{-1}K \cap m_0x^{-1}N_0 \neq \phi$; taking inverses, we see that $KxK \cap xm_0^{-1}N_0 \neq \phi$, so $m_0 \in +M_0$.

Corollary 4.3.2. $1+\sigma(\overline{n}) \asymp 1+\langle \rho_P, H_P(\overline{n}) \rangle$ $(\overline{n} \in \overline{N})$.

Proof. Recall that, for any linear algebraic group, $1+\sigma$ depends upon the faithful representation used in defining σ only up to $\asymp$. Let π be a strongly rational representation with $2\rho_P$ as highest weight and let v_1 denote the highest weight vector. Then the mapping $\overline{n} \mapsto \pi(\overline{n})v_1$ injects $\overline{N}$ into V, since P is the full stability group of $\Omega \cdot v_1$. There is a basis $v_1, \ldots, v_r$ of V, where v_i is of the form $\pi(\overline{n}_i)v_1$ with $\overline{n}_i \in \overline{N}$, $i = 1, \ldots, r$. Since K is compact, there is a constant $C > 1$ such that $C^{-1}|\pi(\mu(\overline{n}))v_1| \leq |\pi(\overline{n})v_1| = |v_1 + \sum_{j>1} \pi_{1j}(\overline{n})v_j| \leq$ $\leq C|\pi(\mu(\overline{n}))v_1|$. We also have

$$|\pi(\mu(\bar{n}))v_i| = |\pi(\mu(\bar{n}\bar{n}_i))v_1| = q^{<\lambda_{2\rho}, H(\bar{n}\bar{n}_i)>}$$

$$= \max_{1< j\leq r} (1, |\pi_{1j}(\bar{n}\bar{n}_i)|),$$

or

$$1+<2\rho, H(\bar{n})> \asymp 1+\log_q(|\pi(\bar{n})|),$$

since $1+<2\rho, H(\bar{n}\bar{n}_i)> \asymp 1+<2\rho, H(\bar{n})>$. Since, furthermore, $1+\log_q|\pi(\bar{n})| \asymp 1+\log_q|\pi($ the coefficients of $\pi(\bar{n}^{-1})$ being polynomials in the coefficients of $\pi(\bar{n})$, and since $1+\sigma(\bar{n}) \asymp 1+\max(\log_q(|\pi(\bar{n})|), \log_q(|\pi(\bar{n}^{-1})|))$, we may conclude that $1+<\rho, H(\bar{n})> \asymp 1+\sigma$

Corollary 4.3.3. For any $\gamma > 0$ the set of all $\bar{n} \in \bar{N}$ such that $<\rho_P, H_P(\bar{n})> \leq \gamma$ is compact.

Proof. Since $1+<\rho_P, H_P(\bar{n})> \asymp 1+\sigma(\bar{n})$, it suffices to show that, for given $\gamma > 0$, the set $S = \{\bar{n} \in \bar{N} | \sigma(\bar{n}) \leq \gamma\}$ is compact. Assuming--as we may--that $\bar{N}$ is a group of unipotent matrices, we see that S may be regarded as a closed subset of some cartesian product of fractional ideals in Ω. This implies that S is compact.

Set $\gamma(a) = \gamma_P(a) = \inf_{\alpha\in\Sigma^0(P,A)} <\alpha, H_P(a)> \quad (a \in A)$.

Lemma 4.3.4. For any $\lambda \in (\mathfrak{a}^*)^+$ there exists a constant $c = c(\lambda) > 0$ such that $<\lambda, H_P(\bar{n}^a)> \leq \max(c, <\lambda, H_P(\bar{n})> - \gamma(a)+c)$ for all $a \in A^+$ and $\bar{n} \in \bar{N}$.

Proof. Without loss of generality let π be a strongly rational representation of G in a vector space V over Ω whose highest weight is λ. Let $v_\lambda = v_1$ be a highest weight vector and let $v_1, \ldots, v_t$ be a base for V with respect to which $\pi|A$ is diagonalized. Then $\pi(\bar{n}^a)v_1 = \pi(a)\pi(\bar{n})v_1\xi(a^{-1}) =$

$v_1 + \sum_{i=2}^{t} \xi_i(a)c_i(\bar{n})v_i$, where $\xi_i(a) = \lambda_i(a)\lambda(a)^{-1}$ and $\lambda_i = \lambda - \Sigma m_{i\ell}\alpha_\ell$ $(m_{i\ell} \geq 0,$

$\alpha_\ell \in \Sigma^0(P,A)$ and $\lambda_i \neq \lambda$ $(i > 1)$). Thus, $|\xi_i(a)| \leq q^{-\gamma(a)} \leq 1$ for $a \in A^+$. Hence, $|\pi(\bar{n}^a)v_1| \leq \max(1, |\pi(\bar{n})v_1|q^{-\gamma(a)})$ and this, combined with Lemma 4.3.1 and the fact that $|\pi(\bar{n}^a)v_1| \gtrsim q^{\langle\lambda, H_P(\bar{n}^a)\rangle}$, implies the lemma.

Taking $\lambda = 2\rho$, we obtain:

<u>Corollary</u> 4.3.5. There exists a positive constant c such that

$$\delta_P(\bar{n}^a)^{\frac{1}{2}} \leq c(1+\delta_P(\bar{n})^{\frac{1}{2}}q^{-\frac{1}{2}\gamma(a)})$$

for all $\bar{n} \in \bar{N}$ and $a \in A^+$.

Indeed this is true for $\bar{n} \in \bar{N}_0$, since $\bar{N}_0 = \bar{N} \cdot \bar{N}_0 \cap M$ and $\delta_P | \bar{N}_0 \cap M = 1$. For later use we formulate a consequence of Corollary 4.3.5:

<u>Corollary</u> 4.3.6. Put $\rho_0 = \rho_{P_0}$. Then there is a positive constant c_0 such that

$$\delta_{P_0}(\bar{n}^a)^{\frac{1}{2}} \leq c_0(1+\delta_{P_0}(\bar{n})^{\frac{1}{2}}q^{-\frac{1}{2}\gamma(a)})$$

for all $\bar{n} \in \bar{N}$ and $a \in A^+$.

Let $(P,A) \succ (P_0, A_0)$ and let $(\bar{P}, A)(\bar{P} = M\bar{N})$ be the opposite p-pair. Let $\rho_0 = \rho_{P_0}$ and $H_0 = H_{P_0}$. Note that

$$\int_{P_0\backslash G} d\bar{x} = \int_{P_0 \cap K\backslash K} d\bar{k} = \int_{\bar{N}} \delta_{P_0}^{-1}(\bar{n})d\bar{n} = \int_{\bar{N}} q^{-2\langle\rho_0, H_0(\bar{n})\rangle} d\bar{n} < \infty.$$

<u>Theorem</u> 4.3.7. Let r_0 be the minimum value of the constant of Theorem 4.2.1. Then

$$\int_{\bar{N}} q^{-\langle\rho_0, H_0(\bar{n})\rangle}(1+\sigma(\bar{n}))^{-r_0-\epsilon} d\bar{n} < \infty$$

for any $\epsilon > 0$.

<u>Proof.</u> Set ${}^*P_0 = P_0 \cap M = M_0 \cdot N_0 \cap M = M_0 \cdot {}^*N_0$. Then $N_0 = {}^*N_0 N$. It follows from Theorem 4.2.1 and Lemma 4.2.2 that there are positive constants c and r_0 such that

$$c(1+\sigma(a))^{r_0} \geq \int_{\bar{N}_0} q^{-\langle \rho_0, H_0(\bar{n}_0^a) + H_0(\bar{n}_0)\rangle} d\bar{n}_0$$

$$= \int_{{}^*\bar{N}_0 \bar{N}} q^{-\langle \rho_0, H_0(\bar{n}^a\, {}^*\bar{n}_0) + H_0(\bar{n}\, {}^*\bar{n}_0)\rangle} d\bar{n}\, d\, {}^*\bar{n}_0$$

for $a \in A^+$. Since ${}^*\bar{n}_0 = {}^*k\, {}^*m_0\, {}^*n_0$ (${}^*k \in K_M$, ${}^*m_0 \in M_0$, ${}^*n_0 \in {}^*N_0$), we have $H_0(\bar{n}\, {}^*\bar{n}_0) = H_0(\bar{n}\, {}^*k\, {}^*m_0) = H_0(\bar{n}\, {}^*k) + H_0({}^*\bar{n}_0)$; similarly, $H_0(\bar{n}^a\, {}^*n_0) = H_0(\bar{n}^a\, {}^*k) + H_0({}^*\bar{n}_0)$. Note that ${}^*k$ commutes with $a \in A$ and normalizes $\bar{N}$. Observing also that $\langle \rho_0, H_0({}^*\bar{n}_0)\rangle = \langle {}^*\rho_0, H_0({}^*\bar{n}_0)\rangle$ (${}^*\rho_0 = \frac{1}{2}\Sigma\alpha$, $\alpha \in \Sigma({}^*P_0, A_0)$) and that $\int_{{}^*\bar{N}_0} q^{-2\langle {}^*\rho_0, H_0({}^*\bar{n}_0)\rangle} d\, {}^*\bar{n}_0$ converges, we have

$$c'\,(1+\sigma(a))^{r_0} \geq \int_{\bar{N}} q^{-\langle \rho_0, H_0(\bar{n}^a) + H_0(\bar{n})\rangle} d\bar{n} \quad (a \in A^+).$$

Now fix $a_0 \in A$ such that $2\gamma = \gamma_P(a_0) > 0$. Then, if $a = a_0^t$, where t is a positive integer, we obtain, using Corollary 4.3.6, the relations

$$q^{\langle \rho_0, H_0(\bar{n}^a)\rangle} \leq c_0 \,(1+q^{\langle \rho_0, H_0(\bar{n})\rangle - \gamma t})$$

i.e.,

$$q^{-\langle \rho_0, H_0(\bar{n}^a) + H_0(\bar{n})\rangle} \geq \frac{c_0^{-1} q^{-\langle \rho_0, H_0(\bar{n})\rangle}}{1+q^{\langle \rho_0, H_0(\bar{n})\rangle - t\gamma}} .$$

It follows that

$$\int_{\bar{N}} q^{-\langle\rho_0, H_0(\bar{n})\rangle}(1+q^{\langle\rho_0, H_0(\bar{n})\rangle - t\gamma})^{-1} d\bar{n} \leq c_1(1+t)^{r_0}.$$

Let $\bar{N}_r = \{\bar{n} \in \bar{N} | \langle\rho_0, H_0(\bar{n})\rangle \leq \gamma 2^r\}$. Then, if $t = 2^r$ and $\bar{n} \in \bar{N}_r$, we have $q^{\langle\rho_0, H_0(\bar{n})\rangle - t\gamma} \leq 1$. It follows that $\frac{1}{2}\int_{\bar{N}_r} q^{-\langle\rho_0, H_0(\bar{n})\rangle} d\bar{n} \leq c_1(1+\gamma 2^r)^{r_0} \leq c_2 2^{rr_0}$.

Letting $\bar{N}(r) = \bar{N}_r - \bar{N}_{r-1}$ $(r \geq 1)$, we have

$$\int_{\bar{N}(r)} q^{-\langle\rho_0, H_0(\bar{n})\rangle}(1+|\langle\rho_0, H_0(\bar{n})\rangle|)^{-r_0-\epsilon} d\bar{n} \leq \frac{c_3 2^{rr_0}}{2^{(r-1)(r_0+\epsilon)}} = c_4 2^{-(r-1)\epsilon}.$$

By Corollary 4.3.3 the set $\bar{N}_\gamma$ of all $\bar{n} \in \bar{N}$ such that $\langle\rho_0, H_0(\bar{n})\rangle \leq \gamma$ is compact, so that $\bar{N} - \bigcup_{r>1} \bar{N}(r)$ is also compact. Using Corollary 4.3.2 to replace $|\langle\rho_0, H_0(\bar{n})\rangle|$ by $\sigma(\bar{n})$, we obtain the present theorem by summing the geometric series $\sum_{r>0} 2^{-(r-1)\epsilon}$.

Let Ξ_M be the function defined in §4.2. We let $\rho = \rho_P$, $H = H_P$, and $H_0 = H_{P_0}$.

<u>Corollary</u> 4.3.8. $\int_{\bar{N}} q^{-\langle\rho, H(\bar{n})\rangle} \Xi_M(\bar{n})(1+\sigma(\bar{n}))^{-r_0-\epsilon} d\bar{n} < \infty$.

<u>Proof.</u> It is clear that, if $\bar{n} = kmn$ $(k \in K,\ m \in M,\ n \in N)$, then $H_0(\bar{n}) = H_0(m) = H(m) + {}^*H_0(m) = H(\bar{n}) + {}^*H_0(m)$. Hence, $\langle\rho_0, H_0(\bar{n})\rangle = \langle\rho, H(\bar{n})\rangle + \langle{}^*\rho_0, {}^*H_0(m)\rangle$. Now let $k \in K_M$. Then k normalizes N, so

$$\begin{aligned}\langle\rho_0, H_0(\bar{n}^k)\rangle &= \langle\rho_0, H_0(\bar{n}k^{-1})\rangle \\ &= \langle\rho, H(\bar{n})\rangle + \langle{}^*\rho_0, {}^*H_0(mk^{-1})\rangle.\end{aligned}$$

It follows that

$$\int_{K_M} q^{-\langle \rho_0, H_0(\bar{n}^k)\rangle} dk = q^{-\langle \rho, H_P(\bar{n})\rangle} \int_{K_M} q^{-\langle {}^*\rho_0, {}^*H_0(mk)\rangle} dk$$

$$= q^{-\langle \rho, H_P(\bar{n})\rangle} \Xi_M(m)$$

$$= q^{-\langle \rho, H_P(\bar{n})\rangle} \Xi_M(\bar{n}) .$$

Assuming that $\sigma(k_1 \times k_2) = \sigma(x)$ $(k_1, k_2 \in K,\ x \in G)$, we have

$$\int_{K_M} q^{-\langle \rho_0, H_0(\bar{n}^k)\rangle} (1+\sigma(\bar{n}^k))^{-r_0-\epsilon} dk$$

$$= q^{-\langle \rho, H_P(\bar{n})\rangle} \Xi_M(\bar{n})(1+\sigma(\bar{n}))^{-r_0-\epsilon}$$

To complete the proof use Theorem 4.3.7, the fact that $k \in K_M$ normalizes $\bar{N}$, and Fubini's theorem.

Corollary 4.3.9. Let $c > 0$ be a constant such that $c + \langle \rho, H_P(\bar{n})\rangle \geq 1$ for all $\bar{n} \in \bar{N}$. Then $\int_{\bar{N}} q^{-\langle \rho, H_P(\bar{n})\rangle} \Xi_M(\bar{n})(c + \langle \rho, H_P(\bar{n})\rangle)^{-r_0-\epsilon} d\bar{n} < \infty$.

Proof. This follows from Corollary 4.3.8 and Corollary 4.3.2.

Corollary 4.3.10. For any $\epsilon > 0$

$$\int_{\bar{N}} \Xi_M(\bar{n}) q^{-(1+\epsilon)\langle \rho, H_P(\bar{n})\rangle} d\bar{n} < \infty.$$

Proof. Immediate from Corollary 4.3.9.

Assume (as always) that $\sigma(k_1 x k_2) = \sigma(x)$ ($k_1, k_2 \in K$ and $x \in G$). Set $\sigma^*_M(m) = \inf_{a \in A} \sigma(ma)$ ($m \in M$). Extend σ^*_M to a function defined on G by setting $\sigma^*_M(kmn) = \sigma^*_M(m)$ ($k \in K$, $m \in M$, $n \in N$).

In the following let $\pi = \pi_{2\rho_0}$ be an irreducible rational representation of G in an Ω-vector space V whose highest weight is $2\rho_0$ ($\rho_0 = \rho_{P_0} = \frac{1}{2}\Sigma\alpha$, $\alpha \in \Sigma(P_0, A_0)$). Fix a base for V with respect to which $\pi|A$ is diagonalized. Let E be the projection, relative to the fixed base, of V on the subspace which transforms under $\pi|A$ as $2\rho_P$. Notice: (1) that $\pi|M$ commutes with E; (2) the highest weight vector lies in the image of E.

Lemma 4.3.11. We can choose $0 < c_1 \leq c_2$ and $d \geq 0$ such that

$$c_1 \frac{\Xi_M(x)}{(1+\sigma^*_M(x))^d} \leq |\pi(x)E|^{-\frac{1}{2}} q^{\langle \rho_P, H_P(x) \rangle} \leq c_2 \Xi_M(x)$$

for all $x \in G$.

Proof. It is obviously sufficient to prove these inequalities for $x \in M$. Put ${}^*P_0 = P_0 \cap M$. Then $M = K_M A^+_{0,M} \omega K_M$, where ω is a finite subset of M_0 ($P_0 = M_0 N_0$) and $A^+_{0,M}$ is the set consisting of all $a \in A_0$ such that $|\alpha(a)| \geq 1$ for all $\alpha \in \Sigma^0({}^*P_0, A_0)$. It follows that $A^+_{0,M} = A^+_0 \cdot A$. Hence $M = (K_M A^+_0 \omega K_M) A$. Write ${}^*\rho_0 = \frac{1}{2}\Sigma\alpha$, α ranging over $\Sigma({}^*P_0, A_0)$.

By Theorem 4.2.1 there exist positive constants c and d such that

$$\Xi_M(a)(1+\sigma^*_M(a))^{-d} \leq c q^{-\langle {}^*\rho_0, H_0(a) \rangle}$$

for all $a \in A^+_{0,M}$. Since $|\pi(a)E| = q^{2\langle \rho_0, H_0(a) \rangle}$ ($a \in A^+_{0,M}$), i.e., $|\pi(a)E|^{-\frac{1}{2}} q^{\langle \rho, H(a) \rangle} = q^{-\langle {}^*\rho_0, H_0(a) \rangle}$ ($a \in A^+_{0,M}$), we see that $\Xi_M(a)(1+\sigma^*_M(a))^{-d} \leq c |\pi(a)E|^{-\frac{1}{2}} q^{\langle \rho, H(a) \rangle}$ ($a \in A^+_{0,M}$). Similarly, knowing that there is a constant $c' > 0$ such that $q^{<{}^*\rho_0, H_0(a)>}\, \Xi_M(a) \geq c'$ $(a \in A^+_{0,M})$, we may conclude that $\Xi_M(a) \geq c' |\pi(a)E|^{-\frac{1}{2}} q^{<\rho, H(a)>}$ for all $a \in A^+_{0,M}$. Since ω is finite and $M = K_M \omega A^+_{0,M} K_M$, the required result follows easily.

Lemma 4.3.12. For $x \in G$ put $M(x) = M \cap KxN$. Let f be a real-valued measurable function on M and put $g(x) = \sup_{m \in M(x)} f(m)$. Then g is measurable on G.

Proof. Let μ^* denote the projection of G on the discrete space $K\backslash G/N$. Since K is open in G, μ^* is continuous. Set $f_0(m) = \sup_{k \in K_M} f(km)$ $(m \in M)$. Then f_0 is a measurable function on $K_M \backslash M \cong K\backslash G/N$. Since $g(x) = f_0(\mu^*(x))$ and μ^* is continuous, g is measurable.

Lemma 4.3.13. Let π be a representation of $A \cdot N$ such that $\pi|A$ is scalar-valued. Then $\pi|N = 1$.

Proof. For any $n \in N$ and $a \in A$ we have $\pi(n^a) = \pi(n)$. It is sufficient to show that the commutator subgroup of $A \cdot N = N$.

Choose $a \in A$ such that $|\alpha(a)| > 1$ for all positive roots α. Then $n \mapsto n^a n^{-1}$ defines an algebraic isomorphism of N onto N. Thus, N lies in the commutator of $A \cdot N$. Since $A \cdot N/N \cong A$, an abelian group, N equals the commutator of $A \cdot N$.

Recall that $\sigma_*(x) = \min_{z \in Z} \sigma(xz)$.

Lemma 4.3.14. Given any compact subset ω of M, we can choose positive constants c and d such that

$$\int_{\bar{N}} \sup_{m \in M(\bar{n})} |\pi(m_0 \bar{n}^a m^{-1})E|^{-\frac{1}{2}} \delta_P(\bar{n})^{-1} d\bar{n} \leq c(1+\sigma_*(a))^d$$

for all $m_0 \in \omega$ and $a \in A^+$ ($M(\bar{n})$ as in Lemma 4.3.12).

Proof. Using Lemma 4.3.11, we have a positive constant c_2 such that $|\pi(x)E|^{-\frac{1}{2}} \leq c_2 \, \Xi_M(x) q^{-\langle \rho, H_P(x) \rangle}$ for all $x \in G$. Therefore, there is a positive constant c_3 such that

$$\int_K |\pi(xk)E|^{-\frac{1}{2}} dk \leq c_2 \int_K \Xi_M(xk) q^{-\langle \rho, H_P(xk) \rangle} dk = c_3 \Xi(x),$$

(Lemma 4.2.4). Without loss of generality we may assume that, for $k \in K_M$, the matrix $E\pi(k) = \pi(k)E$ has integer entries; from Lemma 4.3.13 it follows that $\pi(n)E = E$ for all $n \in N$. Thus, the function $k \mapsto |\pi(xk)E|$ may be regarded as a function defined on $K/K \cap P$. Changing variables, we obtain

$$\int_{\bar{N}} |\pi(x\kappa(\bar{n}))E|^{-\frac{1}{2}} \delta_P(\bar{n})^{-1} d\bar{n} \leq c_3 \, \Xi(x)$$

($\bar{n} = \kappa(\bar{n})\mu(\bar{n})n(\bar{n}) \in KMN$, as usual). Since $\kappa(\bar{n}) \in \bar{n}\mu(\bar{n})^{-1}N$, we have, for $x = m_0 a$,

$$\int_{\bar{N}} |\pi(m_0 a \bar{n} \mu(\bar{n})^{-1})E|^{-\frac{1}{2}} \delta_P(\bar{n})^{-1} d\bar{n} \leq c_3 \, \Xi(m_0 a).$$

Since $\pi(a)E = (2\rho)(a)E$ $(a \in A)$, it follows that

$$\int_{\bar{N}} |\pi(m_0 \bar{n}^a \mu(\bar{n})^{-1})E|^{-\frac{1}{2}} \delta_P(\bar{n})^{-1} d\bar{n} \leq c_3 \, \Xi(m_0 a) q^{\langle \rho, H(a) \rangle}.$$

For $a \in A$ one sees that $\langle \rho, H(a) \rangle = \langle \rho_0, H_0(a) \rangle$, so the required inequality follows from Theorem 4.2.1.

As usual, put $\gamma_P(a) = \inf_{\alpha \in \Sigma^0(P,A)} \langle \alpha, H_P(a) \rangle \quad (a \in A)$.

Lemma 4.3.15. Fix $\epsilon > 0$ and a compact subset ω of M. Then the

$$\lim_{\gamma_P(a)\to\infty} \int_{\bar{N}} |\pi(m\bar{n}^a \mu(\bar{n})^{-1}E|^{-\frac{1}{2}} q^{\epsilon\langle\rho, H_P(\bar{n}^a)-H_P(\bar{n})\rangle} \delta_P(\bar{n})^{-1} d\bar{n}$$

exists uniformly for $m \in \omega$.

Proof. For C a compact subset of $\bar{N}$, write $C' = \bar{N}-C$. We shall show that, for C and $\gamma_P(a)$ sufficiently large, the integral over C' is arbitrarily small for all $m \in \omega$. Observe that, having proved this, we have proved the lemma.

For $a \in A^+$ write $C'_a = \{\bar{n} \in C' \,|\langle\rho, H_P(\bar{n})\rangle \leq \frac{1}{2}\gamma_P(a)\}$. Then C'_a is relatively compact in $\bar{N}$ (Corollary 4. 3. 3). Let $C''_a = C' - C'_a$. Set $f(m, a, \bar{n}) = |\pi(m\bar{n}^a \mu(\bar{n})^{-1}E|^{-\frac{1}{2}} q^{\epsilon\langle\rho, H_P(\bar{n}^a)-H_P(\bar{n})\rangle} \delta_P(\bar{n})^{-1}$. For any measurable set $U \subset \bar{N}$, define $J_{m,a}(U) = \int_U f(m, a, \bar{n}) d\bar{n}$. Clearly, $J_{m,a}(C') = J_{m,a}(C'_a) + J_{m,a}(C''_a)$.

Since ω is compact, there is a constant $c_1 > 0$ such that $|\pi(mx)E| \geq c_1|\pi(x)E|$ for all $m \in \omega$ and $x \in G$. It follows from the definition of E that $\pi(\bar{n}m)E = \pi(m)E + (1-E)\pi(\bar{n}m)E$ for all $\bar{n} \in \bar{N}$ and $m \in M$. Therefore, there exists a constant $c_2 > 0$ such that $|\pi(\bar{n}m)E| \geq c_2|\pi(m)E|$ for all $\bar{n} \in \bar{N}$ and $m \in M$. Thus for any $m \in \omega$, we have $|\pi(m\bar{n}^a \mu(\bar{n})^{-1})E| \geq c_1|\pi(\bar{n}^a \mu(\bar{n})^{-1})E| \geq c_1 c_2|\pi(\mu(\bar{n})^{-1})E|$. Consequently, $|\pi(m\bar{n}^a \mu(\bar{n})^{-1})E|^{-\frac{1}{2}} \leq c_2^{-1}|\pi(\mu(\bar{n})^{-1})E|^{-\frac{1}{2}}$, which, by Lemma 4. 3.11, is dominated by $\Xi_M(\mu(\bar{n}))q^{-\langle\rho, H_P(\bar{n})\rangle}$ for all $\bar{n} \in \bar{N}$. We shall also need the fact, implied by Corollary 4. 3. 5, that there is a constant $c > 0$ such that $q^{\langle\rho, H_P(\bar{n}^a)-H_P(\bar{n})\rangle} \leq 2c \max(q^{-\langle\rho_P, H_P(\bar{n})\rangle}, q^{-\frac{1}{2}\gamma_P(a)})$ for all $\bar{n} \in \bar{N}$ and $a \in A^+$.

We conclude that, with a new constant $c_3 > 0$,

$$J_{m,a}(C'_a) \leq c_3 \int_{C'_a} \Xi_M(\mu(\bar{n}))(2c)^{\epsilon} q^{-(1+\epsilon)\langle\rho, H_P(\bar{n})\rangle} d\bar{n}$$

$$\leq c_4 \int_{C'} \Xi_M(\bar{n}) q^{-(1+\epsilon)\langle\rho, H_P(\bar{n})\rangle} d\bar{n}$$

$(c_4 = c_3(2c)^{\epsilon})$. It follows from Corollary 4.3.10 that $J_{m,a}(C'_a)$ can be made arbitrarily small for all $m \in \omega$ and $a \in A^+$ by taking C sufficiently large.

On the other hand, if $m \in \omega$ and $\bar{n} \in C''_a$, we have, for $c_5 > 0$,

$$f(m, a, \bar{n}) \leq (2c)^{\epsilon} q^{-\frac{\epsilon}{2}\gamma_P(a)} c_5^{-\frac{1}{2}} |\pi(\bar{n}^a \mu(\bar{n})^{-1})E|^{-\frac{1}{2}} \delta_P(\bar{n})^{-1}.$$

Put $c_6 = (2c)^{\epsilon} c_5^{-\frac{1}{2}}$. Then, by Lemma 4.3.14,

$$J_{m,a}(C''_a) \leq c_6 q^{-\frac{\epsilon}{2}\gamma_P(a)} \int_{\bar{N}} |\pi(\bar{n}^a \mu(\bar{n})^{-1})E|^{-\frac{1}{2}} \delta_P(\bar{n})^{-1} d\bar{n}$$

$$\leq c_7 q^{-\frac{\epsilon}{2}\gamma_P(a)} (1+\sigma_*(a))^d \to 0, \quad \gamma_P(a) \to \infty, \text{ and we are done.}$$

Theorem 4.3.16. Fix $\epsilon > 0$ and a compact subset ω of M. Then there exists $d > 0$ such that

$$\lim_{\gamma_P(a)\to\infty} \int_{\bar{N}} \frac{\Xi_M(\bar{n}^{ma} m\mu(\bar{n})^{-1}) q^{-(1-\epsilon)\langle\rho, H_P(\bar{n}^{ma})\rangle - (1+\epsilon)\langle\rho, H_P(\bar{n})\rangle}}{(1+\sigma^*_M(\bar{n}^{ma} m\mu(\bar{n})^{-1}))^d} d\bar{n}$$

exists uniformly for $m \in \omega$.

Proof. Using Lemma 4.3.11, we have $c_1, d > 0$ such that

$$|\pi(m\bar{n}^a \mu(\bar{n})^{-1})E|^{-\frac{1}{2}} \geq c_1 \frac{\Xi_M(m\bar{n}^a \mu(\bar{n})^{-1}) q^{-\langle\rho, H_P(m\bar{n}^a) - H_P(\bar{n})\rangle}}{(1+\sigma^*_M(m\bar{n}^a \mu(\bar{n})^{-1}))^d}$$

for all $m \in \omega$, $\bar{n} \in \bar{N}$, and $a \in A^+$. Together with Lemma 4.3.15 this implies the present theorem.

Corollary 4.3.17.

$$\lim_{\gamma_P(a)\to\infty} \int_{\bar{N}} \frac{\Xi_M(\bar{n}m\mu(\bar{n}^{am^{-1}})^{-1})\, q^{-(1-\epsilon)\langle\rho, H_P(\bar{n}^{am^{-1}})\rangle - (1+\epsilon)\langle\rho, H_P(\bar{n})\rangle}}{(1+\sigma_M^*(\bar{n}m\mu(\bar{n}^{am^{-1}})^{-1}))^d}\, d\bar{n}$$

exists uniformly for $m \in \omega$.

Proof. Note that $\Xi_M(m) = \Xi_M(m^{-1})$ and $\sigma_M^*(m) = \sigma_M^*(m^{-1})$ for all $m \in M$. Therefore,

$$\begin{aligned} \Xi_M(\bar{n}^{ma} m\mu(\bar{n})^{-1}) &= \Xi_M(\mu(\bar{n}^{ma})m\mu(\bar{n})^{-1}) \\ &= \Xi_M(\mu(\bar{n})m^{-1}\mu(\bar{n}^{ma})^{-1}) \\ &= \Xi_M(\bar{n}m^{-1}\mu(\bar{n}^{ma})^{-1}). \end{aligned}$$

Similarly, $\sigma_M^*(\bar{n}^{ma}m\mu(\bar{n})^{-1}) = \sigma_M^*(\bar{n}m^{-1}\mu(\bar{n}^{ma})^{-1})$. To complete the proof replace m by m^{-1} and use Theorem 4.3.16.

Recall that $\rho_0 = \rho_{P_0}$ and $H_0 = H_{P_0}$.

Lemma 4.3.18. For all $\bar{n} \in \bar{N}$ and $m \in M$

$$\sigma_*(\bar{n}m) \asymp \max(\sigma(\bar{n}), \sigma_*(m)).$$

Proof. There exists an embedding of $M\bar{N}$ as the rational points of a product variety. It follows in this case that $\sigma(\bar{n}m) = \max(\sigma(\bar{n}), \sigma(m))$; moreover, $\sigma(\bar{n}) = \sigma_*(\bar{n})$, so $\sigma_*(\bar{n}m) = \max(\sigma(\bar{n}), \sigma_*(\overline{m}))$.

Theorem 4.3.19. Fix $r > r' \geq 0$. Then there exist positive constants c and r_0 such that

$$\delta_P(m)^{\frac{1}{2}} \int_{\bar{N}} \Xi(\bar{n}m)(1+\sigma_*(\bar{n}m))^{-2r_0-r} d\bar{n} \le c\Xi_M(m)(1+\sigma_*(m))^{-r'} ,$$

for all $m \in M$.

Proof. In M we have the minimal p-pair $({}^*P_0, A_0)$, where ${}^*P_0 = P_0 \cap M = M_0 {}^*N_0$. We have the mapping $H_{{}^*P_0} : M \to \mathfrak{a}_0$; we write $M^+_{0,M} = H^{-1}_{M_0}({}^*\mathfrak{a}^+_0)$, where ${}^*\mathfrak{a}^+_0 \subset \mathfrak{a}_0$ denotes the subset which is nonnegative on all $\alpha \in \Sigma^0({}^*P_0, A_0)$. Then $M = K_M(M^+_{0,M}/{}^0M_0)K_M$. It is sufficient to verify the theorem for $m \in M^+_{0,M}$.

For $\bar{n}m \in \bar{N}M$ we have $m_0 \in M^+_0$ such that $\bar{n}m = k_1 m_0 k_2$ $k_1, k_2 \in K)$. Since $\Xi(\bar{n}m) = \Xi(m_0) \prec q^{-\langle\rho_0, H_0(m_0)\rangle}(1+\sigma_*(m_0))^{r_0}$ (Theorem 4.2.1), we have, using Lemma 4.3.18,

$$\Xi(\bar{n}m)(1+\sigma_*(\bar{n}m))^{-2r_0-r-\epsilon} \prec q^{-\langle\rho_0, H_0(m_0)\rangle}(1+\sigma_*(m))^{-r}(1+\sigma(\bar{n}))^{-r_0-\epsilon}$$

$$\prec q^{-\langle\rho_0, H_0(\bar{n}m)\rangle}(1+\sigma_*(m))^{-r}(1+\sigma(\bar{n}))^{-r_0-\epsilon}$$

for any $\epsilon > 0$ and all $\bar{n}m \in \bar{N}M$, $m_0 = m_0(\bar{n}m)$. Therefore, by Theorem 4.2.1, Lemma 4.2.3, Corollary 4.3.2, and Theorem 4.3.7, combined with the fact that $H_0(\bar{n}m) = H_0(\bar{n})+H_0(m)$,

$$\int_{\bar{N}} \Xi(\bar{n}m)(1+\sigma_*(\bar{n}m))^{-2r_0-r-\epsilon} d\bar{n} \prec q^{-\langle\rho_0, H_0(m)\rangle}(1+\sigma_*(m))^{-r}.$$

Recalling that $\rho_0 = {}^*\rho_0 + \rho_P$ and applying Theorem 4.2.1 again, we obtain

$$q^{-\langle\rho_0, H_0(m)\rangle}(1+\sigma_*(m))^{-r} \prec q^{-\langle\rho, H(m)\rangle}\Xi_M(m)(1+\sigma_*(m))^{-r}$$

for all $m \in M^+_{0,M}$. The theorem follows immediately.

In the same way one proves:

Theorem 4.3.20. Fix $r > r' \ge 0$. Then there exist positive constants c and r_0 such that

$$\delta_P(m)^{\frac{1}{2}}\int_{\bar{N}} \Xi(\bar{n}m)(1+\sigma(\bar{n}m))^{-2r_0-r}\, d\bar{n} \lesssim \Xi_M(m)(1+\sigma(m))^{r'}$$

for all $m \in M$.

§4.4. The Schwartz Spaces and Square Integrable Forms.

All preparations having been completed, we set forth in this section the definitions and basic properties of the Schwartz spaces.

Fix a minimal p-pair (P_0, A_0) $(P_0 = M_0N_0)$ of G and an A_0-good maximal compact subgroup K of G. Note that these choices determine the function Ξ of §4.2.

Let K_0 be an open compact subgroup of G. Write $\mathcal{C}_{K_0}(G)[\mathcal{C}_{*K_0}(G)]$ for the set of all $f \in C(G/\!/K_0)$ which satisfy the following condition: For every $r > 0$ there exists a positive constant $C = C_{r,f}$ such that $|f(x)| \leq C\Xi(x)(1+\sigma(x))^{-r}$ $[|f(x)| \leq C\,\Xi(x)(1+\sigma_*(x))^{-r}]$ for all $x \in G$. Set $\mathcal{C}(G) = \bigcup_{K_0} \mathcal{C}_{K_0}(G)[\mathcal{C}_*(G) = \bigcup_{K_0} \mathcal{C}_{*K_0}(G)]$. The spaces $\mathcal{C}(G)$ and $\mathcal{C}_*(G)$ are called the <u>Schwartz spaces</u> of G.

In general if we say <u>the</u> Schwartz space of G we refer to $\mathcal{C}(G)$. To distinguish the two spaces we observe that $\mathcal{C}(G) \subset \mathcal{C}_*(G)$ and call $\mathcal{C}_*(G)$ the "big" or "star" Schwartz space; sometimes we call $\mathcal{C}(G)$ the "little" Schwartz space. If G is semi-simple, then the two spaces coincide. Given $\chi \in \hat{Z}$, we write $\mathcal{C}_*(G,\chi) = \{f \in \mathcal{C}_*(G) \mid f(xz) = \chi(z)f(x)\ \ (x \in G,\ z \in Z)\}$. If (V,τ) is a double representation of K, then we write $\mathcal{C}(G,\tau)$, $\mathcal{C}_*(G,\tau)$, or $\mathcal{C}_*(G,\tau,\chi)$ for the intersection of the appropriate Schwartz space tensored with V and $C(G,\tau)$.

We note that the spaces $\mathcal{C}(G)$ and $\mathcal{C}_*(G)$ are complete locally convex Hausdorff spaces, whose topologies may be defined in terms of the following, respective, semi-norms: For $f \in C(G/\!/K_0)$ and $r > 0$ set

$\nu_r(f) = \sup_{x \in G} |f(x)| \; \Xi^{-1}(x)(1+\sigma(x))^r [\nu_{*r}(f) = \sup_{x \in G} |f(x)| \; \Xi^{-1}(x)(1+\sigma_*(x))^r]$, provided these numbers are finite. One sees that $\mathcal{C}_{K_0}(G)[\mathcal{C}_{*K_0}(G)] =$ $\{f \in C(G/\!/K_0) | \nu_r(f) < \infty [\nu_{*r}(f) < \infty]$ for all $r > 0\}$, that $C_c^\infty(G) \subset \mathcal{C}(G)$, and that $C_c^\infty(G,\chi) \subset \mathcal{C}_*(G,\chi)$ for any $\chi \in \hat{Z}$; observe also that the inclusion maps are continuous and the images are of course dense. (The topology on $C_c^\infty(G)$ or $C_c^\infty(G,\chi)$ is defined as follows: $f_n \to 0$, $n \to \infty$ if and only if $f_n \in C_c(X/\!/K_0)$ or $C_c(XZ/\!/K_0, \chi)$ ($n > n_0$; X compact) and $f_n(x) \to 0$, $n \to \infty$ ($x \in G$).)

Theorem 4.4.1. The inclusion maps below are continuous.

(1) $\mathcal{C}(G) \subset L^2(G)$.

(2) For any $\chi \in \hat{Z}$: $\mathcal{C}_*(G,\chi) \subset L^2(G,\chi)$.

Proof. The containments follow immediately from Lemma 4.2.5.

Let us check continuity in (1); the proof of (2) goes similarly. Let $f_n \to f_0$, $n \to \infty$ with respect to the Schwartz topology $(\{f_n\}_{n=0}^\infty \subset \mathcal{C}(G))$. Then $\nu_r(f_n - f_0) \to 0$, $n \to \infty$ for every $r > 0$, so, for r large enough, we have

$$\int_G |f_n - f_0|^2 dg \leq \nu_r^2(f_n - f_0) \int_G \frac{\Xi^2(x)}{(1+\sigma(x))^{2r}} dx \to 0, \; n \to \infty.$$

Theorem 4.4.2. Both multiplications below are continuous.

(1) $\mathcal{C}(G)$ is an algebra under convolution on G.

(2) For any $\chi \in \hat{Z}$, $\mathcal{C}_*(G,\chi)$ is an algebra under convolution on G/Z.

Proof. We shall check only (1), the proof of (2) being similar.

Let $f_1, f_2 \in \mathcal{C}_{K_0}(G)$. It follows from Theorem 4.4.1 that $f_1 * f_2$ exists as an element of $C(G//K_0)$. We must show that $f_1 * f_2(y) \prec \Xi(y)(1+\sigma(y))^{-r}$ for all $r > 0$. Note that, for all $x, x^{-1}y \in G$, $\sigma(x)+\sigma(x^{-1}y) \geq \sigma(y)$, so $(1+\sigma(x))(1+\sigma(x^{-1}y) \geq 1+\sigma(y)$. It follows that, for every positive r_1 and r_2, we have

$$|f_1(x)f_2(x^{-1}y)| \leq C_{r_1, r_2} \Xi(x)\Xi(x^{-1}y)(1+\sigma(x))^{-r_1}(1+\sigma(x^{-1}y))^{-r_2}$$

$$\leq C_{r_1, r_2} \Xi(x)\Xi(x^{-1}y)(1+\sigma(x))^{r_2-r_1}(1+\sigma(y))^{-r_2}$$

for some $C_{r_1,r_2} > 0$. Take $r_1 - r_2 > r$, where r is given by Lemma 4.2.5(1). Since $\Xi(x) = \Xi(x^{-1})$, $\sigma(k_1 x k_2) = \sigma(x)$ and $\Xi(k_1 x k_2) = \Xi(x)$ $(k_1, k_2 \in K; x \in G)$, and $\int_K \Xi(xky)dk = \Xi(x)\Xi(y)$ $(\int_K dk = 1;$ $x, y \in G)$,

$$|\int_G f_1(x)f_2(x^{-1}y)dx| \leq C_{r_1, r_2} \int_G \Xi(x)\Xi(x^{-1}y)(1+\sigma(x))^{-r}dx(1+\sigma(y))^{-r_2}$$
$$= C_{r_1, r_2} \int_G \int_K \Xi(x)\Xi(x^{-1}ky)(1+\sigma(x))^{-r}dx(1+\sigma(y))^{-r_2}$$
$$= C_{r_1, r_2} \int_G \Xi^2(x)(1+\sigma(x))^{-r}dx\Xi(y)(1+\sigma(y))^{-r_2}.$$
$$\leq C\Xi(y)(1+\sigma(y))^{-r_2},$$

where C is a positive constant. The proof of the continuity of the convolution product is left to the reader.

Let $f \in \mathcal{C}_*(G)$ and let (P, A) $(P = MN)$ be a p-pair of G. Set $f^P(m) = \delta_P^{\frac{1}{2}}(m)\int_N f(mn)dn$ $(m \in M)$.

Theorem 4.4.3. For any $f \in \mathcal{C}_*(G)$ and $m \in M$ the integral defining $f^P(m)$ converges absolutely. The mapping $f \mapsto f^P$ defines a continuous linear mapping from $\mathcal{C}_*(G)$ to $\mathcal{C}_*(M)$; in fact, for every $r > 0$ there exists a constant $c(r, f) > 0$ such that $|f^P(m)| \leq c(r, f)\ \Xi_M(m)(1+\sigma_*(m))^{-r}$ for all $m \in M$. Let (V, τ) be a double representation of K which satisfies the associativity conditions of §1.12. Defining $f \mapsto f^P$ for $f \in \mathcal{C}(G, \tau)$ as above, we obtain a continuous homomorphism of algebras from $\mathcal{C}(G, \tau)$ to $\mathcal{C}(M, \tau_M)$.

Proof. The absolute convergence of the integral, the continuity of the mapping, and the inequality stated above follow from Theorem 4.3.19 (to see that $f^P \in \mathcal{C}_*(M)$ one must observe that $f \in \mathcal{C}_{*K_0}(G)$ implies $f^P \in C(M/\!/K_0 \cap M)$). The final statement follows, by taking limits, from Lemma 2.7.5.

Let $\chi_1, \ldots, \chi_{\ell_0}$ be a set of generators for $X(G)$. Define $\nu(x)$ by setting $q^{\nu(x)} = \max_{1 \le i \le \ell_0} (|\chi_i(x)|, |\chi_i(x^{-1})|)$ $(x \in G)$. For any $T > 0$ set $G_T = \{x \in G \mid \nu(x) \le T\}$. Notice that ${}^0G G_T {}^0G = G_T$ for any $T > 0$ and that the following theorem, in effect, characterizes the intersection of $\mathcal{C}_*(G)$ and $\mathcal{A}(G)$. Let (P, A) $(P = MN)$ be a p-pair and note that $a \underset{P}{\to} \infty$, means that $|\alpha(a)| \to \infty$ for all $\alpha \in \Sigma^0(P, A)$, whereas $a \to \infty$ simply means that there exists $\chi \in X(A)$ such that $|\chi(a)| \to \infty$. Also notice that, if $f \in C(M/\!/M_0)$ (M_0 open and compact), then $f(am) \to 0$, $a \to \infty$, for all $m \in M$ if and only if $f(am) \to 0$, $a \to \infty$, uniformly on compact subsets of M.

Theorem 4.4.4. Let $f \in \mathcal{A}(G)$. Then the following statements are equivalent:

(1) For every $T \ge 0$, $\int_{G_T} |f(x)|^2 dx < \infty$.

(2) For every $T \ge 0$, every standard p-pair (P, A) $(P = MN)$, and every compact subset ω of M, the limit $\|\underset{\sim}{f}_P(ma)\| \to 0$, $a \underset{P}{\to} \infty$ $(a \in G_T)$, exists uniformly on ω.

(3) For every $T \ge 0$, every standard p-pair (P, A) $(P = MN)$, and every compact subset ω of M, the limit $\|\underset{\sim}{f}_P(ma)\| \to 0$, $a \to \infty$ $(a \in G_T \cap A^+(P))$, exists uniformly on ω.

(4) Let $\alpha_1, \ldots, \alpha_\ell$ denote the simple roots for a p-pair (P, A). Let $\chi \in \mathfrak{X}(A)$ and define $\gamma \in \mathfrak{a}^*$ by setting $|\chi(a)| = q^{\langle \gamma, H_P(a) \rangle}$ $(a \in A)$. Then for every semistandard (P, A) and $\chi \in \mathfrak{X}_f(P, A)$ there exist positive constants $c_1, \ldots, c_\ell$ such

that $\gamma + \sum_{i=1}^{\ell} c_i\alpha_i \in \mathfrak{z}^*$, where $\mathfrak{z}^*$ is the dual real Lie algebra of the split component Z of G.

(5) For every $T > 0$ and $r \geq 0$ there exists a positive constant $c = c(T, r)$ such that

$$|f(x)| \leq c \frac{\Xi(x)}{(1+\sigma(x))^r} \quad (x \in G_T).$$

<u>Proof.</u> (1) $\Rightarrow$ (2):

Set $M_0^+(T) = G_T \cap M_0^+$. Then

$$\int_{G_T} |f(x)|^2 dx = \sum_{m \in M_0^+(T)/{}^0M_0} \int_{KmK} |f(x)|^2 dx$$

$$= \sum_{m \in M_0^+(T)/{}^0M_0} [K : K \cap m^{-1}Km] \, \| \underset{\sim}{f}(m) \|^2 \quad \text{(normalize measures)}$$

$$\asymp \sum_{m \in M_0^+(T)/{}^0M_0} \delta_{P_0}(m) \, \| \underset{\sim}{f}(m) \|^2 \quad \text{(Lemma 4.1.1).}$$

If (2) fails for $f \in \mathcal{A}(G)$ and some $(P, A) \succ (P_0, A_0)$, then there is a fixed $m \in M_0^+(T)$, a fixed $\epsilon > 0$, and a sequence $\{a_j\} \subset A^+ \cap G_T$ such that $a_j \underset{P}{\rightarrow} \infty$, $j \rightarrow \infty$, and $\delta_P(ma_j) \| \underset{\sim}{f}(ma_j) \|^2 > \epsilon$ for all positive integers j. Since $\delta_{P_0}(ma) \asymp \delta_P(ma)$ (fixed $m \in M_0$ and variable $a \in A$), we obviously contradict (1).

(2) $\Rightarrow$ (3): Let $f \in \mathcal{A}(G)$. It follows from Corollary 4.1.7 that (2)[(3)] means: For every $(P, A) \succ (P_0, A_0)$, $\chi \in \mathfrak{X}_f(P, A)$, and $a \in A^+(P, t) \cap G_T$, where $t > 1$ [respectively, $t = 1$ and $a^j \rightarrow \infty$, $j \rightarrow \infty$], $|\chi(a^j)| \rightarrow 0$, $j \rightarrow \infty$.

Assume that (3) fails for (f, P, A). Let $a \in A^+(P) \cap G_T$ and assume that $a^j \rightarrow \infty$ and $\chi(a^j) \not\rightarrow 0$, $j \rightarrow \infty$. Let $F \subset \Sigma^0(P, A)$ be such that $\alpha \notin F$ if and only if $|\alpha(a^j)| \rightarrow \infty$, $j \rightarrow \infty$. Let $(P', A') = (P, A)_F$. Then $(P', A') \succ (P_0, A_0)$

and $\chi' = \chi | A' \in \mathfrak{X}_f(P', A')$ (Lemma 3.1.2). Since $a \in \bigcap_{\alpha \in F} \ker \alpha$, it follows that $a^m \in A'$ for some $m > 0$, indeed, $a^m \in A'(P', t)$ with $t > 1$. If (2) is true, then $\chi'(a^{mj}) \to 0$, $j \to \infty$, but this is clearly impossible, so we have a contradiction.

(3) => (4): First, let (P, A) be a standard p-pair. If (4) fails for (P, A, f), then there exists $\chi \in \mathfrak{X}_f(P, A)$ such that $\log_q |\chi(a)| = \sum_{i=1}^{\ell} c_i \langle \alpha_i, H(a) \rangle$ for all $a \in {}^0G \cap A$ with $c_1 \geq 0$. Since $\bigcap_{i=2}^{\ell} \operatorname{Ker} \alpha_i \neq (1)$ in A, we may choose $a_1 \in A \cap {}^0G$ such that $\langle \alpha_1, H(a_1) \rangle > 0$ and $\langle \alpha_j, H(a_1) \rangle = 0$ $(j \geq 2)$. Since $a_1 \in {}^0G \cap A$, we have $\langle \zeta, H(a_1) \rangle = 0$ for all $\zeta \in \mathfrak{z}^*$. We also have $a_1^n \in {}^0G \cap A^+(P)$ for every positive integer n and $a_1^n \to \infty$, $n \to \infty$. By (3), $|\chi(a_1^n)| \to 0$, $n \to \infty$; on the other hand, $|\chi(a_1^n)| = q^{c_1 \langle \alpha_1, H(a_1) \rangle n} \geq 1$ for all n. Contradiction: We must have $c_1 < 0$, so not (4) implies not (3), in the case (P, A) standard.

If (P', A') is semi-standard, then there is a standard p-pair (P, A) and an element $y \in K$ such that $(P', A') = (P^y, A^y)$. We have $\chi \in \mathfrak{X}_f(P, A)$ if and only if $\chi^y \in \mathfrak{X}_f(P^y, A^y)$ (Lemma 3.1.3), so that

$$\log_q |\chi^y(a^y)| = \sum_{i=1}^{\ell} c_i \langle \alpha_i, H(a) \rangle.$$

The roots have obvious transformation properties, too, under conjugation. Choosing $a_1^y \in A^y \cap {}^0G$, we may proceed as before.

(4) => (5): Recall that $G = K\omega_0 A_0^+ K$, where ω_0 is a finite subset of M_0. If not (5), then there exists $T > 0$, $m_0 \in M_0$, and a sequence $\{a_i\}_{i=1}^{\infty} \subset A_0^+$ such that:

1) $-T \leq \langle \zeta, H(a_i) \rangle \leq T$ for all integers i and generators ζ of $\mathfrak{z}^*$;

2) $a_i \to \infty$, $i \to \infty$; and

3) $\|\underset{\sim}{f}(m_0a_i)\,\Xi(m_0a_i)^{-1}(1+\sigma(m_0a_i))^r\| \to \infty$ for some $r > 0$.

Let $F \subset \Sigma^0(P_0, A_0)$ be the complement of the set of all $\alpha \in \Sigma^0(P_0, A_0)$ such that $|\xi_\alpha(a_i)| \to \infty$, $i \to \infty$. Define $(P, A) = (P_0, A_0)_F$. Then $\underset{\sim}{f}_P(m_0a_i) = \delta_P^{\frac{1}{2}}(m_0a_i)\underset{\sim}{f}(m_0a_i) \asymp \delta_{P_0}^{\frac{1}{2}}(m_0a_i)\underset{\sim}{f}(m_0a_i)$ for i sufficiently large. Since $\delta_{P_0}(m_0a_i)^{\frac{1}{2}}\,\Xi(m_0a_i) \geq c_1 > 0$ (Theorem 4.2.1), we see that $\|\underset{\sim}{f}(m_0a_i)\,\Xi(m_0a_i)^{-1}(1+\sigma(m_0a_i))^r\| \leq c_2\|\underset{\sim}{f}_P(m_0a_i)(1+\sigma(m_0a_i))^r\|$. If the left hand side does not remain bounded as $i \to \infty$, then we may conclude (Corollary 4.1.7) that there is at least one element $\chi \in \mathfrak{X}_f(P, A)$ such that $|\chi(a_i)| \not\to 0$, $i \to \infty$. This contradicts (4).

(5) $\Longrightarrow$ (1) is an immediate consequence of Lemma 4.2.5.

Corollary 4.4.5. If $\omega \in \mathcal{E}_2(G)$, then $\mathcal{A}(\omega) \subset \mathcal{C}_*(G)$.

Proof. By Theorem 4.4.4, if $f \in \mathcal{A}(\omega)$, then $|f(x)| \prec \Xi(x)(1+\sigma(x))^{-r}$ for all $x \in G_T$ (fixed $T > 0$) and any $r > 0$. Since $\omega \in \mathcal{E}_2(G)$, $\chi_\omega \in \hat{Z}$, so $|f(xz)| = |f(x)|$ $(x \in G,\ z \in Z)$. Since $\sigma_*(x) \leq \sigma(x)$, it is clear that $|f(x)| \prec \Xi(x)(1+\sigma_*(x))^{-r}$ for all $x \in G$, as required. (In the above proof we have also used the fact that $[G : {}^0GZ] < \infty$, i.e., there exists $T > 0$ such that $G_T \cdot Z = G$.)

Corollary 4.4.6. $\mathcal{E}_2(G) \subset \mathcal{E}'(G)$ (cf. §§2.4 and 3.2).

Proof. This is an immediate consequence of Theorem 4.4.4.

Corollary 4.4.7. Let $(G, Z) \succcurlyeq\!\!\!\!/\ (P, A)(P = MN)$. Let $\omega \in \mathcal{E}_2(G)$ and let $f \in \mathcal{A}(\omega)$. Then $f^P = 0$.

Proof. We have already noted that the above hypotheses imply that $f \in \mathcal{C}_*(G)$ (Corollary 4.4.5). It follows (Theorem 4.4.3) that, for every $m \in M$, the integral defining $f^P(m)$ is absolutely convergent; the function $f^P \in \mathcal{C}_*(M)$.

We want first to show that f^P is an A-finite function. Noting that $(\rho(mn)f)^P = \delta_P(m)^{\frac{1}{2}}\rho_M(m)f^P$ $(m \in M,\ n \in N)$, we observe that $f \mapsto f^P$ is a P-module homomorphism from $\mathcal{A}(\omega)$ into $\mathcal{C}_*(M)$ which factors through $\mathcal{A}(\omega)/\mathcal{A}(\omega)(N)$. It follows from Corollary 2.3.6 that f^P generates an admissible submodule of $\mathcal{C}_*(M)$, that $f^P \in \mathcal{A}(M)$. This implies that f^P is an A-finite function.

We want to show that $f^P = 0$. Let $\Phi = (\phi_1, \ldots, \phi_d)$ span the subspace of $\mathcal{A}(M)$ generated by f^P under translation by elements of A. Let $\rho(a)\Phi = \Phi\chi(a)$. Fix $r \geq 1$ and $c(r,f) > 0$ such that $|f^P(m)| \leq c(r,f)\ \Xi_M(m)(1+\sigma_*(m))^{-r}$ for all $m \in M$ (Theorem 4.4.3). Then $|\Phi(m)| \leq c\ \Xi_M(m)(1+\sigma_*(m))^{-r}$ for all $m \in M$, too. Recalling that $\Xi_M(ma) = \Xi_M(m)$ $(m \in M,\ a \in A)$, we have $|\Phi(ma)| \leq c\ \Xi_M(m)(1+\sigma_*(ma))^{-r} \leq c'\ \Xi_M(m)(1+\sigma_*(a))^{-r}$. This implies that $|\chi(a)| \leq c'(1+\sigma_*(a))^{-r}$, or that $|1| = |\chi(a)\chi^{-1}(a)| \leq |\chi(a)|\,|\chi(a^{-1})| \leq c'^2(1+\sigma_*(a))^{-2r}$ for all $a \in A$. This is impossible.

Remark. The preceding illustrates the principle that, if A/Z is not compact, then an A-finite function cannot converge to zero at ∞--there are no A-finite L^2 functions on A.

The following is a reformulation of the preceding corollary.

Corollary 4.4.8. Let $f \in \mathcal{A}(G) \cap \mathcal{C}_*(G)$. Then f is a cusp form (i.e., $f^P = 0$ for all P).

Remarks.

1) Let $\mathcal{C}_*(G, Z)$ denote the set of Z-finite functions lying in $\mathcal{C}_*(G)$. Then $\mathcal{C}_*(G, Z) = \oplus \sum_{\chi \in \hat{Z}} \mathcal{C}_*(G, \chi)$.

2) If $f \in \mathcal{C}(G)$ and $\chi \in \hat{Z}$, then $\int_Z f(xz)\chi(z^{-1})dz$ exists and defines an element of $\mathcal{C}_*(G, \chi)$.

§4.5. Tempered Representations and the Weak Constant Term.

A distribution T on G is called a tempered distribution if T extends to a continuous functional on $\mathcal{C}(G)$. Clearly, such an extension, if it exists, is unique.

To a locally integrable function h on G we attach a distribution T_h by setting $T_h(f) = \int_G h(x)f(x)dx$ for $f \in C_c^\infty(G)$. We call h a tempered function if the associated distribution T_h is a tempered distribution.

Fix a minimal p-pair (P_0, A_0) of G and an A_0-good maximal compact subgroup K of G.

Let $h \in C(G)$. We say that h satisfies the weak inequality if there exist positive constants c and r such that $|h(x)| \leq c\, \Xi(x)(1+\sigma(x))^r$ for all $x \in G$. Let $h \in C(G/\!/K_0)$ for some open compact subgroup K_0 of G. Then it follows easily from Theorem 4.2.1 that h satisfies the weak inequality if and only if $jh = \underset{\sim}{h}$ satisfies the following condition: There exist positive constants c' and r' such that

$$\delta_{P_0}(m)^{\frac{1}{2}} \|\underset{\sim}{h}(m)\| \leq c'(1+\sigma(m))^{r'}$$

for all $m \in M_0^+$.

Theorem 4.5.1. (1) If $h \in C(G)$ satisfies the weak inequality, then h is a tempered function.

(2) Let K_0 be an open compact subgroup of G. If $h \in C(G/\!/K_0)$ is a tempered function, then h satisfies the weak inequality.

Proof. (1) Assume that there exist positive constants c_1 and r_1 such that $|h(x)| \leq c_1 \; \Xi(x)(1+\sigma(x))^{r_1}$ for all $x \in G$. Then, obviously, the distribution T_h extends to $\mathcal{C}(G)$: Given $f \in \mathcal{C}(G)$, we may choose a constant r_2 arbitrarily large and a positive constant c_2 such that

$$\int_G |f(x)h(x)|\,dx \leq c_2 \int_G \Xi^2(x)(1+\sigma(x))^{-r_2} dx < \infty.$$

To show that T_h is a continuous functional on $\mathcal{C}(G)$ set $\mu_{r_1}(f) = \int_G |f(x)| \; \Xi(x)(1+\sigma(x))^{r_1} dx$ $(f \in C_c^\infty(G))$. Since T_h is obviously continuous with respect to μ_{r_1}, it is sufficient to show that μ_{r_1} is a semi-norm on $C_c^\infty(G)$ which is continuous with respect to the semi-norms ν_r defined previously (§4.4). However, it is clear that μ_{r_1} is a semi-norm and that, if $\nu_r(f_n) \to 0$, $n \to \infty$, for all r ($f_n \in C_c^\infty(G)$ for all n), then $\mu_{r_1}(f_n) \to 0$, $n \to \infty$, too.

(2) Since h is tempered, we have a constant $c_r > 0$ such that $|\int_G h(x)f(x)dx| \leq c_r \nu_r(f) = c_r \sup_{x \in G} |f(x)| \; \Xi^{-1}(x)(1+\sigma(x))^r$ for any $r > 0$ and $f \in C_c^\infty(G)$. In particular, taking for f the characteristic function of the double coset $K_0 \, x \, K_0$, we obtain

$$\left|\int_{K_0 x K_0} h(x)dx\right| = \mu(K_0 \, x \, K_0)|h(x)| \leq c(1+\sigma(x))^r \; \Xi^{-1}(x).$$

It is sufficient to show that

$$|h(x)| \prec \frac{(1+\sigma(x))^r}{\Xi(x)\mu(K x K)} \prec (1+\sigma(x))^r \; \Xi(x).$$

The first domination follows from the equivalence $\mu(K_0 \times K_0) \asymp \mu(K \times K)$, i.e., $\mu(K \times K)[K : K_0]^{-2} \leq \mu(K_0 \times K_0) \leq \mu(K \times K)$. To prove the second we shall check that $1 \prec \Xi^2(x)\mu(K \times K)$. It suffices to check this for $x = m_0 \in M_0^+$. In this case we have

$$\Xi^2(m_0)\mu(K \times K) \asymp \Xi^2(m_0)\delta_P(m_0) \succ 1,$$

as follows from Lemma 4.1.1 and Theorem 4.2.1.

Let π be an admissible representation of G. We call π <u>tempered</u> if every $f \in \mathcal{A}(\pi)$ is tempered. Clearly, if π is tempered, then so are $\tilde{\pi}$, π^*, and $\bar{\pi}$.

<u>Lemma</u> 4.5.2. (1) If π is tempered, then the character Θ_π of π is tempered. (2) If π is completely reducible and Θ_π is tempered, then π is tempered.

<u>Proof.</u> (1) Let π act in a vector space V. Let $E_{K_0} \in C_c(G/\!/K_0)$ be the identity element, for K_0 an open compact subgroup of G. Set $\Theta_{\pi, K_0}(x) = \mathrm{tr}(\pi(E_{K_0})\pi(x)\pi(E_{K_0}))$. Then Θ_{π, K_0} satisfies the weak inequality and represents $\Theta_\pi | C_c(G/\!/K_0)$. Since, if $K_1 \subset K_0$, Θ_{π, K_1} defines on $C_c(G/\!/K_0)$ the same functional as Θ_{π, K_0}, we see that Θ_π has a well-defined extension to $\mathcal{C}_{K_0}(G)$ for every K_0.

(2) It is sufficient to consider the case in which π is irreducible. Since Θ_{π, K_0} as defined above is a tempered function, since the property of being a tempered function is stable under convolution by elements of $C_c^\infty(G)$, and since $\mathcal{A}(\pi)$ is a simple double-module over $C_c^\infty(G)$, it follows that π is tempered.

We write ${}_{w}\mathcal{A}(G)$ or $\mathcal{A}_T(G)$ for the set of tempered elements of $\mathcal{A}(G)$. Then ${}_{w}\mathcal{A}(G)$ is a double module over $\mathcal{Z}(G)$ and a submodule of $\mathcal{A}(G)$ with respect to $C_c^\infty(G)$.

<u>Lemma</u> 4.5.3. Let $f \in \mathcal{A}(G)$. The following statements are equivalent:

(1) $f \in {}_{w}\mathcal{A}(G)$.

(2) Given any p-pair $(P, A) \leqq (G, Z)$ and $\chi \in \mathfrak{X}_f(P, A)$, we have $|\chi(a)| \leq 1$ for all $a \in A^+$.

(3) $\mathfrak{X}_f(G, Z) \subset \hat{Z}$ and, given any proper p-pair (P, A) and $\chi \in \mathfrak{X}_f(P, A)$, either $\chi \in \hat{A}$ or $\chi(a) \to 0$, $a \underset{P}{\to} \infty$.

<u>Proof</u>. (3) $\Rightarrow$ (2): If not (2), then there exists a standard p-pair (P, A) and $b \in A^+$ such that $|\chi(b)| > 1$ (cf. Lemma 3.1.3). Let (P, A) be maximal with this property. Then, obviously, $(P, A) = (G, Z)$ if and only if $\mathfrak{X}_f(G, Z) \not\subset \hat{Z}$, so we may assume that $(P, A) \neq (G, Z)$ and $b \in A^+ \cap {}^0G$. In this case, obviously, $\chi \notin \hat{A}$. Assume that $\chi(a) \to 0$, $a \underset{P}{\to} \infty$. Then there is a non-empty set $F \subset \Sigma^0(P, A)$ (simple roots) such that $|\alpha(b)| = 1$ $(\alpha \in F)$ and $|\alpha(b)| > 1$ $(\alpha \in \Sigma^0 - F)$. Let $(P', A') = (P, A)_F$. Without loss of generality $([\bigcap_{\alpha \in F} \ker \alpha : A'] < \infty)$, assume that $b \in A'$. By Lemma 3.1.2 we know that $\chi' = \chi | A' \in \mathfrak{X}_f(P', A')$. Moreover, $b^m \underset{P'}{\to} \infty$, $m \to \infty$ and $\chi'(b^m) \not\to 0$, $m \to \infty$, so we have shown that not (2) implies not (3).

(2) $\Rightarrow$ (1): In view of Theorem 4.5.1 it suffices to show that, assuming (2), we can find positive constants c and r such that $\|\delta_{P_0}(m)^{\frac{1}{2}} \underset{\sim}{f}(m)\| \leq c(1+\sigma(m))^r$ *for* all $m \in M_0^+$. This certainly holds for $m \in Z\omega$, where ω is an arbitrary compact subset of M_0^+. Note that there, obviously, exists a compact set $\omega \subset M_0^+$ with the following property: There is a constant $c > 0$ such that, for every $m \in M_0^+ - Z\omega$

there is at least one standard p-pair (P, A) for which $\|\delta_{P_0}(m)^{\frac{1}{2}}\underset{\sim}{f}(m)\| \leq c\|\underset{\sim}{f}_P(m)\|$. Since $\|\underset{\sim}{f}_P(m)\| \leq \sum_{\chi \in \mathcal{X}_f(P,A)} \|\underset{\sim}{f}_{P,\chi}(m)\| \leq c'(1+\sigma(m))^{r'}$ for all $m \in M_0^+$ (a constant $c' > 0$), we have proved that (2) $\Longrightarrow$ (1).

(1) $\Longrightarrow$ (3): Obviously, $f \in {}_w\mathcal{A}(G)$ combined with the fact that $\Xi(zx) = \Xi(x)$ for all $x \in G$ and $z \in Z$ implies that $\mathcal{X}_f(G, Z) \subset \hat{Z}$. Assume not (3). Then there exists a standard p-pair (P, A) and $\chi \in \mathcal{X}_f(P, A)$ such that neither $\chi \in \hat{A}$ nor $\chi(a) \to 0$, $a \underset{P}{\to} \infty$. It follows that there exists $b \in A^+(t)$ $(t > 1)$ such that $|\chi(b)| > 1$ (i.e., note that $A^+(t) \cup A^+(t)^{-1}$ generates A). If (1), we have $\|\chi^{-1}(b^n)\underset{\sim}{f}_P(mb^n)\| \leq c|\chi^{-1}(b^n)|(1+\sigma(mb^n))^r \to 0$, $n \to \infty$. This contradicts $|\chi^{-1}(b^n)|\ \|\underset{\sim}{f}_{P,\chi}(mb^n)\| \not\to 0$, $n \to \infty$ (Corollary 4.1.7). Thus, not (3) implies not (1).

Corollary 4.5.4. Let $f \in {}_w\mathcal{A}(G)$ and let $(P, A)(P = MN)$ be a p-pair of G. Then $f_P \in {}_w\mathcal{A}(M)$ if and only if $\mathcal{X}_f(P, A) \subset \hat{A}$.

Proof. Both necessity and sufficiency follow immediately from Lemma 4.5.3 combined with the transitivity of the constant term (Theorem 2.7.2).

Theorem 4.5.5. Let $f \in {}_w\mathcal{A}(G)$ and let $(P, A)(P = MN)$ be a p-pair. There is a unique element ${}_wf_P \in {}_w\mathcal{A}(M)$ such that $\lim_{a \underset{P}{\to} \infty} |\delta_P(ma)^{\frac{1}{2}}f(ma) - {}_wf_P(ma)| = 0$ for every $m \in M$.

Proof. Consider $f_P = \Sigma f_{P,\chi}$ $(\chi \in \mathcal{X}_f(P, A))$ (cf. §§2.6 and 3.1). Set ${}_wf_P = \Sigma f_{P,\chi}$ $(\chi \in \mathcal{X}_f(P, A) \cap \hat{A})$. It follows from Corollary 4.1.7 and Lemma 4.5.3 that $\lim_{a \underset{P}{\to} \infty} |\delta_P(ma)f(ma) - {}_wf_P(ma)|$ exists and equals zero. The uniqueness of ${}_wf_P$

follows from Corollary 4.1.6.

For $f \in {}_{w}\mathcal{A}(G)$ we have ${}_{w}f_{P} = \Sigma f_{P,\chi}$ (summation over all $\chi \in \mathfrak{X}_f(P, A) \cap \hat{A}$). We write ${}_{w}\mathfrak{X}_f(P, A)$ for $\mathfrak{X}_f(P, A) \cap \hat{A}$ and call ${}_{w}\mathfrak{X}_f(P, A)$ the set of <u>weak exponents</u> of f. Similarly, if π is an irreducible admissible tempered representation of G, then $\mathcal{A}(\pi) \subset {}_{w}\mathcal{A}(G)$. For any nonzero $f \in \mathcal{A}(\pi)$ we may set ${}_{w}\mathfrak{X}_f(P, A) = {}_{w}\mathfrak{X}_{\pi}(P, A)$. It follows from Theorem 3.1.1 that ${}_{w}\mathfrak{X}_{\pi}(P, A)$ is an invariant of π. We define the <u>multiplicity</u> of a weak exponent $\chi \in \mathfrak{X}_f(P, A)$ to be its multiplicity as an exponent; similarly we know what the term <u>simple weak exponent</u> means.

We write ${}_{w}\mathfrak{X}_f(P, A)$ for the set of all tempered elements $\omega \in \mathfrak{X}_f(P, A)$; we call such an ω a <u>weak or tempered class exponent of</u> f. Given π and f as before, we set ${}_{w}\mathfrak{X}_f(P, A) = {}_{w}\mathfrak{X}_{\pi}(P, A)$ and call this set the set of <u>weak or tempered class exponents of</u> π. Clearly, ${}_{w}\mathfrak{X}_{\pi}(P, A)$ is also an invariant of π.

<u>Lemma</u> 4.5.6. Let π be an irreducible admissible tempered representation of G in a vector space V. Let $(P, A)(P = MN)$ be a p-pair and let $\omega \in \mathfrak{X}_{\pi}(P, A)$. Then $\omega \in {}_{w}\mathfrak{X}_{\pi}(P, A)$ if and only if the central exponent $\chi_{\omega} \in {}_{w}\mathfrak{X}_{\pi}(P, A)$.

<u>Proof.</u> We know that, if $\omega \in {}_{w}\mathfrak{X}_{\pi}(P, A)$, then $\chi_{\omega} \in \mathfrak{X}_{\pi}(P, A) \cap \hat{A} = {}_{w}\mathfrak{X}_{\pi}(P, A)$. To prove the converse take $0 \neq h \in \mathcal{A}(\omega)$; it is sufficient to show that $h \in {}_{w}\mathcal{A}(M)$. By Theorem 3.2.4, $h = f_P$ for some $f \in \mathcal{A}(\pi)$. It now follows from Corollary 4.5.4 that $h \in {}_{w}\mathcal{A}(M)$, so $\omega \in {}_{w}\mathfrak{X}_{\pi}(P, A)$.

<u>Corollary</u> 4.5.7. Let π be an irreducible admissible tempered representation of G. Let (P, A) be a p-pair of G and $\chi \in {}_{w}\mathfrak{X}_{\pi}(P, A)$. Then there exists

$\sigma \in \omega \in {}_w\mathfrak{X}_\pi(P, A)$ such that $\chi_\omega = \chi$ and $\pi \subset \mathrm{Ind}_{\overline{P}}^G(\delta_{\overline{P}}^{\frac{1}{2}}\sigma)$.

Proof. This follows from Lemma 4.5.6 and the standard argument of Jacquet, i.e., that used in the proof of Theorem 2.4.1.

Lemma 4.5.8. Let π be an irreducible admissible tempered representation of G. Let $(P', A') \succ (P, A)$ be p-pairs of G. If $\chi \in {}_w\mathfrak{X}_\pi(P, A)$, then $\chi|A' \in {}_w\mathfrak{X}_\pi(P', A')$.

Proof. This is an immediate consequence of Lemmas 3.1.2 and 4.5.6.

Corollary 4.5.9. Let $(P', A') \succ (P, A)$ $(P' = M'N')$. Set $({}^*P, {}^*A) = (P \cap M', A)$, a p-pair of M'. Let $f \in {}_w\mathcal{A}(G)$. Then ${}_wf_P = {}_w({}_wf_{P'})_{*P}$.

Proof. We have $(f_{P'})_{*P} = f_P$ (Theorem 2.7.2). In fact, $f_{P'} = \Sigma f_{P', \eta}$ $(\eta \in \mathfrak{X}_f(P', A'))$ and $(f_{P'})_{*P} = \Sigma (f_{P', \eta})_{*P}$ and $(f_{P', \eta})_{*P} = \Sigma f_{P, \chi}$, $\chi \in \mathfrak{X}_f(P, A)$ such that $\chi|A' = \eta$. It follows that ${}_w({}_wf_{P'})_{*P} = {}_w(\Sigma f_{P', \eta})_{*P}$ $(\eta \in \mathfrak{X}_f(P', A') \cap \hat{A}') = \Sigma f_{P, \chi} (\chi \in \hat{A} \cap \mathfrak{X}_f(P, A))$.

Recall that $\mathcal{A}(G, \eta) = \{f \in \mathcal{A}(G) \mid f(gz) = \eta(z) f(g)\}$.

Theorem 4.5.10. Let $f \in \mathcal{A}(G, \eta)$, with $\eta \in \hat{Z}$. The following statements are equivalent.

(1) $\int_{G/Z} |f(x)|^2 dx^* < \infty$.

(2) The function $f \in {}_w\mathcal{A}(G)$ and, for every p-pair $(P, A) \neq (G, Z)$, ${}_wf_P = 0$.

(3) Let $\alpha_1, \ldots, \alpha_\ell$ denote the simple roots for a p-pair (P, A) and let $\chi \in \mathfrak{X}(A)$. Define $\gamma \in \mathfrak{a}^*$ by setting $|\chi(a)| = q^{\langle \gamma, H_P(a) \rangle}$ $(a \in A)$. For every semi-standard p-pair and $\chi \in \mathfrak{X}_f(P, A)$ there exist positive constants $c_1, \ldots, c_\ell$ such that

$\gamma = -\sum_{i=1}^{\ell} c_i \alpha_i$.

(4) For every $r > 0$ there exists a constant $c = c(r) > 0$ such that

$$|f(x)| \leq c \frac{\Xi(x)}{(1+\sigma_*(x))^r} \quad (x \in G).$$

Proof. This is just a reformulation of Theorem 4.4.4.

Let π be an irreducible admissible tempered representation of G and let (P, A) be a p-pair of G. Then (P, A) is called weakly π-minimal if ${}_W\mathfrak{X}_\pi(P, A) \neq \phi$ and (P, A) is minimal with this property. Weakly π-minimal p-pairs, obviously, always exist. An element $\chi \in {}_W\mathfrak{X}_\pi(P, A)$ is called a weakly π-critical exponent if: For every $(P', A') \underset{\neq}{\leq} (P, A)$ and every $\chi' \in {}_W\mathfrak{X}_\pi(P', A')$ such that $\chi' | A = \chi$ we have $d(\chi') < d(\chi)$. If $\mathfrak{X}_\pi(P, A)$ contains a weakly π-critical exponent, then (P, A) is called a weakly π-critical p-pair. If (P, A) is weakly π-minimal, then (P, A) is weakly π-critical. Note that the converse is true too (Corollary 4.5.12(3)).

Corollary 4.5.11. Let π be an irreducible admissible tempered representation of G. Then there exists a standard p-pair (P, A) $(P = MN)$ and $\sigma \in \omega \in \mathcal{E}_2(M)$ such that $\pi \subset \mathrm{Ind}_{\overline{P}}^{G}(\delta_{\overline{P}}^{\frac{1}{2}} \sigma)$. More precisely, if (P, A) is a standard weakly π-critical p-pair and $\chi \in {}_W\mathfrak{X}_\pi(P, A)$ is a weakly π-critical exponent, then we may choose $\omega \in \mathcal{E}_2(M)$ with central exponent χ.

Proof. Let (P, A) be a standard weakly π-critical p-pair and χ a weakly π-critical exponent (e.g., take (P, A) weakly π-minimal). Applying Corollary 4.5.7, we have $\pi \subset \mathrm{Ind}_{\overline{P}}^{G}(\delta_{\overline{P}}^{\frac{1}{2}} \sigma)$, where $\sigma \in \omega \in {}_W\mathfrak{X}_\pi(P, A)$ and $\chi_\omega = \chi$. Applying Corollary 4.5.9,

Theorem 4.5.10, Theorem 3.2.4, and arguing as in Corollary 2.4.2, we obtain $\omega \in \mathcal{E}_2($

Corollary 4.5.12. Let π be an irreducible admissible tempered representation of G. Assume that there exists a p-pair (P, A) $(P = MN)$ such that $\mathfrak{X}_\pi(P, A)$ contains $\omega \in \mathcal{E}_2(M)$. Let $\chi_\omega = \chi$ be the central exponent of ω. Then, for any p-pair (P_1, A_1) $(P_1 = M_1N_1)$:

(1) ${}_W\mathfrak{X}_\pi(P_1, A_1) \subset \{\chi_1 \in \hat{A}_1 | \chi_1 = \chi \circ s$ for some $s \in W(A|A_1)\}$. In particular, if $W(A|A_1) = \phi$, then ${}_W\mathfrak{X}_\pi(P_1, A_1) = \phi$.

(2) If $A_1 \sim A$, then

$${}_W\mathfrak{X}_\pi(P_1, A_1) \subset \{\omega_1 \in \mathcal{E}_2(M_1) | \omega_1 = \omega^s \text{ for some } s \in W(A|A_1)\}.$$

(3) If (P_1, A_1) is a weakly π-critical p-pair, then (P_1, A_1) is weakly π-minimal.

Proof. The proof is essentially the same--replacing Corollary 2.4.2 by Corollary 4.5.11--as that of Lemma 3.3.3 and is omitted.

Corollary 4.5.13. If π is an irreducible admissible tempered representation of G, then π is unitary.

Proof. Apply Corollary 1.7.9 and Corollary 4.5.11.

Notation: Write ${}_W\mathcal{E}(G)$ or $\mathcal{E}_T(G)$ for the set of classes of irreducible admissible tempered representations of G.

Corollary 4.5.14. Let π as above, (P, A) and $(\bar{P}, A)$ opposite p-pairs of G. Then ${}_W\mathfrak{X}_\pi(P, A) = {}_W\mathfrak{X}_\pi(\bar{P}, A)$ and ${}_W\mathfrak{X}_\pi(P, A) = {}_W\mathfrak{X}_\pi(\bar{P}, A)$.

Proof. Since π is unitary we may apply Corollary 3.2.6. Since $\omega[\chi] \in {}_W\mathfrak{X}_\pi(P, A)$ $[{}_W\mathfrak{X}_\pi(P, A)]$ satisfies $\omega^*[\chi^*] = \omega[\chi]$, the present corollary follows immediately.

Lemma 4.5.15. Let (P, A) $(P = MN)$ be a p-pair of G and let $\omega \in {}_w\mathcal{E}(M)$. Then $\pi = \mathrm{Ind}_P^G(\delta_P^{\frac{1}{2}}\sigma)$ is a completely reducible tempered representation of G.

Proof. The complete reducibility follows from the fact that π is unitary (Corollary 4.5.13). According to Lemma 4.5.2, in order to show that π is tempered, it is sufficient to show that the character Θ_π of π is a tempered distribution.

Let $f \in C_c(G/\!/K_0)$, K_0 an open compact subgroup of G. Then, it follows from Theorem 1.13.2 that $\Theta_\pi(f) = \theta_\sigma(\bar{f}^P)$, where θ_σ denotes the character of σ and $\bar{f}(x) = \int_K f(kxk^{-1})dk$. Given $f_0 \in \mathcal{C}_{K_0}(G)$, set $\Theta_\pi(f_0) = \theta_\sigma(\bar{f}_0^P)$. It follows almost immediately from Theorem 4.4.3 and the fact that σ is tempered that we obtain in this way a continuous extension of Θ_π.

Let (V, τ) be a smooth double representation of K which satisfies associativity conditions (cf. §1.12). Note that ${}_w\mathcal{A}(G, \tau)$ is a double module over $\mathcal{C}(G, \tau)$; indeed, for any admissible tempered representation π of G, $\mathcal{A}(\pi, \tau)$ is a submodule. Clearly, $(\mathcal{C}_{K_0} \otimes V) \cap \mathcal{C}(G, \tau)$ stabilizes the subspace $\mathcal{A}(\pi, \tau) \cap (C(G/\!/K_0) \otimes V)$ for any open compact subgroup K_0 of G.

Lemma 4.5.16. Let $f \in \mathcal{C}(G, \tau)$ and $\psi \in {}_w\mathcal{A}(G, \tau)$. Then ${}_w(f*\psi)_P = f^P \underset{M}{*} {}_w\psi_P$ and ${}_w(\psi*f)_P = {}_w\psi_P \underset{M}{*} f^{\bar{P}}$ for any p-pair (P, A) $(P = MN)$ and its opposite $(\bar{P}, A)$ $(\bar{P} = M\bar{N})$.

Proof. We shall prove only the first equality, as the second goes similarly. Let K_0 be an open compact subgroup of G and assume (without loss of generality) that $f \in (\mathcal{C}_{K_0}(G) \otimes V_0) \cap C(G, \tau)$ and $\psi \in (({}_w\mathcal{A}(G) \cap C(G/\!/K_0)) \otimes V_0) \cap C(G, \tau)$, where V_0 is finite-dimensional. We remark that $V_0 \cdot V_0$ is finite-dimensional and that $(\mathcal{C}_{K_0}(G) \otimes V_0) \cap \mathcal{C}(G,\tau) * \psi$ is a finite-dimensional subspace of ${}_w\mathcal{A}(G,\tau)$. Let $\{X_n\}_{n=1}^{\infty}$ be an increasing sequence of compact open subsets of G such that $K_0 X_n K_0 = X_n$ and $\bigcup_{n=1}^{\infty} X_n = G$. Set $f_n = f|X_n$. Then $f_n \in C_c(G,\tau) \cap (C(G/\!\!/K_0) \cap V_0)$ for all n and $f_n \to f$, $n \to \infty$ in $\mathcal{C}(G,\tau)$. It follows from Theorem 2.7.6 that $(f_n * \psi)_P = f_n^P \underset{M}{*} \psi_P$, and this implies that ${}_w(f_n * \psi)_P = {}_w(f_n^P * \psi_P) = f_n^P * {}_w\psi_P$. We know that $f_n^P \to f^P$, $n \to \infty$ in $\mathcal{C}(M,\tau_M)$. Therefore, $(f_n^P * {}_w\psi_P) \to (f^P * {}_w\psi_P)$ uniformly on compact subsets of M.

We shall show that ${}_w(f_n * \psi)_P(m) \to {}_w(f * \psi)_P(m)$ $(m \in M)$ also uniformly on compact subsets of M. To see this note that, for m fixed or in a compact subset of M, we may choose t large enough so that

$$\delta_P^{\frac{1}{2}}(ma)\int_G (f_n(x)-f(x))\psi(x^{-1}ma)dx = ((f_n-f) * \psi)_P(ma)$$

for all $a \in A^+(t)$ and n, by our remark above. It is sufficient to show that $\|((f_n-f) * \psi)_P(ma)\| \to 0$, $n \to \infty$, since certainly this implies $\|{}_w((f_n-f) * \psi)_P(ma)\| \to 0$, $n \to \infty$. We have (for positive constants c, c', d, r, and r_0 and t sufficiently large)

$$\|\delta_P^{\frac{1}{2}}(ma)\int_G (f_n(x)-f(x))\psi(x^{-1}ma)dx\| \leq c\delta_P^{\frac{1}{2}}(ma)\nu_r(f_n-f)\int_G \frac{\Xi(x)}{(1+\sigma(x))^r}\Xi(x^{-1}ma)(1+\sigma(x^{-1}ma))^d$$

$$\leq c'\nu_r(f_n-f)\int_G \frac{\Xi^2(x)}{(1+\sigma(x))^{r-d}}dx(1+\sigma(ma))^{d+r_0}.$$

The implication we require follows immediately, as $\nu_r(f_n-f) \to 0$, $n \to \infty$.

Write ${}^0\mathcal{C}(G)[{}^0\mathcal{C}_*(G)]$ for the set of all $f \in \mathcal{C}(G)[\mathcal{C}_*(G)]$ such that $f^P = 0$ for all p-pairs (P,A) of G. We also assign obvious meanings to notations such as ${}^0\mathcal{C}(G:U)$, ${}^0\mathcal{C}(G,\tau)$, etc.

Write ${}_w^0\mathcal{A}(G)$ for ${}_w\mathcal{A}(G) \cap \mathcal{C}_*(G) = \mathcal{A}(G) \cap \mathcal{C}_*(G)$ and call ${}_w^0\mathcal{A}(G)$

the space of __cusp forms on__ G. Similarly, define ${}_{w}{}^{0}\mathcal{A}(G,\tau)$ and ${}_{w}{}^{0}\mathcal{A}(G,\chi)$ $(\chi \in \hat{Z})$, etc. One sees that ${}_{w}{}^{0}\mathcal{A}(G) = \oplus\, {}_{w}{}^{0}\mathcal{A}(G,\chi)$.

We write $\mathcal{E}_{sp}(G) = \mathcal{E}_2(G) - {}^{0}\mathcal{E}(G)$, by definition, the set of classes of __special representations of__ G. Let $C(\pi) \in \mathcal{E}_{\mathbb{C}}(G)$, $\chi_\pi = \chi$. If $\mathcal{A}(\pi) \subset {}_{w}{}^{0}\mathcal{A}(G,\chi) - {}^{0}\mathcal{A}(G,\chi)$, then π is a special representation of G.

__Corollary__ 4.5.17. Assume $\chi \in \hat{Z}$. Let $f \in {}^{0}\mathcal{C}(G)$ and $\psi \in {}_{w}\mathcal{A}(G,\chi)$ or $f \in \mathcal{C}(G)$ and $\psi \in {}_{w}{}^{0}\mathcal{A}(G,\chi)$. Then both $f*\psi$ and $\psi*f \in {}_{w}{}^{0}\mathcal{A}(G,\chi)$.

__Proof.__ Immediate from Lemma 4.5.16.

Let (P,A) $(P = MN)$ and $(\overline{P},A)$ $(\overline{P} = M\overline{N})$ be opposite semistandard p-pairs of G.

__Theorem__ 4.5.18. Let π be an irreducible admissible tempered representation of G in a vector space V. Let $\sigma \in \omega \in \mathcal{E}_2(M)$. The following are equivalent:

(1) $\omega \in {}_{w}\mathfrak{X}_{\pi}(P,A)$.

(2) $\delta_{\overline{P}}^{\frac{1}{2}}\omega$ occurs as a quotient in $V/V(\overline{N})$.

(3) π occurs as a summand in the completely reducible representations $\mathrm{Ind}_P^G(\delta_P^{\frac{1}{2}}\sigma)$ and $\mathrm{Ind}_{\overline{P}}^G(\delta_{\overline{P}}^{\frac{1}{2}}\sigma)$.

__Proof.__ The proof differs from the proof of Theorem 3.3.1 in that one must assume that the module (W,μ) which corresponds to that of Theorem 3.3.1 is tempered. It is easy to see that the analogue of the module $\overline{V}_\chi$ of Theorem 3.3.1 is tempered. Otherwise this theorem and Theorem 3.3.1 have essentially identical proofs.

§4.6. The General Maass-Selberg Relations.

Let A_0 be a maximal split torus of G and let K be an A_0-good maximal compact subgroup of G. Let (V, τ) be a smooth unitary double representation of K which satisfies associativity conditions (cf. §1.12). Let A be a standard torus of G and set $M = Z_G(A)$.

Theorem 4.6.1. Let $\omega \in \mathcal{E}_2(M)$ be unramified in G (cf. §2.5). Then, if $\sigma \in \omega^s$ $(s \in W(G/A))$ and $P \in \mathcal{P}(A)$, the induced representation $\pi_{P,\omega^s} = \mathrm{Ind}_P^G(\delta_P^{\frac{1}{2}}\sigma)$ is an irreducible unitary tempered representation of G. Moreover, the class $C_P^G(\omega^s)$ of π_{P,ω^s} is the same for all choices of P and s. (Notation: $C_P^G(\omega^s) = C_M^G(\omega)$).

Proof. The first part follows from Theorem 2.5.9, Corollary 4.5.13, and Lemma 4.5.15. The proof of Theorem 3.3.2 proves the second part--one must substitute $\mathcal{E}_2$ for ${}^0\mathcal{E}$ and Theorem 4.5.18 for Theorem 3.3.1.

Corollary 4.6.2. Let (P, A) $(P = MN)$ and (P_1, A_1) $(P_1 = M_1N_1)$ be p-pairs of G. Let $\omega \in \mathcal{E}_2(M)$ and assume that ω is unramified. Let $\pi \in C_M^G(\omega)$. Then:

(1) ${}_w\mathfrak{X}_\pi(P, A) = \{\omega^s \mid s \in W(G/A)\}$.

(2) ${}_w\mathfrak{X}_\pi(P_1, A_1) = \{\chi \circ s \mid s \in W(A \mid A_1)\}$.

In particular, $W(A \mid A_1) = \phi$ implies ${}_w\mathfrak{X}_\pi(P_1, A_1) = \phi$.

Proof. Statement (1) follows from Theorems 4.5.18 and 4.6.1 and Corollary 4.5.12. Statement (2) follows from (1), Corollary 4.5.12, and Lemma 4.5.8.

A class $\omega \in \mathcal{E}_2(M)$ is called simple if: (1) ω is unramified in G; and (2) for every $P \in P(A)$ and every $\eta \in {}_w\mathfrak{X}_\pi(P, A)$ $(C(\pi) = C_M^G(\omega))$ the exponent

η is simple (cf. §§4.5 and 3.1). It is proved in Chapter 5 that every unramified $\omega \in \mathcal{E}_2(M)$ is simple. Since at this stage we still do not know that any simple classes exist, the theorem we shall soon state will remain without content until later.

Let $\omega \in \mathcal{E}_2(M)$ be a simple class and let $f \in \mathcal{A}(\pi, \tau)$ $(C(\pi) = C_M^G(\omega))$. Fix a Haar measure on M/A. It follows from results analogous to those proved in §3.4 that ${}_w f_P \in \bigoplus_{s \in W(A)} \mathcal{A}(\omega^s, \tau_M)$. Write $f_{P,s}$ for the component of ${}_w f_P \in \mathcal{A}(\omega^s, \tau_M)$. It makes sense to define

$$\|f_{P,s}\|^2 = \int_{M/A} \|f_{P,s}\|_V^2 dm^*.$$

Theorem 4.6.3. (General Maass-Selberg relations). Assume that $\omega \in \mathcal{E}_2(M)$ is simple with respect to G. If $f \in \mathcal{A}(C_M^G(\omega), \tau)$, then

$$\|f_{P_1, s_1}\| = \|f_{P_2, s_2}\|$$

for all P_1 and $P_2 \in \mathcal{P}(A)$ and s_1 and $s_2 \in W(G/A)$.

To prove Theorem 4.6.2 one can simply go through §§3.4 and 3.5, substituting terminology and, in a few places, using results from §4.5 in place of the more elementary results employed previously.

Let us indicate the substitutions:

$$\mathcal{A} \longrightarrow {}_w\mathcal{A}$$

$${}^0\mathcal{A} \longrightarrow {}_w{}^0\mathcal{A}$$

$$C_c \longrightarrow \mathcal{C}$$

$${}^0C_c \longrightarrow {}^0\mathcal{C}$$

$$\mathfrak{X}_\pi \longrightarrow {}_w\mathfrak{X}_\pi$$

$\mathfrak{X}_\pi \longrightarrow {}_w\mathfrak{X}_\pi$

${}^0\mathcal{E} \longrightarrow \mathcal{E}_2$

supercuspidal $\longrightarrow$ square integrable

admissible $\longrightarrow$ tempered

$f_P \longrightarrow {}_w f_P$

π-critical $\longrightarrow$ weakly π-critical

π-minimal $\longrightarrow$ weakly π-minimal

Corollary 2.4.2 $\longrightarrow$ Corollary 4.5.11

Lemma 2.7.3 $\longrightarrow$ obvious unformulated analogue

Lemma 2.7.6 $\longrightarrow$ Lemma 4.5.16

Theorem 2.8.1 $\longrightarrow$ Theorem 4.5.10

Theorem 3.3.1 $\longrightarrow$ Theorem 4.5.18

Theorem 3.3.2 $\longrightarrow$ Theorem 4.6.1

Lemma 3.3.3 $\longrightarrow$ Corollary 4.5.12

Corollary 3.3.4 $\longrightarrow$ Corollary 4.6.2

One needs to define negligibility in terms of the above substitutions. We leave it to the reader to verify that, with these substitutions, §§3.4 and 3.5 prove Theorem 4.6.3.

§4.7. The Steinberg Character of G and an Application.

Fix a minimal p-pair (P_0, A_0) of G and an A_0-good maximal compact subgroup K of G. In this context the spherical function Ξ of §4.2 is defined (in any case Ξ is unique up to $\underset{\frown}{\smile}$).

For any $x \in G$ define $D_G(x) = D(x)$ by setting

$$\det(t+1-\mathrm{Ad}(x))_{\mathfrak{g}/\mathfrak{z}} = D(x)t^{\ell} + \dots .$$

In this expression ℓ denotes the dimension of Γ/Z, where Γ is any Cartan subgroup and Z is the split component of G ($\mathfrak{g}$ and $\mathfrak{z}$ denote, respectively, the Lie algebras of G and Z). An element $x \in G$ for which $D(x) \neq 0$ is called a <u>regular element of</u> G.

In order that $x \in G$ be regular it is necessary and sufficient that x lie in one and only one Cartan subgroup of G. If x is an element of the Cartan subgroup Γ of G, then $D(x) = D_{G/\Gamma}(x) = \det(1-\mathrm{Ad}(x))_{\mathfrak{g}/\underline{\Gamma}}$ ($\underline{\Gamma}$ = Lie algebra of Γ). More generally, for $M_1 \supset M_2$ reductive groups with Lie algebras $\mathfrak{m}_1 \supset \mathfrak{m}_2$, we write $D_{M_1/M_2}(m) = \det(1-\mathrm{Ad}(m))_{\mathfrak{m}_1/\mathfrak{m}_2}$ $(m \in M_2)$.

We write $G'[\Gamma']$ for the set of regular elements of G [of a Cartan subgroup Γ of G]. The set $G'[\Gamma']$ is an open dense subset of $G[\Gamma]$. Choosing a set $\{\Gamma\}$ of representatives for the conjugacy classes of Cartan subgroups of G, we may write $G' = \bigcup_{\Gamma \in \{\Gamma\}} (\Gamma')^G$.

A Cartan subgroup Γ of G is called <u>elliptic</u> if Γ/Z is compact. A regular element $x \in G$ is called an <u>elliptic element</u> if x lies in an elliptic Cartan subgroup. We write G_e for the set of elliptic elements of G. The set $G_e = \bigcup_{\Gamma \text{ elliptic}} \Gamma'$, an open subset of G.

Let φ_e denote the characteristic function of the set G_e. The main result of this section is:

<u>Theorem</u> 4.7.1. $\int_K \varphi_e(xk)dk \preccurlyeq \Xi(x) \quad (x \in G)$.

The organization of this section is as follows. First, as consequences and applications of Theorem 4.7.1, we shall prove Corollaries 4.7.1-4.7.4. Then we state and prove Lemma 4.7.5, which shows that Theorem 4.7.1 follows from the existence of a certain function defined on the regular elements of G and possessing four specified properties. After proving the fundamental Lemma 4.7.6 (cf. also [13a]), which concerns the characters of induced representations of G, we shall prove that plus or minus the character of the Steinberg representation of G possesses the four properties required in Lemma 4.7.5.

For $f \in C_c^\infty(G)$ set

$$F_f^{G/\Gamma}(\gamma) = F_f^{\Gamma}(\gamma) = F_f(\gamma) = |D(\gamma)|^{\frac{1}{2}} \int_{G/\Gamma} f(x\gamma x^{-1}) dx^*,$$

where $\gamma \in \Gamma' \subset \Gamma$ for some Cartan subgroup Γ of G and dx^* is a measure chosen on the homogeneous space G/Γ such that $dx = d\gamma dx^*$ (dx and $d\gamma$, respectively, Haar measures on G and on Γ).

Let $\{\Gamma\}$ denote a set of representatives for the conjugacy classes of Cartan subgroups of G. We shall assume "Weyl's integration formula": For $f \in C_c^\infty(G)$,

$$\int_G f(x) dx = \sum_{\{\Gamma\}} [W(G/\Gamma)]^{-1} \int_\Gamma |D(\gamma)|^{\frac{1}{2}} F_f^{G/\Gamma}(\gamma) d\gamma$$

$(W(G/\Gamma) = N_G(\Gamma)/\Gamma = \widetilde{\Gamma}/\Gamma)$. (The main step in the proof is Lemma 22, p. 58 of [7c].)

Corollary 4.7.2. Let $f \in \mathcal{C}(G)$. Then $\int_{G_e} |f(x)| dx < \infty$.

Proof. Since $f \in \mathcal{C}(G)$, there is, for any $r > 0$, a constant c_r such that $|f(x)| \leq c_r \, \Xi(x)(1+\sigma(x))^{-r}$. Therefore

$$\int_{G_e} |f(x)|dx \le c_r \int_G \frac{\Xi(x)}{(1+\sigma(x))^r} \varphi_e(x)dx$$

$$\le c_r \int_G \int_K \frac{\Xi(x)}{(1+\sigma(x))^r} \varphi_e(xk)dkdx$$

$$\le c \int_G \frac{\Xi(x)^2}{(1+\sigma(x))^r} dx < \infty \quad \text{for } r \text{ sufficiently large.}$$

Corollary 4.7.3. Let Γ be an elliptic Cartan subgroup of G. Let $f \in \mathcal{C}(G)$. Then $F_f(\gamma)$ is defined for almost all $\gamma \in \Gamma'$ and $|D^{\frac{1}{2}}(\gamma)|F_f(\gamma)$ is a summable function on Γ.

Proof. Set $G_\Gamma = (\Gamma')^G$, a subset of G_e. Corollary 4.7.2 implies that $\int_{G_\Gamma} |f(x)|dx$ exists. Fubini's theorem implies the present corollary, inasmuch as

$$\int_{G_\Gamma} |f(x)|dx = \int_{\Gamma'} \int_{G/\Gamma} |D(\gamma)| \; |f(x\gamma x^{-1})| dx^* d\gamma$$

$$= \int_{\Gamma'} |D(\gamma)|^{\frac{1}{2}} F_f^{G/\Gamma}(\gamma) d\gamma \; .$$

Corollary 4.7.4. Let $\omega \in \mathcal{E}_2(G)$. Then the character Θ_ω of ω is represented on G_e by a locally summable function.

Proof. Let f be a diagonal matrix coefficient for ω. Then $f \in \mathcal{C}_*(G)$ (Corollary 4.4.5) and, if Γ is an elliptic Cartan subgroup of G, then, up to a constant,

$$\int_{G/\Gamma} f(x\gamma x^{-1})dx^* = \Theta(\gamma) \quad (\gamma \in (\Gamma')^G)$$

represents Θ_ω on an open subset of $(\Gamma')^G$ (cf. Theorem 9, p. 43 of [7c]).

Remark. We shall see in the next section that, if $\operatorname{char} \Omega = 0$, then Θ_ω is represented on G' by a locally constant function.

Given Γ a Cartan subgroup of G, we write A_Γ for the split component of Γ and $M_\Gamma = Z_G(A_\Gamma)$. Since Γ/A_Γ is compact, Γ is an elliptic Cartan subgroup of M_Γ. It is also clear that A_Γ is a special torus of G, i.e., that A_Γ is the split component of M_Γ.

The function Θ of the following lemma will turn out to be, up to a sign, the character of the Steinberg representation of G.

Lemma 4.7.5. Suppose there exists a measurable function Θ on G and a constant $c_0 \geq 0$ such that the following four conditions hold:

1) $\Theta(xyx^{-1}) = \Theta(y)$ for all $x \in G$ and $y \in G'$.

2) For any Cartan subgroup Γ of G $|D_{G/M_\Gamma}(\gamma)|^{\frac{1}{2}} |\Theta(\gamma)| \leq c_0$ for all $\gamma \in \Gamma'$.

3) $\int_K \Theta(xk)dk = 0$.

4) $\Theta|G_e = 1$.

Then there exists a constant $c > 0$ such that $\int_K \varphi_e(xk)dk \leq c\ \Xi(x)$ for all $x \in G$.

Proof. For any pair (Γ, A_Γ) there exists $y \in G$ such that $(A_\Gamma)^y \subset A_0$. Up to conjugacy we may assume $A_\Gamma = A \subset A_0$ and $Z_G(A_\Gamma) = M_\Gamma = M$. We may also assume that $(P, A) \succ (P_0, A_0)(P = MN)$.

We have, for $f \in C_c^\infty(G)$ and $\gamma \in \Gamma'$,

$$F_f^{G/\Gamma}(\gamma) = |D(\gamma)|^{\frac{1}{2}} \int_{G/\Gamma} f(x\gamma x^{-1})dx^* \quad (dx = d\gamma\, dx^*)$$

$$= |D(\gamma)|^{\frac{1}{2}} \int_{K \times N \times M/\Gamma} f(\gamma^{knm})dkdndm^*$$

$$(dx = dkd_r p = dkdndm)$$

$$= |D(\gamma)|^{\frac{1}{2}} \int_{N \times M/\Gamma} \bar{f}(\gamma^{nm})dndm^*,$$

where $\overline{f}(x) = \int_K f(x^k)dk,$

$$= |D_M(\gamma)|^{\frac{1}{2}} \delta_P(\gamma)^{\frac{1}{2}} \int_{M/\Gamma} \int_N \overline{f}(\gamma^m n)dndm^*$$

(cf. [7c], p. 58 for proof of this change of variables formula; also use Lemma 4.7.7).

Now we set $g_f(m) = (\overline{f})^P(m) = \delta_P(m)^{\frac{1}{2}} \int_N \overline{f}(mn)dn$ and obtain $F_f^{G/\Gamma}(\gamma) = F_{g_f}^{M/\Gamma}(\gamma) \quad (\gamma \in \Gamma')$.

Since Γ/A is compact, Γ is elliptic in M.

Next we pick a set of representatives for the finitely many conjugacy classes of standard tori in G. For A such a representative let C_A denote a set of representatives for the conjugacy classes of Cartan subgroups Γ of G with split component $A_\Gamma = A$. Using Weyl's integration formula, we have (for $f \in C_c^\infty(G)$; from now on we will also assume f to be positive-valued)

$$\int_G f(x)|\Theta(x)|dx = \sum_{\{\Gamma\}} [W(G/\Gamma)]^{-1} \int_\Gamma |\Theta(\gamma)| \; |D_G(\gamma)|^{\frac{1}{2}} F_f^{G/\Gamma}(\gamma)d\gamma,$$

where $\{\Gamma\} = \bigcup C_A$. Fixing a representative A and $\Gamma \in C_A$, we have

$$\int_\Gamma |\Theta(\gamma)| \; |D_G(\gamma)|^{\frac{1}{2}} F_f^{G/\Gamma}(\gamma)d\gamma \leq c_0 \int_\Gamma |D_M(\gamma)|^{\frac{1}{2}} F_{g_f}^{M/\Gamma}(\gamma)d\gamma$$

by hypothesis 2). Thus,

$$\sum_{\Gamma \in C_A} [W(G/A)]^{-1} \int_\Gamma |\Theta(\gamma)| \; |D_G(\gamma)|^{\frac{1}{2}} F_f^{G/\Gamma}(\gamma)d\gamma$$

$$\leq c_1 \sum_{\Gamma \in C_A} [W(M/\Gamma)]^{-1} \int_\Gamma |D_M(\gamma)|^{\frac{1}{2}} F_{g_f}^{M/\Gamma}(\gamma)d\gamma$$

$$\leq c_1 \int_{M_e} g_f(m)dm < \infty,$$

where $c_1 > 0$ is a constant. We note that the next-to-last inequality results from the fact that the elements of the set C_A, being non-conjugate in G, are also non-conjugate in M.

Since, by 3), $\int_K \Theta(xk)dk = 0$, we have, for $f \in C_c(G/\!/K)$,

$$0 = \Theta(f) = \int_G \Theta(x)f(x)dx$$

$$= \int_{G_e} fdx + \sum_{\{\Gamma\} \text{ nonelliptic}} [W(G/\Gamma)]^{-1} \int_{\Gamma'} |D_G(\gamma)|^{\frac{1}{2}} \Theta(\gamma) F_f^{G/\Gamma}(\gamma)d\gamma.$$

It follows that

$$\int_{G_e} fdx \leq c_1 \sum_{\{A\}} \sum_{\Gamma \in C_A} [W(M/\Gamma)]^{-1} \int_{\Gamma'} |D_M(\gamma)|^{\frac{1}{2}} F_{g_f}^{M/\Gamma}(\gamma)d\gamma$$

$$\leq c_2 \sum_{\{A\}} \int_{M_{A,e}} g_f(m)dm$$

$$= c_2 \sum_{\{A\}} \int_{M_A} g_f(m)\phi_{A,e}(m)dm,$$

where $M_A = Z_G(A)$ and $\phi_{A,e}$ is the characteristic function of the elliptic set in M_A.

Since $f \in C_c(G/\!/K)$, it is clear that $g_f \in C_c(M/\!/K_M)$, $M = Z_G(A)$. Therefore, using induction on the semisimple rank of G, we may assume that

$$\int_M g_f(m)\phi_{M,e}(m)dm = \int_M g_f(m) \int_{K_M} \phi_{M,e}(mk)dk_M$$

$$\leq C_M \int_M g_f(m)\, \Xi_M(m)dm.$$

Having extended Ξ_M to a function defined on G by setting $\Xi_M(kmn) = \Xi_M(m)$, we may write

$$\int_M g_f(m)\phi_{M,e}(m)dm \leq C_M \int_{MN} \delta_P(m)^{\frac{1}{2}} f(mn)\, \Xi_M(m)dmdn$$

$$= C_M \int_G \delta_P(x)^{-\frac{1}{2}} f(x)\, \Xi_M(x)dx.$$

Summing, we have

$$\int_G f\phi_e dx \le \sum_{\{A\}} C_{M_A} \int_G \delta_P(x)^{-\frac{1}{2}} f(x)\, \Xi_{M_A}(x)dx.$$

Putting in an extra integration over K and using Lemma 4.2.4, we conclude that $\int_G f\phi_e dx \le \sum_{\{A\}} C_{M_A} \int_G f(x)\Xi(x)dx \le c\int_G f(x)\Xi(x)dx$. In other words, setting $\Phi(x) = \int_K \phi_e(xk)dk$, we have shown that $\int_G f\Phi dx \le c\int_G f\, \Xi\, dx$ for all positive $f \in C_c(G/\!/K)$. This implies that $\Phi(x) \le c\, \Xi(x)$ for all $x \in G$.

In the following A is a fixed standard torus of G and $M = Z_G(A)$.

Lemma 4.7.6. Let θ be a locally summable class function on M. Given $f \in C_c^\infty(G)$ set $g_f(m) = \int_N\int_K \delta_P(m)^{\frac{1}{2}} f(kmnk^{-1})dkdn$. Define a distribution Θ on G by setting

$$\Theta(f) = \int_M \theta(m)g_f(m)dm.$$

Then the distribution Θ is represented by a locally summable class function on G, whose values are determined by the values of θ on Cartan subgroups of M. Specifically, $|D_G(\gamma)|^{\frac{1}{2}}\Theta(\gamma) = \sum_{s\in W(A_\Gamma|A)} |D_{M^s}(\gamma)|^{\frac{1}{2}}\theta^s(\gamma)$ $(\gamma \in \Gamma')$, where $\theta^s(\gamma) = \theta(y^{-1}\gamma y)$ with $y = y(s) \in K$ and Γ is any Cartan subgroup of M.

Proof. Let $Y = Y(M,\Gamma) = \{y \in G | y^{-1}\Gamma y \subset M\}$. Let $\widetilde{\Gamma}$ denote the normalizer of Γ. Then, clearly, $Y = \widetilde{\Gamma}YM$. For any $y \in Y$ define $s_y \in W(A_\Gamma|A)$ by setting $a^{s_y} = yay^{-1} \in A_\Gamma$ $(a \in A)$. It is clear that $y \mapsto s_y$ defines a surjective mapping of Y/M on $W(A_\Gamma|A)$. One sees that $W(G/\Gamma)s_y$ is the image of $\widetilde{\Gamma}yM$ in $W(A_\Gamma|A)$ $(y \in Y)$. In particular, it is clear that Y/M and $\Gamma\backslash Y/M$ are finite sets.

Let C_Γ be the set of all Cartan subgroups of M which are conjugate to Γ in G. Then M operates on C_Γ. Let $y_1, \ldots, y_r$ be a complete set of representatives for $\widetilde{\Gamma}\backslash Y/M$. Set $\Gamma_i = y_i^{-1}\Gamma y_i$ $(i = 1, \ldots, r)$. It is routine to check that $C_\Gamma/M \overset{1-1}{\leftrightarrow} \{\Gamma_1, \ldots, \Gamma_r\}$.

Let $y_1, \ldots, y_r$, as before, be a complete set of representatives for $\widetilde{\Gamma}\backslash Y/M$. Set $s_i = s_{y_i} \in W(A_\Gamma|A)$. We may write $W(A_\Gamma|A) = \coprod_{i=1}^{r} W(G/\Gamma)s_i$. The subgroup $W(M^{s_i}/\Gamma)$ is the stabilizer of s_i in $W(G/\Gamma)$. Therefore,

$$W(G/\Gamma)s_i = \{t \cdot s_i | t \in W(G/\Gamma)/W(M^{s_i}/\Gamma)\}$$

$$= \{t^{-1} \cdot s_i | t \in W(M^{s_i}/\Gamma)\backslash W(G/\Gamma)\}.$$

For $\gamma \in \Gamma'$ consider

$$\sum_{s \in W(A_\Gamma|A)} |D_{M^s}(\gamma)|^{\frac{1}{2}} \theta^s(\gamma) = \sum_{i=1}^{r} \sum_{t \in W(M^{s_i}/\Gamma)\backslash W(G/\Gamma)} |D_{M^{s_i}}(\gamma^t)|^{\frac{1}{2}} \theta^{s_i}(\gamma^t).$$

Recall that $M^{s_i} = y_i M y_i^{-1}$ and $D_{M^{s_i}}(\gamma^t) = D_M(y_i^{-1}\gamma^t y_i) = D_M((\gamma_i)^{t_i})$, where $\gamma_i = y_i^{-1}\gamma y_i \in \Gamma_i$ and $t_i = y_i^{-1} t y_i \in W(G/\Gamma_i)$. We also have $\theta^{s_i}(\gamma^t) = \theta((\gamma_i)^{t_i})$. Thus

$$\sum_{s \in W(A_\Gamma|A)} |D_{M^s}(\gamma)|^{\frac{1}{2}} \theta^s(\gamma) = \sum_{i=1}^{r} \sum_{t \in W(M^{s_i}/\Gamma)\backslash W(G/\Gamma)} |D_M(\gamma_i^t)|^{\frac{1}{2}} \theta(\gamma_i^t).$$

We want to verify that the above function represents Θ.

For this note first that $F_f^{G/\Gamma_i}(\gamma_i) = F_f^{G/\Gamma}(\gamma)$. Assuming suitable normalization of measures, we obtain

$$[W(G/\Gamma)]^{-1}\int_\Gamma F_f^{G/\Gamma}(\gamma) \sum_{s\in W(A_\Gamma|A)} |D_{M^s}(\gamma)|^{\frac{1}{2}}\theta^s(\gamma)d\gamma$$

$$= [W(G/\Gamma)]^{-1}\sum_{i=1}^{r}[W(M/\Gamma_i)]^{-1} \sum_{t\in W(G/\Gamma_i)} \int_{\Gamma_i} F_f^{G/\Gamma_i}(\gamma_i)|D_M(\gamma_i^t)|\theta(\gamma_i^t)d\gamma_i$$

$$= \sum_{i=1}^{r}[W(M/\Gamma_i)]^{-1}\int_{\Gamma_i} F_f^{G/\Gamma_i}(\gamma_i)|D_M(\gamma_i)|^{\frac{1}{2}}\theta(\gamma_i)d\gamma_i$$

$$= \sum_{i=1}^{r}[W(M/\Gamma_i)]^{-1}\int_{\Gamma_i} F_{g_f}^{M/\Gamma_i}(\gamma)|D_M(\gamma)|^{\frac{1}{2}}\theta(\gamma)d\gamma$$

$$= \int_{(\Gamma')^G\cap M} g_f(m)\theta(m)dm.$$ Thus, we have shown that

$$\sum_{\{\Gamma\}} [W(G/\Gamma)]^{-1}\int_\Gamma F_f^{G/\Gamma}(\gamma) \sum_{s\in W(A_\Gamma|A)} |D_{M^s}(\gamma)|^{\frac{1}{2}}\theta^s(\gamma)d\gamma$$

$$= \int_M g_f(m)\theta(m)dm$$

$$= \Theta(f).$$

In the following lemma we write $\mathfrak{g}$ for the Lie algebra of G, $\mathfrak{n}$, $\mathfrak{m}$, and $\bar{\mathfrak{n}}$, respectively, for the subalgebras corresponding to N, M, and $\bar{N}$. The vector space $\mathfrak{g}$ is the direct sum $\mathfrak{g} = \mathfrak{n}+\mathfrak{m}+\bar{\mathfrak{n}}$.

Lemma 4.7.7. For $\gamma \in \Gamma \subset M \subset G$, $|D_{G/M}(\gamma)|^{\frac{1}{2}} = \delta_P(\gamma)^{\frac{1}{2}}|\det(1-\mathrm{Ad}(\gamma^{-1}))_{\mathfrak{n}}|$.

Proof. To compute $D_{G/M}$ we extend Ω to a splitting field Ω' for Γ and regard each of the spaces $\mathfrak{n}$, $\mathfrak{m}$, and $\bar{\mathfrak{n}}$ as vector spaces over Ω'. We may assume an order on $\Sigma(\underset{\sim}{G}, \underset{\sim}{\Gamma})$ such that $\mathfrak{n}[\bar{\mathfrak{n}}]$ is a sum of positive [negative] root spaces for Γ. It follows that

$$D_{G/M}(\gamma) = \det(1-\mathrm{Ad}(\gamma)_{\mathfrak{g}/\mathfrak{m}})$$

$$= \det(1-\mathrm{Ad}(\gamma))_{\mathfrak{n}} \det(1-\mathrm{Ad}(\gamma))_{\bar{\mathfrak{n}}}$$

$$= \det \mathrm{Ad}(\gamma)_{\mathfrak{n}} (-1)^{\dim \mathfrak{n}} \det(1-\mathrm{Ad}(\gamma^{-1}))_{\mathfrak{n}}^{2},$$

since $\det(1-\mathrm{Ad}(\gamma))_{\mathfrak{n}} = \det(1-\mathrm{Ad}(\gamma^{-1}))_{\bar{\mathfrak{n}}}$. Noting that $|\det \mathrm{Ad}(\gamma)_{\mathfrak{n}}| = \delta_P(\gamma)$, we obtain the lemma.

For $(P, A) \succ (P_0, A_0)$ $(P = MN)$ let 1_P denote the trivial representation of the parabolic subgroup P. Set $\pi_P = \mathrm{Ind}_P^G 1_P$. Then π_P is an admissible representation of G. Write Θ_P for the character of π_P. It follows immediately from Lemma 4.7.6 that the distribution Θ_P is represented by a function, also to be denoted Θ_P, defined on the regular elements of G. Define the distribution--and the function--

$$\Theta = \sum_{(P,A) \succ (P_0, A_0)} (-1)^{\dim A_0/A} \Theta_P .$$

One can show (and we shall not) that $(-1)^{\dim A_0} \Theta$ is the character of a unitary representation belonging to the discrete series of G, the Steinberg representation of G (cf. [5a]).

Our goal is to verify that the function $\Theta(x)$ satisfies all the hypotheses of Lemma 4.7.5. This will conclude the proof of Theorem 4.7.1. Since Θ is a linear combination of characters, the function Θ is clearly a class function on G' i.e., hypothesis 1) is true. It is easy to see that each representation $\pi_P|K$ contains 1_K exactly once; indeed $\int_K \Theta_P(xk)dk$ is independent of P. Therefore, since

$$\sum_{(P,A) \succ (P_0, A_0)} (-1)^{\dim A_0/A} = \sum_{(P,A) \succ (P_0, A_0)} (-1)^{\mathrm{prk}\, P} =$$

$$\sum_{F \subset \Sigma(P_0, A_0)} (-1)^{[F]} = (1-1)^{\dim A_0/Z} = 0 \text{ for } A_0 \neq Z, \text{ we may con-}$$

clude that $\int_K \Theta(xk)dk = 0$, i.e., 3) is true, provided $\text{rank}_\Omega G > 0$. To see that 4) holds, that $\Theta|G_e = 1$ it suffices to observe that $\Theta_P|G_e = 1$, $P = G$, and 0 otherwise. This follows from the fact that $\Theta_P(f) = \int_P\int_K f(p^k)dkd_\ell p$ $(f \in C_c^\infty(G))$ and the fact that $G_e \cap P = \phi$ $(P \neq G)$. We must verify that Θ satisfies hypothesis 2) of Lemma 4.7.5.

Let Γ be a Cartan subgroup of G, let A_Γ be its split component, and let $M_\Gamma = Z_G(A_\Gamma)$.

Lemma 4.7.8. Set $\Phi_\Gamma(\gamma) = |D_{G/M_\Gamma}(\gamma)|^{\frac{1}{2}}\Theta(\gamma)$ $(\gamma \in \Gamma')$. Then

$$\Phi_\Gamma(\gamma) = \sum_{(P,A) \succ (P_0, A_0)} (-1)^{\text{prk}\, P} \sum_{s \in W(A_\Gamma|A)} \delta_{P^s}(\gamma)^{-\frac{1}{2}} |D_{M^s/M_\Gamma}(\gamma)|^{\frac{1}{2}}.$$

Proof. The function θ_P associated with Θ_P in the formula of Lemma 4.7.6 is $\delta_P^{-\frac{1}{2}}$. Thus, we have $|D_G(\gamma)|^{\frac{1}{2}}\Theta_P(\gamma) = \sum_{s \in W(A_\Gamma|A)} \delta_P^s(\gamma)^{-\frac{1}{2}} |D_{M^s}(\gamma)|^{\frac{1}{2}}$. Note that for all $s \in W(A_\Gamma|A)$, we have $\Gamma \subset M_\Gamma \subset M^s$. The lemma follows from the relations $|D_G(\gamma)| = |D_{G/M_\Gamma}(\gamma)|\,|D_{M_\Gamma}(\gamma)|$, $|D_{M^s}(\gamma)| = |D_{M^s/M_\Gamma}(\gamma)|\,|D_{M_\Gamma}(\gamma)|$, and $\delta_{P^s}(\gamma) = \delta_P(\gamma^{s^{-1}}) = \delta_P^s(\gamma)$.

Now assume that the split component A_Γ of Γ lies in A_0. Write $Q(\Gamma) = \{(P,A) | A \subset A_\Gamma\}$. Let $T = \{(P,A,s) | (P,A) \succ (P_0, A_0)$ and $s \in W(A_\Gamma|A)\}$. Define a mapping $\phi : T \to Q(\Gamma)$ by setting $\phi(P,A,s) = (P,A)^s$. It is routine to check that ϕ is a bijection. Observing that $(P,A) \in Q(\Gamma)$, i.e., $(P,A,1) \in T$, if and only if there exists $(P_1, A_1, s_1) \in T$ such that $(\overline{P}, A) = (P_1, A_1)^{s_1}$, we may write

$$\Phi_\Gamma(\gamma) = \sum_{(P,A,s)\in T} (-1)^{\mathrm{prk}\,P} \delta_{P^s}(\gamma)^{-\frac{1}{2}} |D_{M^s/M_\Gamma}(\gamma)|^{\frac{1}{2}}$$

$$= \sum_{(P,A)\in Q(\Gamma)} (-1)^{\mathrm{prk}\,P} \delta_P(\gamma)^{+\frac{1}{2}} |D_{M/M_\Gamma}(\gamma)|^{\frac{1}{2}} \quad (\gamma \in \Gamma' \subset M_\Gamma).$$

Since $[Q(\Gamma)] < \infty$, the following lemma implies hypothesis 2) of Lemma 4.7.5 and completes the proof of Theorem 4.7.1.

Lemma 4.7.9. For any Cartan subgroup Γ and $\gamma \in \Gamma'$, $|\Phi(\gamma)| \leq [Q(\Gamma)]$.

Proof. Note that $[\Gamma : A_\Gamma \cdot {}^0\Gamma] < \infty$, where ${}^0\Gamma = \bigcap_{\chi \in X(\Gamma)} \ker|\chi|$. Fix $u \in A_\Gamma$ and let $\Gamma(u) = \{\gamma \in \Gamma \mid \gamma^n \in u \cdot {}^0\Gamma$ for some integer $n = n(\gamma) \geq 1\}$. It is enough to prove the above inequality for $\gamma \in \Gamma(u) \cap \Gamma'$. Choose an order on the set of roots $\Sigma(G, A_\Gamma)$ such that $|\alpha(u)| > 1$ implies $\alpha > 0$ $(\alpha \in \Sigma(G, A_\Gamma))$. Let (P_Γ, A_Γ) $(P_\Gamma = M_\Gamma N_\Gamma)$ be the p-pair of G such that $\Sigma(P_\Gamma, A_\Gamma)$ comprises the set of positive roots in $\Sigma(G, A_\Gamma)$. Set $F_u = \{\alpha \in \Sigma^0(P_\Gamma, A_\Gamma) \mid |\alpha(u)| = 1\}$. Then $(P_u, A_u) = (P_\Gamma, A_\Gamma)_{F_u} \succ (P_\Gamma, A_\Gamma)$. Write $P_u = M_u N_u$ for the Levi decomposition associated to (P_u, A_u).

We may regard $\Sigma(P_u, A_u)$ as a subset, via restriction, of $\Sigma(P_\Gamma, A_\Gamma)$. Under this identification $|\alpha(u)| > 1$ for $\alpha \in \Sigma(P_\Gamma, A_\Gamma)$ if and only if $\alpha \in \Sigma(P_u, A_u)$. Put another way, $\langle \alpha, H_{M_u}(u) \rangle > 0$ for all $\alpha \in \Sigma(P_u, A_u)$. (Notice that u is not necessarily an element of A_u, so we cannot speak of $\alpha(u)$ for $\alpha \in X(A_u)$ without resorting to some subterfuge.)

Let $(P, A) \in Q(\Gamma)$, $P = MN$. Define ${}^*P = M \cap P_u$ and ${}^*A = AA_u$. To check that $({}^*P, {}^*A)$ is a p-pair of M observe first that $A_u A$ lies in the center of $M \cap M_u$. By considering the roots of A_u in M it is possible to check easily hat $A_u A$ is the maximal split torus in the center of $M \cap M_u$, i.e., look at the roots in the Lie algebras $\mathfrak{m} = \mathfrak{n}_u \cap \mathfrak{m} + \mathfrak{m} \cap \mathfrak{m}_u + \bar{\mathfrak{n}}_u \cap \mathfrak{m}$. It is also obvious that $M \cap P_u = M \cap M_u \cdot N_u \cap M$ is a parabolic subgroup of M. Note that, since both $A_u \subset A_\Gamma$ and $A \subset A_\Gamma$, ${}^*A = AA_u \subset A_\Gamma$.

Recall that there is a one-one correspondence between p-pairs $({}^*P, {}^*A)$ of M and p-pairs (P', A') of G such that $(P, A) \succ (P', A')$ (§0.3). Let $Q(u) = \{(P', A') \mid A_u \subset A' \subset A_\Gamma\}$. Then, clearly, $Q(u) \subset Q(\Gamma)$. For $(P', A') \in Q(u)$ set $F^+(P', A') = \{\alpha \in \Sigma(P', A') \mid \langle \alpha, H_{P'}(u) \rangle > 0\}$. Set $T(u) = \{(P', A', F) \mid (P', A') \in Q(u),\ F \subset F^+(P', A')\}$.

Lemma 4.7.10. The mapping $(P', A', F) \mapsto (P', A')_F$ defines a bijection of $T(u)$ with $Q(\Gamma)$.

Proof. First, it is clear that, for any $F \subset \Sigma^0(P', A')$, $(P', A')_F = (P, A)$ with $A \subset A' \subset A_\Gamma$, so $(P, A) \in Q(\Gamma)$. Let us prove surjectivity. Given $(P, A) \in Q(\Gamma)$, $P = MN$, define $({}^*P, {}^*A) = (P_u \cap M, AA_u)$, a p-pair of M. Set $(P', A') = (P_u \cap M \cdot N, A \cdot A_u) = (M_u \cap M \cdot N_u \cap M \cdot N, AA_u)$. Clearly, $A_u \subset A' \subset A_\Gamma$ and $(P, A) \succ (P', A')$. To prove that the above mapping is surjective it suffices to check that $(P, A) = (P', A')_F$ with $F \subset F^+(P', A')$. However, since A commutes with $N_u \cap M$ and A_u normalizes $N_u \cap M$, we need only check that the roots of A_u in $N_u \cap M$ are positive. This follows from the identification $\Sigma(P_u, A_u) \subset \Sigma(P_\Gamma, A_\Gamma)$.

Now assume that $(P, A) = (P_1, A_1)_{F_1} = (P_2, A_2)_{F_2}$ with $(P_i, A_i, F_i) \in T(u)$ $(i = 1, 2)$. Since the restriction mapping $F^+(P_i, A_i) \to \Sigma(P_u, A_u)$ is, as one easily sees, injective, there is a subset $S \subset \Sigma(P_u, A_u)$ such that $(\mathfrak{A} \cap \mathfrak{A}_u)^0 =$

$(\bigcap_{\alpha \in S} \ker \alpha \cap \underset{\sim}{A})^0 = (\bigcap_{\alpha \in F_i} \ker \alpha \cap \underset{\sim}{A}_u)^0$. Obviously all three of the sets S, F_1, and F_2 may be identified with the same subset of $\Sigma(P_\Gamma, A_\Gamma)$ and it follows from this both that $(P_1, A_1) = (P_2, A_2)$ and that $F_1 = F_2$.

Now consider

$$\Phi(\gamma) = \sum_{(P,A)\in Q(\Gamma)} (-1)^{\operatorname{prk} P} \delta_P(\gamma)^{\frac{1}{2}} |D_{M/M_\Gamma}(\gamma)|^{\frac{1}{2}}$$

$$= \sum_{(P',A')\in Q(u)} \sum_{F \subset F^+(P',A')} (-1)^{\operatorname{prk} P' - [F]} \delta_{P(F)}(\gamma)^{\frac{1}{2}} |D_{M(F)/M_\Gamma}(\gamma)|^{\frac{1}{2}},$$

where $P(F) = M(F)N(F)$ and $(P(F), A(F)) = (P', A')_F$. We have $A_\Gamma \supset A' \supset A_u A(F) \supset A(F)$, so $M_\Gamma \subset M' \subset M(F)$. One sees also that $D_{M(F)/M_\Gamma}(\gamma) = D_{M(F)/M'}(\gamma) D_{M'/M_\Gamma}(\gamma)$.

Let us compute $D_{M(F)/M'}(\gamma)$. It follows from Lemma 4.7.7 that $|D_{M(F)/M'}(\gamma)|^{\frac{1}{2}} = \delta_{{}^*P}(\gamma)^{\frac{1}{2}} |\det(1 - \operatorname{Ad}(\gamma^{-1}))_{{}^*\mathfrak{n}}|^{\frac{1}{2}}$ $({}^*P = M(F) \cap P' = M' \cap M(F) \cdot N' \cap M(F))$. Since $({}^*P, {}^*A)$ is the p-pair of M' which corresponds to $F \subset F^+(P', A')$, $\langle \alpha, H_{{}^*P}(\gamma)\rangle > 0$ for all $\alpha \in \Sigma({}^*P, {}^*A)$ $(\gamma \in \Gamma(u))$. Therefore, $|\det(1 - \operatorname{Ad}(\gamma^{-1}))_{{}^*\mathfrak{n}}| = 1$. We obtain the result

$$\delta_{P(F)}(\gamma)^{\frac{1}{2}} |D_{M(F)/M_\Gamma}(\gamma)|^{\frac{1}{2}} = \delta_{P(F)}(\gamma)^{\frac{1}{2}} \delta_{{}^*P}(\gamma)^{\frac{1}{2}} |D_{M'/M_\Gamma}(\gamma)|^{\frac{1}{2}}$$

$$= \delta_{P'}(\gamma)^{\frac{1}{2}} |D_{M'/M_\Gamma}(\gamma)|^{\frac{1}{2}}.$$

Substitution yields the relations

$$\Phi(\gamma) = \sum_{(P',A')\in Q(u)} (-1)^{\operatorname{prk} P'} \delta_{P'}(\gamma)^{\frac{1}{2}} |D_{M'/M_\Gamma}(\gamma)|^{\frac{1}{2}} \sum_{F \subset F^+(P',A')} (-1)^{[F]}$$

$$= \sum_{\substack{(P', A') \in Q(u) \\ F^+(P', A') = \phi}} (-1)^{\operatorname{prk} P'} \delta_{P'}(\gamma)^{\frac{1}{2}} |D_{M'/M_\Gamma}(\gamma)|^{\frac{1}{2}},$$

since $\sum_{F \subset F^+(P', A')} (-1)^{[F]} = (1-1)^{[F^+(P', A')]} = 0$ whenever $F^+(P', A') \neq \phi$.

On the other hand, $F^+(P', A') = \phi$ if and only if $\langle \alpha, H_{M'}(u) \rangle \leq 0$ for all $\alpha \in \Sigma(P', A')$. This means that $\delta_{P'}(\gamma) \leq 1$ for all $\gamma \in \Gamma(u)$. Moreover, since $M_u \supset M' \supset M_\Gamma$, we have $|\beta(\gamma)| = 1$ for all $\gamma \in \Gamma(u)$ and all roots $\beta \in \Sigma(\underset{\sim}{M}', \underset{\sim}{\Gamma}) - \Sigma(\underset{\sim}{M}_\Gamma, \underset{\sim}{\Gamma})$ (β defined over some splitting field for $\underset{\sim}{\Gamma}$). Thus, $|D_{M'/M_\Gamma}(\gamma)| = |\prod(1 - \beta(\gamma))| \leq 1$. We may, therefore, conclude that

$|\Phi(\gamma)| \leq [Q(u)] \leq [Q(\Gamma)]$.

§4.8. Howe's Theorem and Consequences.

For this section we assume $\operatorname{char} \Omega = 0$.

Let K_1 and K_2 be open compact subgroups of G. Let ω be a subset of G. Let $\underline{d}_i \in \mathcal{E}(K_i)$ $(i = 1, 2)$ and let $\xi_{\underline{d}_i}$ denote the character of $\underline{d}_i$. We say that ω intertwines $\underline{d}_1$ with $\underline{d}_2$ if there exists $f \in C(G)$ such that $\omega \cap \operatorname{Supp}(\xi_{\underline{d}_1} * f * \xi_{\underline{d}_2}) \neq \phi$.

The purpose of this section is to present a proof, modulo Howe's unpublished result Theorem 4.8.6, of the following theorem due to Roger Howe. We also derive some consequences.

Theorem 4.8.1. Let Γ be a Cartan subgroup of G with Γ' the set of regular elements (cf. §4.7) of Γ. Let ω be a compact subset of Γ'. Then there exists a compact open subgroup K_1 of G with the following property. Fix an open compact subgroup K_2 of G and an element $\underline{d}_2 \in \mathcal{E}(K_2)$. Let F denote the set of all $\underline{d}_1 \in \mathcal{E}(K_1)$ such that

1) G intertwines $\underline{d}_1$ with $\underline{d}_2$.

2) ω intertwines $\underline{d}_1$ with itself.

Then F is a finite set.

Remarks. (1) Let ω be a compact subset of Γ. Then $\omega \subset \Gamma'$ if and only if there exists a constant $c(\omega) > 0$ such that $|\beta(\gamma)-1| \geq c(\omega)$ for all $\gamma \in \omega$ and all roots $\beta \in \Sigma(\underset{\sim}{G}, \underset{\sim}{\Gamma})$.

(2) Let K_1 be an open subgroup of G which satisfies the conditions of Theorem 4.8.1. Then it is obvious that any open subgroup K_1' of K_1 also satisfies the same conditions.

Before proceeding to the proof of Theorem 4.8.1 we give two interesting consequences.

Recall that a distribution T on an open subset $X \subset G$ is said to be represented by a function F_T on X if, for every $f \in C_c^\infty(X)$,

$$T(f) = \int_X f(x) F_T(x) dx.$$

Corollary 4.8.2. Let π be an admissible representation of G on V and suppose that V is a G-module of finite type under π. Then the character of π is represented on G' by a locally constant function.

Proof. Let Γ be a Cartan subgroup of G and let ω be a compact open subset of the set of regular elements Γ' of Γ. It is sufficient to show that there is an open subset of G which contains ω on which the character of π is represented by a locally constant function.

Since V is finitely generated, there is an open compact subgroup K_2 of G such that each element of a finite generating set for V is K_2-invariant. Let K_1 be an open compact subgroup which is so small that the set F which consists of all $\underline{d} \in \mathcal{E}(K_1)$ such that G intertwines $\underline{d}$ with 1_{K_2} and ω intertwines $\underline{d}$ with $\underline{d}$ is finite (existence of infinitely many subgroups K_1 follows from Theorem 4.8.1). Let E_F be the idempotent in $C^\infty(K_1)$ associated with F. If $\theta_F(x)$ denotes the trace of the operator $\pi(E_F)\pi(x)\pi(E_F)$, then $\theta_F(x)$ is a locally constant function on G. Since, if $\underline{d} \in \mathcal{E}(K_1)$ occurs in $\pi|K_1$, then G must intertwine $\underline{d}$ with 1_{K_2}, we see that, for all $F \subset F' \subset \mathcal{E}(K_1)$, the function $\theta_{F'}|\omega = \theta_F$. Moreover, since $\theta_{F'}(k^{-1}xk) = \theta_{F'}(x)$ for all $k \in K_1$ and $x \in G$, it is clear that $\theta_{F'} = \theta_F$ on an open neighborhood of ω. Now an elementary computation shows that $\operatorname{tr} \pi(f) = \int_G \theta_{F'}(x)f(x)dx$ for any $f \in E_{F'} * C_c^\infty(G)$. It is therefore clear that θ_F represents the character of π on an open set containing ω.

Corollary 4.8.3. Let Γ be a Cartan subgroup of G. Let K_0 be an open compact subgroup of G. Fix $\gamma_0 \in \Gamma'$, the set of regular elements of Γ. Then there exists a neighborhood ω of γ_0 in Γ' such that $F_f^{G/\Gamma}(\gamma) = F_f^{G/\Gamma}(\gamma_0)$ for all $f \in C_c(G/K_0)$ and $\gamma \in \omega$. In other words, for any compact subset ω of Γ', the mapping $f \mapsto F_f|\omega^G$ of $C_c(G/K_0)$ to functions supported in ω^G has a finite-dimensional range.

Proof. Let K be an open compact subgroup of G. Let $\mathcal{E}_0(K) = \{\underline{d} \in \mathcal{E}(K) | G$ intertwines $\underline{d}$ with $1_{K_0}\}$. Fix a compact subset ω of Γ'. Let $\mathcal{E}_\omega(K) = \{\underline{d} \in \mathcal{E}(K) | \omega$ intertwines $\underline{d}$ with itself$\}$. We shall first consider the following statements:

1) If $f \in C(G/K_0)$, then $\xi_{\underline{d}} * f = 0$ unless $\underline{d} \in \mathcal{E}_0(K)$.

2) Let $g \in C(G)$ and set $\bar{g}(x) = \int_K g(x^k)dk$. Then, for any $\underline{d} \in \mathcal{E}(K)$, $\xi_{\underline{d}} * \bar{g} = 0$ on ω unless $\underline{d} \in \mathcal{E}_\omega(K)$.

3) Fix $x, y \in G$, $f \in C(G/K_0)$, $\gamma \in \omega$, and $\underline{d} \in \mathcal{E}(K)$. Then

$$\int_{G \times K} \xi_{\underline{d}}(\gamma^k z) f(yz^{-1}x) dz dk = 0$$

unless $\underline{d} \in \mathcal{E}_0(K) \cap \mathcal{E}_\omega(K)$.

Assertions 1) and 2) being obvious, let us prove 3). Set $g(z) = f(yzx)$. Then the above integral is $\overline{(\xi_{\underline{d}} * g)}(\gamma)$. Note that $\overline{\xi_{\underline{d}} * g} = \xi_{\underline{d}} * \bar{g}$. Therefore, 2) implies that, if $\underline{d} \notin \mathcal{E}_\omega(K)$, then $\xi_{\underline{d}} * \bar{g} = 0$ on ω. Note that $\lambda(y)f \in C(G/K_0)$. Therefore, 1) implies that, if $\underline{d} \notin \mathcal{E}_0(K)$, then $0 = \xi_{\underline{d}} * \lambda(y)f = \rho(x)(\xi_{\underline{d}} * \lambda(y)f) = \xi_{\underline{d}} * g$. It follows that $\overline{\xi_{\underline{d}} * g} = 0$ on ω unless $\underline{d} \in \mathcal{E}_\omega(K) \cap \mathcal{E}_0(K)$.

We may write $\bar{g}|\omega = (\sum_{\underline{d} \in \mathcal{E}_0(K) \cap \mathcal{E}_\omega(K)} d(\underline{d}) \xi_{\underline{d}} * \bar{g})|\omega$. Since ω is compact, it follows from Theorem 4.8.1 that $\mathcal{E}_0(K) \cap \mathcal{E}_\omega(K)$ is finite. Let $K_F = \bigcap_{\underline{d} \in F} \operatorname{Ker} \underline{d}$. Then K_F is a normal subgroup of finite index in K, so $\xi_{\underline{d}} \in C(K /\!/ K_F) \subset C_c(G /\!/ K_F)$ $(\underline{d} \in F)$. It follows that $\bar{g}|\omega = (\sum_{\underline{d} \in F} d(\underline{d}) \xi_{\underline{d}} * \bar{g})|\omega = \sum_{\underline{d} \in F} \bar{g}_{\underline{d}}$ on ω, with $g_{\underline{d}} \in C(G /\!/ K_F)$. Thus if $\gamma_0 \in \omega$ and $\gamma \in (\gamma_0 K_F) \cap \omega$, then $\bar{g}(\gamma_0) = \bar{g}(\gamma)$. We may conclude that, for $\gamma_0 \in \omega$ and $\gamma \in (\gamma_0 K_F) \cap \omega$, $\int_K f(y\gamma^k x)dk = \int_K f(y\gamma_0^k x)dk$ for any pair $(y, x) \in G \times G$. Therefore, $\int_{G/\Gamma} f(x\gamma x^{-1})dx^* = \int_{G/\Gamma} f(x\gamma_0 x^{-1})dx^*$ for γ and γ_0 as before. Since $|D(\gamma)|^{\frac{1}{2}}$ is a locally constant function on Γ', F_f is locally constant on Γ'. More precisely, we have shown that, given $\gamma_0 \in \Gamma'$, we may choose a compact neighborhood

$_{K_0}$ of γ_0 in Γ' such that, for all $f \in C_c(G/K_0)$, $F_f|\omega_{K_0}$ is constant.

Corollary 4.8.4. Let $\mathcal{C}_{K_0}(G)$ denote $C(G/\!/K_0) \cap \mathcal{C}(G)$. The conclusions of Corollary 4.8.3 hold for $\mathcal{C}_{K_0}(G)$.

Proof. Let us consider first the crucial case in which Γ is elliptic. Fix $\gamma_0 \in \Gamma'$ and assume that $f \in \mathcal{C}_{K_0}(G)$. Choose an open neighborhood ω of γ_0 in Γ' such that:

(1) $|D(\gamma)| = |D(\gamma_0)|$ for all $\gamma \in \omega$, and

(2) $\int_{G/\Gamma} g(x\gamma x^{-1})dx^* = \int_{G/\Gamma} g(x\gamma_0 x^{-1})dx^*$ for all $\gamma \in \omega$ and $g \in C_c(G/K_0)$.

Let $\Omega_1 \subset \Omega_2 \subset \ldots \subset \Omega_i \subset \ldots$ be a sequence of open compact subsets of G such that $\Omega_i K_0 = \Omega_i$ and $\bigcup_{i=1}^{\infty} \Omega_i = G$. Let ϕ_i denote the characteristic function of Ω_i. Set $f_i = \phi_i f \in C_c(G/K_0)$. Then, by (1) and (2) above, $F_{f_i}(\gamma) = F_{f_i}(\gamma_0)$ for all i. It follows immediately from the existence of the limit almost everywhere on ω (Corollary 4.7.3) that, in fact, $\lim_{n \to \infty} F_{f_i}(\gamma)$ exists, $= F_{f_i}(\gamma_0)$ for all $\gamma \in \omega$.

Now let Γ be any Cartan subgroup of G with $A_\Gamma \supsetneq Z$ its split component. Then Γ is an elliptic Cartan subgroup of $M_\Gamma = Z_G(A_\Gamma)$. Up to conjugacy we may assume that $A_\Gamma = A \subset A_0$, that (P, A) $(P = MN)$ is a standard p-pair. Let K be an A_0-good maximal compact subgroup of G. Arguing as in the proof of Lemma 4.7.5, we set $\bar{f}(x) = \int_K f(x^k)dk$ $(x \in G)$ and obtain

$$F_f^{G/\Gamma}(\gamma) = |D(\gamma)|^{\frac{1}{2}} \int_{G/\Gamma} f(x\gamma x^{-1})dx^*$$

$$= |D(\gamma)|^{\frac{1}{2}} \int_{K \times N \times M/\Gamma} f(\gamma^{knm})dkdndm$$

$$= |D_M(\gamma)|^{\frac{1}{2}} \int_{M/\Gamma} \bar{f}^P(\gamma^m)dm^*,$$

for any $\gamma \in \Gamma'$, which implies that the general case reduces to the case Γ elliptic.

Corollary 4.8.5. Let Θ_ω be the character of a class $\omega \in \mathcal{E}_2(G)$. Then Θ_ω is represented on G_e by a locally constant and locally summable function.

Proof. Immediate from Corollary 4.8.4 and Corollary 4.7.4.

Conjecture of Howe. Let Γ be an elliptic Cartan subgroup of G. Let ω be an arbitrary compact subset of Γ. For $f \in C_c^\infty(G)$ let $j(f) = F_f|\omega'$, where $\omega' = \omega \cap \Gamma'$. If K_0 is an open compact subgroup of G, then $\dim j(C_c(G/K_0)) < \infty$

This conjecture implies that every irreducible unitary representation of G is admissible. The corollaries 4.8.3 and 4.8.4 verify Howe's conjecture in the special case $\omega \subset \Gamma'$.

Let us next establish what we need to prove Theorem 4.8.1.

Assume that $G \subset GL(n, \Omega)$ and that the Lie algebra $\mathfrak{g} \subset gl(n, \Omega)$. We call any open compact subgroup L of $\mathfrak{g}$ a lattice. Fix a nontrivial additive character χ of Ω. Let B be a nondegenerate symmetric bilinear form on $\mathfrak{g}$ which is invariant under the adjoint representation Ad of G (e.g., the Killing form on $[\mathfrak{g}, \mathfrak{g}]$ extended to all of $\mathfrak{g}$). Given a lattice L of $\mathfrak{g}$, write $L^* = \{X \in \mathfrak{g} \mid \chi(B(X, Y)) = 1 \text{ for all } Y \in L\}$. Then L^* is a lattice of $\mathfrak{g}$ too.

Let π be a uniformizing element of Ω, so that $|\pi| = q^{-1}$, where q is the module of Ω. Let $q = p^f$, where p is a prime element of $\mathbb{Z}$. If p is odd [even], let $\beta \geq e(p-1)^{-1}$ [$\beta \geq 2e+1$], where e is the ramification index of Ω over $\mathbb{Q}_p$. Assume that $[L, L] \subset \pi^\beta L$ (if we choose L small enough, this condition is automatically satisfied).

Let $\lambda \mapsto \exp(\lambda)$ denote the exponential mapping from $\mathfrak{g}$ into G. Let

$L = L_0$ and write $K_0 = \exp(L_0)$. Then K_0 is an open compact subgroup of G. For $\nu \geq 0$ we set $L_\nu = \pi^\nu L$ and $K_\nu = \exp(L_\nu)$. Then, for each ν, L_ν is a lattice and K_ν is an open compact subgroup of G. We have $K_0 \supset \ldots \supset K_\nu \supset \ldots$ and $\bigcap_{\nu \geq 0} K_\nu = (1)$. It is clear that $L_\nu^* = \pi^{-\nu} L^*$, that K_{ν_1} normalizes K_{ν_2}, $\nu_1 < \nu_2$.

For the case $p = 2$, we need to introduce some special terminology. We set $K_\nu^{\frac{1}{2}} = \exp(\frac{1}{2}L_\nu)$. Of course, $\frac{1}{2}L_\nu \supset L_\nu$, so $K_\nu^{\frac{1}{2}} \supset K_\nu$; it is clear that $K_\nu^{\frac{1}{2}}$ normalizes K_ν and acts on $\mathcal{E}(K_\nu)$. We write $\mathcal{E}^{\frac{1}{2}}(K_\nu)$ for the set of equivalence classes under the action of $K_\nu^{\frac{1}{2}}$ on $\mathcal{E}(K_\nu)$. In Theorem 4.8.6 replace $\mathcal{E}(K_\nu)$ by $\mathcal{E}^{\frac{1}{2}}(K_\nu)$ when $p = 2$.

For the proof of Theorem 4.8.1 we shall need the "main theorem of Kirillov theory", for whose proof we refer the reader to [8a].

Theorem 4.8.6. For any $\nu \geq 0$ there is a one-one correspondence $\underline{d} \mapsto O_{\underline{d}}$ between classes $\underline{d} \in \mathcal{E}(K_\nu)$ and $\mathrm{Ad}(K_\nu)$ orbits in $\mathfrak{g}/L_\nu^*$ such that: Suppose $\underline{d}_i \in \mathcal{E}(K_\nu)$ $(i = 1,2)$ and $x \in G$. Then x intertwines $\underline{d}_1$ and $\underline{d}_2$ if and only if $O_{\underline{d}_1} \cap \mathrm{Ad}(x)O_{\underline{d}_2} \neq \phi$.

In the sequel we write O^x for $\mathrm{Ad}(x)O$.

Fix a base $X_1, \ldots, X_r$ for $\mathfrak{g}$ and define a p-adic Banach space structure on $\mathfrak{g}$, i.e., if $X = \sum_{i=1}^{r} c_i X_i \in \mathfrak{g}$, we set $|X| = \max_{1 \leq i \leq r} |c_i|$. Assume that $L = \{Y \in \mathfrak{g} \mid |Y| \leq 1\}$.

Lemma 4.8.7. Fix $\nu \geq 0$. If $x \in G$ intertwines $\underline{d} \in \mathcal{E}(K_\nu)$ with itself, then there is an element $Y \in O_{\underline{d}}$ such that $|Y^x - Y| \leq \max(q^{-(\beta+\nu)}|Y|, q^\nu \delta)$, where $\delta = \sup_{\lambda^* \in L^*} |\lambda^*|$.

Proof. Fix $\nu \geq 0$ and $x \in G$ and suppose that x intertwines $\underline{d} \in \mathcal{E}(K_\nu)$ with itself. This means that $O_{\underline{d}} \cap (O_{\underline{d}})^x \neq \phi$ (Theorem 4.8.6). Let $Y \in (O_{\underline{d}})^{x^{-1}} \cap O_{\underline{d}}$, so that $Y^x \in O_{\underline{d}}$. It follows from Theorem 4.8.6 that $Y^x = Y^k + \lambda$, where $k \in K_\nu$, $\lambda \in L_\nu^*$; thus, $Y^x - Y = Y^k - Y + \lambda$. Let $k = \exp(\lambda_0)$, $\lambda_0 \in L_\nu$. Then $|ad(\lambda_0)Z| \leq q^{-(\beta+\nu)}|Z|$ $(Z \in L_\nu)$. Noting that $Y^k - Y = \sum_{m \geq 1} (ad(\lambda_0))^m Y/m!$, that $v_p(m!) = \sum_{j\geq 1} [m/p^j] < m(p-1)^{-1}$, and that $|(ad(\lambda_0))^m Y/m!| \leq q^{-m(\beta+\nu-e(p-1)^{-1})}|Y|$ $(m \geq 2)$, we deduce that $|Y^k - Y| \leq q^{-(\beta+\nu)}|Y|$, which implies that $|Y^x - Y| \leq \max(|Y^k - Y|, |\lambda|) \leq \max(q^{-(\beta+\nu)}|Y|, q^\nu \delta)$.

Now let B be a Cartan subgroup of G and let $\mathfrak{b}$ be the Lie algebra of B. Then $\mathfrak{g} = \mathfrak{b} \oplus [\mathfrak{b}, \mathfrak{g}]$. Let P denote the projection of $\mathfrak{g}$ on $[\mathfrak{b}, \mathfrak{g}]$. Let ω be a compact subset of the set B' of regular elements of B. Then there exists a positive constant $\gamma(\omega)$ such that $|X^b - X| \geq \gamma(\omega)|PX|$ for all $X \in \mathfrak{g}$ and $b \in \omega$. If ω intertwines $\underline{d} \in \mathcal{E}(K_\nu)$ with itself, then there exists $Y \in O_{\underline{d}}$ and $b \in \omega$ such that $\gamma(\omega)|PY| \leq |Y^b - Y| \leq \max(q^{-(\beta+\nu)}|Y|, q^\nu \delta)$, δ as in Lemma 4.8.7.

We have proved:

Corollary 4.8.8. The set $\omega \subset B'$ cannot intertwine $\underline{d} \in \mathcal{E}(K_\nu)$ with itself unless there exists $Y \in O_{\underline{d}}$ such that $\gamma(\omega)|PY| \leq \max(q^{-(\beta+\nu)}|Y|, q^{\nu\delta})$.

Lemma 4.8.9. Let C_1 and C_2 be compact subsets of $\mathfrak{g}$. Then there exists a constant $c > 0$ and a compact subset C_3 of $\mathfrak{g}$ such that $|PX| \geq c|X|$ for all $X \in (C_1^G + C_2) \cap {}^cC_3$ ($C_1^G = \bigcup_{x \in G} C_1^x$ and ${}^cC_3$ denotes the complement of C_3).

Proof. Extend the ground field Ω so that B and $\mathfrak{b}$ are split (a finite separable extension suffices, so we may assume the Banach space norm extends too). In this situation $\mathfrak{g} = \bar{\mathfrak{n}} \oplus \mathcal{V} \oplus \mathfrak{n}$. Since B and G are now split, there is a B-good maximal compact subgroup K of G such that $G = KB^+K$, B^+ denoting the closure of a positive Weyl chamber. Clearly, $C_1^{B^+K} \subset C' + \mathfrak{n} + \bar{\mathfrak{n}}$, where C' is a compact subset of $\mathcal{V}$, so $C_1^G \subset (C')^K + \mathcal{N}$, where $\mathcal{N}$ denotes the set of all nilpotent elements of $\mathfrak{g}$. Thus, $C_1^G + C_2 \subset C'' + \mathcal{N}$, where C'' is a compact subset of $\mathcal{V}$.

Suppose the lemma is false. Then there is a sequence $\{X_i\} \subset C'' + \mathcal{N}$ such that $|X_i| \to \infty$, $i \to \infty$, and such that $|PX_i|\ |X_i|^{-1} \to 0$, $i \to \infty$. Let $S = \{X \in \mathfrak{g} \mid |X| = 1\}$. Then $X_i = t_iY_i + Z_i$, where $Y_i \in \mathcal{N} \cap S$ and $Z_i \in C''$. Since C'' is compact, we see that, for large i, $|X_i| = t_i$, so $|t_i| \to \infty$, $i \to \infty$. Thus, for large i, $|PX_i|\ |X_i|^{-1} = |P(Y_i + t_i^{-1}Z_i)|$. Clearly, $t_i^{-1}Z_i \to 0$, $i \to \infty$. Since $\mathcal{N} \cap S$ is compact, we may choose a subsequence of $\{X_i\}$ such that $Y_i \to Y_0$, $i \to \infty$. Furthermore, since $|PY_i| \to 0$, $i \to \infty$, we have $|PY_0| = 0$. But $\mathfrak{g} = \mathcal{V} \oplus [\mathcal{V}, \mathfrak{g}]$, so $PY_0 = 0$ implies that $Y_0 \in \mathcal{V}$. Since $\mathcal{V} \cap (\mathcal{N} \cap S) = \phi$, we have found a contradiction, so Lemma 4.8.9 is proved.

Now let us prove Howe's Theorem. It is sufficient to consider $\underline{d}_0 \in \mathcal{E}(K_0)$. Note that $O_{\underline{d}_0}$ is a compact subset of $\mathfrak{g}$. Fix $\nu \geq 0$ such that $c\gamma(\omega) > q^{-(\beta+\nu)}$. Let F_ν be the set of all $\underline{d} \in \mathcal{E}(K_\nu)$ such that:

1) G intertwines $\underline{d}$ with $\underline{d}_0$, and
2) ω intertwines $\underline{d}$ with $\underline{d}$.

To complete the proof of Howe's Theorem we shall show that F_ν is a finite set.

For this we apply Lemma 4.8.9. Let $C_1 = O_{\underline{d}_0}$, $C_2 = L_\nu^*$, and choose a compact subset C of $\mathfrak{g}$ such that:

1) $|PX| \geq c|X|$ for $X \in (O_{\underline{d}_0}^G + L_\nu^*) \cap {}^c C$.

2) $C = C^{K_\nu} + L_\nu^*$.

3) $|X| \leq q^{\beta+2\nu}\delta$ implies $X \in C$ $(X \in \mathfrak{g})$.

Now let $\underline{d} \in F_\nu$. We claim that $C \cap O_{\underline{d}} \neq \phi$. Otherwise, since G intertwines $\underline{d}$ and $\underline{d}_0$, there exists $Y_0 \in O_{\underline{d}} \cap (O_{\underline{d}_0})^G$ such that $Y_0 \notin C$. Since ω intertwines $\underline{d}$ with itself, Corollary 4.8.8 implies the existence of $Y \in O_{\underline{d}}$ such that $\gamma(\omega)|PY| \leq \max(q^\nu\delta, q^{-(\beta+\nu)}|Y|)$. However, both Y and Y_0 lie in $O_{\underline{d}}$, so $Y = Y_0^k + \lambda$ $(k \in K_\nu, \lambda \in L_\nu^*)$. By 2), $Y \notin C$. Since $Y \in ((O_{\underline{d}_0})^G + L_\nu^*) \cap {}^c C$, condition 1) implies that $|PY| \geq c|Y|$.

Combining inequalities yields $c\gamma(\omega)|Y| \leq \max(q^\nu\delta, q^{-(\beta+\nu)}|Y|)$. Since $Y \notin C$, we have, by 3), $|Y| > q^{\beta+2\nu}\delta > 0$. If $c\gamma(\omega)|Y| \leq q^\nu\delta$, then $\gamma(\omega)c \leq q^{-(\beta+\nu)}$. Since ν was chosen so large that $\gamma(\omega)c > q^{-(\beta+\nu)}$, this yields a contradiction, so we must have $c\gamma(\omega)|Y| \leq \max(q^\nu\delta, q^{-(\beta+\nu)}|Y|) = q^{-(\beta+\nu)}|Y|$. However, this implies the same contradiction. It follows that, if $\underline{d} \in F_\nu$, then $O_{\underline{d}} \cap C \neq \phi$ implies $O_{\underline{d}} \subset C$. Since $O_{\underline{d}}$ is open and C compact, only finitely many $\underline{d}$ have the property $O_{\underline{d}} \cap C \neq \phi$.

Chapter 5. The Eisenstein Integral and Applications.

We fix, for the whole chapter, a maximal split torus A_0 of G and an A_0-good maximal compact subgroup K of G.

Let A be an A_0-standard torus, $M = Z_G(A)$, and $\omega \in \mathcal{E}_2(M)$. If ω is unramified, then we have seen that the theory of induced representations provides us with a class $C_M^G(\omega) \in \mathcal{E}(G)$. Let (V, τ) be a smooth unitary double representation of K. Then the map $f \mapsto f_P = \Sigma f_{P,s}$ from $\mathcal{A}(C(\omega), \tau)$ to $\bigoplus_{s \in W(A)} \mathcal{A}(\omega^s, \tau_M)$ gives rise to the Maass-Selberg relations. The Eisenstein integral, $\psi \mapsto E(P:\psi)$ $(P \in \mathcal{P}(A))$, maps $\mathcal{A}(\omega, \tau_M)$ to $\mathcal{A}(C(\omega), \tau)$. The functional equations result by combining the two.

The development of the first half of this chapter proceeds in two stages. In the first or formal stage we assume the existence of simple classes in $\mathcal{E}_2(M)$ and derive the functional equations without making any use of the analytic structure on the spaces of admissible representations. In the second stage we bring in the analytic structure, prove that all unramified classes in $\mathcal{E}_2(M)$ are simple, and interpret the functional equations as relations between analytic functions. The last half of the chapter presents various applications to the harmonic analysis of G, including the Plancherel's formula.

In brief, the first section of the chapter collects various technical facts to be applied later. The second and third sections contain, respectively, the formal and analytic theories of the Eisenstein integral and its functional equations. The fourth section applies the theory to an analysis of the composition series of an induced representation. The fifth and final section introduces wave packets, proves a Plancherel's formula, and gives an explicit characterization of the commuting algebra of an induced representation; the concluding subsection of these notes contains a proof that, if $\dim A_0/Z = 1$, then only tempered representations occur in the Plancherel's formula for G.

In [8c] Roger Howe proved for $GL_n(\Omega)$ that the formal degrees of supercuspidal representations are integers. This implies that all irreducible unitary representations of $GL_n(\Omega)$ are admissible. In unpublished work, Harish-Chandra generalized Howe's result to arbitrary G, char $\Omega = 0$. More recently, Bernstein ([1a]) has given a short proof, not depending upon any restriction of the characteristic of Ω, that all irreducible unitary representations of G are admissible.

In [7f] it is announced that only tempered representations occur in the Plancherel's formula for G.

§5.1. <u>On the Matrix Coefficients of Admissible Representations and Their Constant Terms.</u>

Fix a minimal p-pair (P_0, A_0). Let $\{K_j\}_{j=1}^{\infty}$ be a sequence of open compact subgroups of G such that, for any p-pair $(P, A) \succ (P_0, A_0)$ and index j, we have $K_j = \bar{N}_j M_j N_j$ with obvious notations (cf. §2.1). Note that, for $j_1 \leq j_2$, the group K_{j_2} is a normal subgroup of K_{j_1}.

Let π be an admissible representation of G on a vector space $\mathcal{H}$. For the most part we regard the index j as fixed. Normalize the measure dk n K_j and define the projection operator $E_j = \int_{K_j} \pi(k)dk$. Set $E_j \mathcal{H} = \mathcal{H}_j$ and define $\phi(x) = \phi_j(x) = E_j \pi(x) E_j$. Then $\phi_j \in \mathcal{A}(G : \mathrm{End}(\mathcal{H}_j))$.

In the following assume that all Haar measures are normalized.

Recall that $\phi_{(P)}(m) = \delta_P(m)^{\frac{1}{2}} \phi(m)$.

Lemma 5.1.1.

1) $\int_{K_j} \phi(xky)dk = \phi(x)\phi(y) \quad (x, y \in G)$.

2) If $\bar{N}_j^{m_1} \subset \bar{N}_j$ and $N_j^{m_2^{-1}} \subset N_j$, then $\int_{M_j} \phi(m_1 m m_2)dm = \phi(m_1)\phi(m_2) \quad (m_1, m_2 \in M)$.

3) If $\bar{N}_j^{m_1} \subset \bar{N}_j [N_j^{m_2^{-1}} \subset N_j]$, then $\int_{M_j} \phi_P(m_1 m m_2)dm = \phi_{(P)}(m_1)\phi_P(m_2)$

$$[\phi_P(m_1)\phi_{(P)}(m_2)] \quad (m_1, m_2 \in M).$$

4) $\int_{M_j} \phi_P(m_1 m m_2)dm = \phi_P(m_1)\phi_P(m_2) \quad (m_1, m_2 \in M)$.

5) If $\bar{N}_j^m \subset \bar{N}_j [N_j^{m^{-1}} \subset N_j]$, then $\phi_P(m) = \phi_{(P)}(m)\phi_P(1) =$

$\phi_P(m)\phi_P(1) [= \phi_P(1)\phi_{(P)}(m)] \quad (m \in M)$.

6) $\phi_P(ma) = \phi_P(m)\phi_P(a) = \phi_P(a)\phi_P(m) \quad (m \in M,\ a \in A)$.

Proof.

1) Immediate from the relation $\int_{K_j} \pi(xky)dk = \pi(x)E_j\pi(y)$.

2) Since $\phi(m_1 k m_2) = \phi(m_1 \bar{n} m n m_2) = \phi(\bar{n}^{m_1} m_1 m m_2 n^{m_2^{-1}})$, it is clear that 1), combined with the normalization of measures, implies 2).

3) For t sufficiently large and $a \in A^+(t)$ it is clear that m_1 and $m_2' = am_2$ satisfy the hypotheses of 2). It follows that $\int_{M_j} \phi_{(P)}(m_1 m m_2')dm =$

$\phi_{(P)}(m_1)\phi_{(P)}(m_2')$. Choose t so large that $\phi_{(P)}(m_1 m m_2') = \phi_P(m_1 m m_2')$ for all $m \in M_j$ and $\phi_{(P)}(m_2') = \phi_P(m_2')$. Use the *A-finiteness* of ϕ_P and Lemma 2.6.2 to justify setting $a = 1$ or $m_2' = m_2$. The proof of the bracketed relation goes similarly.

4) Use A-finiteness and Lemma 2.6.2 to deduce 4) from 3).

5) and 6) are both immediate from 3) and 4).

Corollary 5.1.2. Choose $t_j \geq 1$ so large that $\bar{N}_j^a \subset \bar{N}_j$ and $N_j^{a^{-1}} \subset N_j$ for all $a \in A^+(t_j)$. Then $\phi(a_1 a_2) = \phi(a_1)\phi(a_2)$ for all a_1 and $a_2 \in A^+(t_j)$. The number t_j depends upon K_j but not upon π.

Now choose K_j such that $\mathcal{H}_j \neq (0)$ and let $d = d_j = \dim \mathcal{H}_j$. Write

$$W_j = \bigcup_{a \in A^+(t_j)} \ker \phi(a).$$

Lemma 5.1.3. For any $a \in A$, $W_j = \ker \phi_P(a)$. In particular, W_j is a subspace of $\mathcal{H}_j$.

Proof. It follows from Lemma 5.1.1, 6) that $\phi_P(a_2) = \phi_P(a_2 a_1^{-1})\phi_P(a_1)$ for all $a_1, a_2 \in A$. Thus, $\ker \phi_P(a_1) = \ker \phi_P(a_2)$. Since $\ker \phi(a) = \ker \phi_{(P)}(a) = \ker \phi_P(a) = W_j$, $a \in A^+(t)$ for some t, the lemma follows.

Corollary 5.1.4.

1) For every $a \in A^+(t_j)$, $\ker \phi(a^d) = W_j$.

2) $1_{\mathcal{H}_j} - \phi_P(1)$ projects $\mathcal{H}_j$ on W_j and commutes with $\phi(a)$ for all $a \in A^+(t_j)$.

Proof.

1) For $a \in A^+(t_j)$ and n sufficiently large $\ker \phi(a^n) = \ker \phi_P(a^n) = W_j$. How-

ever, for $a \in A^+(t_j)$, $\phi(a^n) = \phi(a)^n$ and $d = \dim \mathcal{H}_j$, so $\ker\phi(a^d) = W_j$.

2) Clearly, $1 - \phi_P(1)$ is a projection operator on $\mathcal{H}_j$ with image W_j. Lemma 5.1.1, 5) implies that $1 - \phi_P(1)$ commutes with $\phi(a)$ for all $a \in A^+(t_j)$.

Notice that the number t of the following lemma depends upon $\mathcal{H}$, but not otherwise upon π.

Lemma 5.1.5. Fix a compact subset ω of M and an element $a_0 \in A^+(t_j)$. Choose $t \geq 1$ such that $\bar{N}_j^{ma} \subset \bar{N}_j$ and $N_j^{m^{-1}a^{-1}} \subset N_j$ for all $m \in \omega a_0^{-d}$ and $a \in A^+(t)$. Then $\phi_{(P)}(ma) = \phi_P(ma)$ for all $m \in \omega$ and $a \in A^+(t)$.

Proof. For $m \in \omega$ and $a \in A^+(t)$ we write $ma = m_1 a_1$, where $m_1 = ma_0^{-d}$ and $a_1 = aa_0^d$. Since $\phi_P(ma) = \phi_P(m_1 a_1) = \phi_{(P)}(m_1 a_1)\phi_P(1)$ (Lemma 5.1.1, 5)), it follows that $\phi_P(ma) - \phi_{(P)}(ma) = \phi_{(P)}(m_1 a_1)(1-\phi_P(1))$. But $\phi(m_1 a_1) = \phi(m_1 a a_0^d) = \phi(m_1 a)\phi(a_0)^d$, so $\phi(m_1 a_1)(1-\phi_P(1)) = \phi(m_1 a)\phi(a_0)^d(1-\phi_P(1)) = 0$. Therefore, $\phi_P(ma) = \phi_{(P)}(ma)$ $(m \in \omega,\ a \in A^+(t))$.

Lemma 5.1.6. Fix $a_0 \in A^+(t_j)$ and set $\bar{U} = (\bar{N}_j)^{a_0^{-d}}$. Then $\int_{\bar{U}} \pi(\bar{n})d\bar{n}(1-\phi_P(1)) = 0$

Proof. Since $(1-\phi_P(1))\mathcal{H}_j = W_j$, it is sufficient to show that $\int_{\bar{U}} \pi(\bar{n})d\bar{n} \cdot W_j = (0)$. Since $W_j = \ker\phi(a^d)$ $(a \in A^+(t_j)$; cf. Corollary 5.1.4, 1)), it follows that $\phi(a_0^d)w = 0$ $(w \in W_j)$ if and only if $E_j\pi(a_0^d)w = 0$. However,

$$E_j\pi(a_0^d)w = \int_{\bar{N}_j M_j N_j} \pi(\bar{n}mna_0^d)d\bar{n}dmdn \cdot w$$

$$= \pi(a_0^d)\int \pi(\bar{n}^{a_0^{-d}} mn^{a_0^{-d}})d\bar{n}dmdn \cdot w$$

$$= \pi(a_0^d)\int_{\bar{N}_j} \pi(\bar{n}^{a_0^{-d}})d\bar{n}\, E_j w$$

$$= \pi(a_0^d) \int_{\bar{U}} \pi(\bar{n}) d\bar{n} \cdot w.$$

Since $\pi(a_0)$ is invertible, $\int_{\bar{U}} \pi(\bar{n}) d\bar{n} \cdot w = 0$.

Lemma 5.1.7. $W_j = \mathcal{H}_j \cap \mathcal{H}(\bar{N})$.

Proof. Lemma 5.1.6 implies that $W_j \subset \mathcal{H}_j \cap \mathcal{H}(\bar{N})$. Let $v \in \mathcal{H}_j \cap \mathcal{H}(\bar{N})$ and choose an open compact subgroup $\bar{U} \subset \bar{N}$ such that $\int_{\bar{U}} \pi(\bar{n}) d\bar{n} \cdot v = 0$. Choose $a \in A^+(t)$, $t \geq 1$, such that $\bar{U}^a \subset \bar{N}_j$. Then $0 = \pi(a) \int_{\bar{U}} \pi(\bar{n}) d\bar{n} \cdot v =$ $\int_{\bar{U}} \pi(\bar{n}^a) d\bar{n} \cdot \pi(a) v$, and this implies, clearly, that $E_j \pi(a) v = \phi(a) v = 0$. Therefore, $v \in \ker \phi_P(a) = W_j$ (t sufficiently large).

Write. ${}^tE_j = \int_{K_j} \tilde{\pi}(k) dk = \int_{K_j} {}^t\pi(k) dk$ and set $\tilde{\mathcal{H}}_j = {}^tE_j \tilde{\mathcal{H}} = (\mathcal{H}_j)^{\sim}$.

Clearly, $\psi(x) = {}^t\phi(x^{-1}) = {}^tE_j \tilde{\pi}(x) {}^tE_j$ bears the same relation to $\tilde{\pi}$ as ϕ bears to π

Corollary 5.1.8. Set $\tilde{W}_j = \ker \psi_{\bar{P}}(1)$. Then $\tilde{W}_j = \tilde{\mathcal{H}}_j \cap \tilde{\mathcal{H}}(N)$. Moreover, for any $\tilde{v} \in \tilde{\mathcal{H}}$ and $v \in \mathcal{H}$, $\langle \tilde{v}, \phi_P(1) v \rangle = \langle \psi_{\bar{P}}(1) \tilde{v}, v \rangle$.

Proof. Only the last statement requires some justification. Observe that, if $T \in \mathrm{End}^0(\mathcal{H})$, then ${}^tT \in \mathrm{End}^0(\tilde{\mathcal{H}})$ and $f_T(x) = \mathrm{tr}(T\pi(x)) = \mathrm{tr}({}^tT\tilde{\pi}(x^{-1})) = f_{{}^tT}(x^{-1})$. It follows that $f_{T,P}(m) = f_{{}^tT,\bar{P}}(m^{-1})$ for any $m \in M$. On the other hand, given $v \in \mathcal{H}_j$ and $\tilde{v} \in \tilde{\mathcal{H}}_j$, we have $T = \tilde{v} \otimes v \in \mathrm{End}^0(\mathcal{H})$ and $\langle \tilde{v}, \phi_P(1) v \rangle = f_{T,P}(1) =$ $f_{{}^tT,\bar{P}}(1) = \langle \psi_{\bar{P}}(1) \tilde{v}, v \rangle$. This proves the corollary.

We may interpret the preceding as follows. The mapping

$_P : T \mapsto f_{T,P}(1)$ is a linear functional on $\mathrm{End}^0(\mathcal{H})$. Using the identification $\mathrm{End}^0(\mathcal{H}) = \mathcal{H} \otimes \widetilde{\mathcal{H}}$ (cf. §1.11), we define a bilinear form on $\widetilde{\mathcal{H}} \times \mathcal{H}$:

$$<\tilde{v}, v>_P = J_P(v \otimes \tilde{v}) \qquad ((v, \tilde{v}) \in \mathcal{H} \times \widetilde{\mathcal{H}}).$$

For $v \in \mathcal{H}_j$ and $\tilde{v} \in \widetilde{\mathcal{H}}_j$ we have $<\tilde{v}, v>_P = <\tilde{v}, \phi_P(1)v> = <\psi_{\bar{P}}(1)\tilde{v}, v>$.

Now let $p : \mathcal{H} \to \mathcal{H}/\mathcal{H}(\bar{N})$ and $p^* : \widetilde{\mathcal{H}} \to \widetilde{\mathcal{H}}/\widetilde{\mathcal{H}}(N)$ be the canonical maps. Let $U = \mathcal{H}/\mathcal{H}(\bar{N})$, $U^* = \widetilde{\mathcal{H}}/\widetilde{\mathcal{H}}(N)$, $U_j = p(\mathcal{H}_j)$, and $U_j^* = p^*(\widetilde{\mathcal{H}}_j)$. Set

$$<p^*(\tilde{v}), p(v)> = <\tilde{v}, v>_P \qquad ((v, \tilde{v}) \in \mathcal{H} \times \widetilde{\mathcal{H}})$$

and obtain a bilinear form on $U^* \times U$. It follows from Lemma 5.1.7 and Corollary 5.1.8 that both this bilinear form and its restriction to $U_j^* \times U_j$ are nondegenerate bilinear forms. Write $U_j^\perp = \{u \in U \mid <u^*, u> = 0 \text{ for all } u^* \in U_j^*\}$. Then $U = U_j + U_j^\perp$ (direct sum). Let $F_j : U \to U_j$ be the corresponding projection operator.

Let $\bar{\pi}$ denote the representation of M on U which satisfies $\bar{\pi}(m)p(v) = p(\pi(m)v)$ for all $v \in \mathcal{H}$ and $m \in M$.

Lemma 5.1.9. As before, let $\phi_j = E_j\pi(x)E_j$ $(x \in G;\ 1 \leq j \in \mathbb{Z})$. Let $(v, \tilde{v}) \in \mathcal{H}_{j_0} \times \widetilde{\mathcal{H}}_{j_0}$. Then

$$\delta_P(m)^{\frac{1}{2}} <p^*(\tilde{v}), F_j\bar{\pi}(m)F_jp(v)> = <\tilde{v}, \phi_{j_0,P}(m)v>$$

for all $j \geq j_0$.

Proof. Choose $t \geq 1$ such that, for all $a \in A^+(t)$ and all $j \geq 1$, $N_j^{m^{-1}a^{-1}} \subseteq N_j$ (it is an easy consequence of the proof of Theorem 2.1.1 that this is possible). Fix $a \in A^+(t)$. Choose $j_1 \geq j_0$ such that $\pi(ma)v \in \mathcal{H}_{j_1}$. Then, for $j \geq j_1$, we have

$$\delta_P(ma)^{\frac{1}{2}} <p^*(\tilde{v}), F_j\bar{\pi}(ma)F_jp(v)> = \delta_P(ma)^{\frac{1}{2}} <p^*(\tilde{v}), p(\pi(ma)v)>$$

$$= \delta_P(ma)^{\frac{1}{2}} \langle \tilde{v}, \pi(ma)v \rangle_P$$

$$= \langle \tilde{v}, \phi_{j_1, P}(1)\phi_{j_1(P)}(ma)v \rangle$$

$$= \langle \tilde{v}, \phi_{j_1, P}(ma)v \rangle \quad \text{(Lemma 5.1.1, 5))}$$

$$= \langle {}^tE_{j_0}\tilde{v}, \phi_{j_1, P}(ma)E_{j_0}v \rangle$$

$$= \langle \tilde{v}, \phi_{j_0, P}(ma)v \rangle.$$

Since the final equality holds for all $a \in A^+(t)$, we may use Lemma 2.6.2 and the A-finiteness of both sides to conclude that the lemma is true.

Corollary 5.1.10. Let $\theta_{\bar{\pi}}$ denote the character of the admissible representation $\bar{\pi}$ (cf. §2.3). Let $\alpha \in C_c(M/\!/M_{j_0})$. Then there exists an index $j_1 \geq j_0$ such that, if $j \geq j_1$,

$$\theta_{\bar{\pi}}(\alpha) = \int_M \alpha(m)\delta_P(m)^{-\frac{1}{2}} \operatorname{tr}(\phi_{j,P}(m))dm.$$

Proof. Let $U_{M_{j_0}}$ denote the space of $\bar{\pi}(M_{j_0})$-fixed vectors in U. Choose j_1 such that $j \geq j_1$ implies $F_j U_{M_{j_0}} = U_{M_{j_0}}$ (this is possible because $\dim U_{M_{j_0}} < \infty$ and $U = \bigcup_{j \geq 1} U_j$). Since K_{j_0} normalizes K_j whenever $j \geq j_0$, it is clear that U_j and $U_j^{\perp}$ are M_{j_0}-modules. It follows that $\theta_{\bar{\pi}}(\alpha) = \int_M \alpha(m)\operatorname{tr}(F_j\bar{\pi}(m)F_j)dm$ $(j \geq j_1)$. The operator $F_j\bar{\pi}(m)F_j$ corresponds to a finite matrix with entries of the form $\langle p^*(\tilde{v}), F_j\bar{\pi}(m)F_j \cdot p(v)\rangle$ with $(v, \tilde{v}) \in \mathcal{H}_j \times \tilde{\mathcal{H}}_j$. The corollary follows immediately from Lemma 5.1.9 and Schur's orthogonality relations.

We conclude the section with a lemma to be applied in §5.5.

Let $f \in \mathcal{A}(G)$. For the next lemma we want to extend f_P to a function defined on all of G. Set $f_P(x) = \delta_P(x)^{\frac{1}{2}}(\lambda(x^{-1})f)_P(1)$. The new function, as follows from Lemma 2.7.1, has the same value as previously for $m \in M$. For $x \in G$ fixed and $a \in A^+(t)$, t large, we have $\delta_P(xa)^{\frac{1}{2}}f(xa) = f_P(xa)$. In general, $f_P(kmn) = (\lambda(k^{-1})f)_P(m)$ for $k \in K$, $m \in M$, and $n \in N$. If $f \in \mathcal{A}(G, \tau)$ for some double representation (V, τ), then $f_P(kmn) = \tau(k)f_P(m)$.

Lemma 5.1.11. Let ω be a compact subset of M. For any $f \in \mathcal{A}(G)$, there exists an open compact subgroup $\bar{U} = \bar{U}(f, \omega) \subset \bar{N}$ such that, if $\bar{U} \subset X \subset \bar{N}$ and X is a compact group, then

$$\int_X f(\bar{n}m)d\bar{n} = \int_X \delta_P(\bar{n}m)^{-\frac{1}{2}} f_P(\bar{n}m)d\bar{n}$$

for any $m \in \omega$. For fixed ω and a fixed open compact subgroup K_0 of G the group $\bar{U}$ may be chosen independent of $f \in \mathcal{A}(G) \cap C(G /\!/ K_0)$.

Proof. Since any $f \in \mathcal{A}(G)$ may be regarded as a matrix coefficient of an admissible representation, it suffices to prove this lemma for the function ϕ defined previously in this section.

Using Lemma 5.1.1, 1) and the normalization of measures, we have

$$\phi(\bar{n}m)\phi(a) = \int_{K_j} \phi(\bar{n}mk_j a)dk_j$$

$$= \int_{\bar{N}_j M_j N_j} \phi(\bar{n}m \cdot \bar{n}_j m_j n_j \cdot a)d\bar{n}_j dm_j dn_j \; .$$

If $a \in A^+(t)$ for t sufficiently large, then $n_j^{a^{-1}} \in N_j$. Take $\bar{U}_1$, an open compact subgroup of $\bar{N}$, such that $\bar{U}_1 \supset \bar{N}_j^m$ for all $m \in \omega$. Then

$$\int_{\bar{U}_1} \phi(\bar{n}m)\phi(a)d\bar{n} = \int_{\bar{U}_1} \phi(\bar{n}ma)d\bar{n}$$

for all $m \in \omega$. Clearly, we may even choose t such that

$$\int_{\bar{U}_1} \phi(\bar{n}m)\phi_P(a)d\bar{n} = \int_{\bar{U}_1} \delta_P(\bar{n}m)^{-\frac{1}{2}}\phi_P(\bar{n}ma)d\bar{n}$$

for all $m \in \omega$. Using A-finiteness and Lemma 2.6.2 we have

$$\int_{\bar{U}_1} \phi(\bar{n}m)\phi_P(1)d\bar{n} = \int_{\bar{U}_1} \delta_P(\bar{n}m)^{-\frac{1}{2}}\phi_P(\bar{n}m)d\bar{n} \quad (m \in \omega).$$

Choose $\bar{U}$ so that $\bar{U} \supset \bar{U}_1$ and $\int_{\bar{U}} \pi(\bar{n})(1-\phi_P(1))E_j d\bar{n} = 0$ (use Lemma 5.1.6). It follows that $\int_{\bar{U}} \phi(\bar{n}m)d\bar{n} = \int_{\bar{U}} \delta_P(\bar{n}m)^{-\frac{1}{2}}\phi_P(\bar{n}m)d\bar{n}$ for all $m \in \omega$.

§5.2. The Eisenstein Integral and Its Functional Equations: The Formal Theory

Let A be an A_0-standard torus of G, let $M = Z_G(A)$, and let $\mathcal{P}(A)$ denote the set of parabolic subgroups of G with A as split component.

§5.2.1. Preliminaries: The Context

Let σ be an admissible representation of M in a vector space U, $\tilde{\sigma}$ in $\tilde{U}$ its contragredient. Write $\sigma_0 = \sigma|K_M[\tilde{\sigma}_0 = \tilde{\sigma}|K_M]$ $(K_M = K \cap M)$ and set $\pi_0 = \mathrm{Ind}_{K_M}^K \sigma_0[\tilde{\pi}_0 = \mathrm{Ind}_{K_M}^K \tilde{\sigma}_0]$. Let $\mathcal{H}[\tilde{\mathcal{H}}] = \{h \in C^\infty(K:U) | h(mk) = \sigma_0(m)h(k)$ $[h(mk) = \tilde{\sigma}_0(m)h(k)]$ $(k \in K,\ m \in K_M)\}$ be the representation space for $\pi_0[\tilde{\pi}_0]$. Fix pre-hilbert space structures $(\ ,\ )_U$ and $(\ ,\ )_{\tilde{U}}$ on U and $\tilde{U}$ such that σ_0 and $\tilde{\sigma}_0$ are unitary. Set $(h_1,h_2)_{\mathcal{H}} = \int_K (h_1(k),h_2(k))_U dk$ $(h_i \in \mathcal{H})$ and $(\tilde{h}_1,\tilde{h}_2)_{\tilde{\mathcal{H}}} = \int_K (\tilde{h}_1(k),\tilde{h}_2(k))_{\tilde{U}} dk$ $(\tilde{h}_i \in \tilde{\mathcal{H}})$. These scalar produces give $\mathcal{H}$ and $\tilde{\mathcal{H}}$ pre-hilbert space structures with respect to which π_0 and $\tilde{\pi}_0$ are unitary. For any $T \in \mathrm{End}^0(\mathcal{H})$ we write T^* for the adjoint of T with respect to this structure.

Given $P \in \mathcal{P}(A)$ $(P = MN)$, define a projection operator $F_P : \mathcal{H} \to \mathcal{H}(P)$ $[F_P : \tilde{\mathcal{H}} \to \tilde{\mathcal{H}}(P)]$ by setting $(F_P h)(k) = \int_{N \cap K} h(nk)dn$ $(h \in \mathcal{H}$ $[h \in \tilde{\mathcal{H}}])$. In the subspace $\mathcal{H}(P)$ of $\mathcal{H}[\tilde{\mathcal{H}}(P)$ of $\tilde{\mathcal{H}}]$ we have the representation $\pi_{0,P} = \mathrm{Ind}_{P \cap K}^{K} \sigma_0 [\tilde{\pi}_{0,P}]$ of K. We extend $h \in \mathcal{H}(P)[\tilde{\mathcal{H}}(P)]$ to a function defined on G by setting $h(pg) = \delta_P^{\frac{1}{2}}(p)\sigma(p)h(g)[h(pg) = \delta_P^{\frac{1}{2}}(p)\tilde{\sigma}(p)h(g)]$ $(p \in P,\ g \in G$-$\sigma(mn) = \sigma(m)$ for $m \in M,\ n \in N)$. In this way we realize the representation $\pi_P = \mathrm{Ind}_P^G(\delta_P^{\frac{1}{2}}\sigma)[\tilde{\pi}_P]$ in the space $\mathcal{H}(P)[\tilde{\mathcal{H}}(P)]$-$(\pi_P(x)h)(g) = h(gx)$, as usual.

Let (V, τ) denote an arbitrary smooth unitary double representation of K which has an involution and satisfies associativity conditions (§1.12). For every $P \in \mathcal{P}(A)$ we have left and right projection operators defined on the space V : $F_P(\tau)v = \int_{N \cap K} \tau(n)v dn$ and $vF_P(\tau) = \int_{N \cap K} v\tau(n)dn$. Write $V_{P_2|P_1}$ or $V(P_2|P_1) = F_{P_2}(\tau)VF_{P_1}(\tau)$ $(P_1, P_2 \in \mathcal{P}(A))$, $V_P = V(P|P)$. Write τ_M for the double representation $\tau|K_M$ of K_M on V and note that, for any P_1, P_2, the subspace $V(P_2|P_1)$ is τ_M-stable. Let $\tau_{P_2|P_1}$ denote τ_M on the space $V(P_2|P_1)$; write τ_P for $\tau_{P|P}$. Clearly, multiplication maps $V(P_3|P_2) \times V(P_2|P_1) \to V(P_3|P_1)$. In particular, V_P is always a smooth unitary double representation algebra which satisfies associativity conditions.

The space $\mathcal{T} = \mathrm{End}^0(\mathcal{H})$ forms with respect to pre and post multiplication by π_0 a particularly important example of a double representation with the above properties (cf. §§1.11-1.12). In particular, $\|T\|^2 = \mathrm{tr}(T^* T)$ $(T \in \mathcal{T})$. For any $P_1, P_2 \in \mathcal{P}(A)$ we have $\mathcal{T}(P_1|P_2) = F_{P_1}\mathcal{T}F_{P_2}$. The space $\mathcal{T}(P)$ is a K-stable subalgebra for any $P \in \mathcal{P}(A)$. Note that $\pi_P(x) = F_P\pi_P(x) = \pi_P(x)F_P$ $(x \in G)$. This implies that, to any pair $P_1, P_2 \in \mathcal{P}(A)$, there corresponds a natural double $C_c^\infty(G)$-module structure both on $\mathcal{T}$ and on $\mathcal{T}(P_1|P_2)$. For later use we "throw this structure to the opposite side" by defining $\alpha \underset{P_2}{*} T = T\pi_{P_2}(\alpha')$ and $T \underset{P_1}{*} \alpha = \pi_{P_1}(\alpha')T$ ($T \in \mathcal{T}$ and $\alpha'(x) = \alpha(x^{-1})$, $\alpha \in C_c^\infty(G)$).

Given $T = \sum_{i=1}^{r} h_i \otimes h_i^\sim \in \mathcal{H} \otimes \mathcal{H}^\sim$ we set

$(Th)(k_2) = \sum_{i=1}^{r} h_i(k_2)\int_K \langle h_i^\sim(k_1), h(k_1)\rangle dk_1$ for any $h \in \mathcal{H}$. This defines an isomorphism between $\mathcal{H} \otimes \mathcal{H}^\sim$ and $\mathcal{T}$ (cf. §1.11) and we frequently identify these two spaces. For any $P_1, P_2 \in \mathcal{P}(A)$ there is a natural correspondence between $\mathcal{H}(P_1) \otimes \mathcal{H}^\sim(P_2)$ and $\mathcal{T}(P_1|P_2)$.

Throughout §5.2 we write V^0 for $C^\infty(K \times K)$ and consider two distinct double representations of K on V^0. Let $v \in V^0$ and $(\ell_1, \ell_2) \in K \times K$. Then:

1) $\tau^0(\ell_1)v\tau^0(\ell_2) : (k_1 : k_2) \mapsto v(k_1\ell_1 : \ell_2 k_2)$ and

2) ${}^0\tau(\ell_1)v\,{}^0\tau(\ell_2) : (k_1 : k_2) \mapsto v(\ell_1^{-1}k_1 : k_2\ell_2^{-1})$.

To each $T \in \mathcal{T}$ we associate a unique smooth function $\kappa_T : K \times K \to \mathrm{End}^0$ such that:

1) $\kappa_T(m_2k_2 : k_1m_1) = \sigma_0(m_2)\kappa_T(k_2 : k_1)\sigma_0(m_1)$ and

2) $(Th)(k_2) = \int_K \kappa_T(k_2 : k_1)h(k_1^{-1})dk_1$

($m_i \in K_M$, $k_i \in K$, $i = 1, 2$; $h \in \mathcal{H}$). If $T = \Sigma h_i \otimes h_i^\sim$, then $\kappa_T(k_2 : k_1) = \Sigma h_i(k_2) \otimes h_i^\sim(k_1^{-1}) \in U \otimes U^\sim$. We frequently regard κ_T as an element of $V^0 \otimes \mathrm{End}^0(U)$. Indeed, it is obvious that every element of $V^0 \otimes \mathrm{End}^0(U)$ which satisfies 1) corresponds to κ_T for some $T \in \mathcal{T}$. The algebra structure on $\mathcal{T}$

compatible with that on $V^0 \otimes \mathrm{End}^0(U)$: For $T_1, T_2 \in \mathcal{T}$ we have

$$\kappa_{T_1 \cdot T_2}(k_2 : k_1) = \int_K \kappa_{T_1}(k_2 : \ell)\kappa_{T_2}(\ell^{-1} : k_1)d\ell$$

$$= (\kappa_{T_1} \cdot \kappa_{T_2})(k_2 : k_1).$$

is easy to see that

$$\kappa_{\pi_0(\ell_2)T\pi_0(\ell_1)}(k_2 : k_1) = \kappa_T(k_2\ell_2 : \ell_1 k_1)$$

$$= (\tau^0(\ell_2) \otimes 1)\kappa_T(k_2 : k_1)(\tau^0(\ell_1) \otimes 1)$$

$$= \tau^0(\ell_2)\kappa_T(k_2 : k_1)\tau^0(\ell_1).$$

emma 5.2.1.1. Let $T \in \mathcal{T}$ have kernel $\kappa_T \in V^0 \otimes \mathrm{End}^0(U)$. Then $\mathrm{tr}(T) =$ $_K \mathrm{tr}(\kappa_T(k : k^{-1}))dk$.

roof. It suffices to consider $\mathcal{T} = h_0 \otimes h_0^{\sim} \in \mathcal{H} \otimes \mathcal{H}^{\sim}$. In this case

$$\cdot(T) = h_0^{\sim}(h_0)$$

$$= \int_K <h_0^{\sim}(k), h_0(k)> dk$$

$$= \int_K \mathrm{tr}(\kappa_T(k : k^{-1}))dk.$$

We note that, for $T \in \mathcal{T}$, the kernel which corresponds to the adjoint * of T is

$$\kappa_{T^*}(k_2 : k_1) = \kappa_T(k_1^{-1} : k_2^{-1})^*.$$

ecall that $v^*(k_2 : k_1) = \bar{v}(k_1^{-1} : k_2^{-1})$; here $\kappa_T \in V^0 \otimes \mathrm{End}^0(U)$, so $(v_i \otimes R_i)^* = \Sigma v_i^* \otimes R_i^*$ for $v_i \otimes R_i \in V^0 \otimes \mathrm{End}^0(U)$.)

Now for any $T \in \mathcal{T}$ we define $\psi_T \in C(M : V^0)$ by setting

$$\psi_T(k_1 : m : k_2) = tr(\sigma(m)\kappa_T(k_2 : k_1))$$

$(m \in M,\ (k_1, k_2) \in K \times K)$. Observe that

$$\psi_{\pi_0(\ell_2)T\pi_0(\ell_1)}(k_1 : m : k_2) = tr(\sigma(m)\kappa_T(k_2\ell_2 : \ell_1 k_1))$$

$$= \psi_T(\ell_1 k_1 : m : k_2\ell_2)$$

$$= {}^0\tau(\ell_1^{-1})\psi_T(k_1 : m : k_2)^0\tau(\ell_2^{-1}).$$

On the other hand, given $P_1, P_2 \in \mathcal{P}(A)$, we have $\tau^0(F_{P_2})\psi_T\tau^0(F_{P_1}) = \psi_{F_{P_1}TF_{P_2}}$.

Thus, if $T \in \mathcal{T}(P_1|P_2)$, then $\psi_T \in C(M : V^0_{P_2|P_1})$

As usual, let $\mathcal{A}(\sigma, \tau_M)$ denote $(\mathcal{A}(\sigma) \otimes V) \cap C(M, \tau_M)$ and $\mathcal{A}(\sigma, \tau_{P_1|P_2}) = (\mathcal{A}(\sigma) \otimes V_{P_1|P_2}) \cap C(M, \tau_M)$ for (V, τ) any double representation of

<u>Lemma</u> 5.2.1.2. For any $T \in \mathcal{T}$ we have $\psi_T \in \mathcal{A}(\sigma, \tau^0_M)$. If $T \in \mathcal{T}(P_2|P_1)$, then $\psi_T \in \mathcal{A}(\sigma, \tau^0_{P_1|P_2})$.

<u>Proof</u>. It suffices to prove the first assertion. Since

$$\psi_T(k_1 : m_1 m m_2 : k_2) = tr(\sigma(m_1 m m_2)\kappa_T(k_2 : k_1))$$
$$= tr(\sigma(m)\kappa_T(m_2 k_2 : k_1 m_1))$$
$$= \psi_T(k_1 m_1 : m : m_2 k_2)$$

$((k_1, k_2) \in K^2,\ (m_1, m_2) \in K_M^2$, and $m \in M)$, it is clear that $\psi_T \in C(M, \tau^0_M)$. To see that $\psi_T \in \mathcal{A}(\sigma) \otimes V^0$ recall that $\kappa_T \in V^0 \otimes \mathrm{End}^0(U)$. Thus,

$$\psi_T(k_1 : m : k_2) = \sum_{j=1}^{r} v_j(k_2 : k_1) tr(\sigma(m)U_j)$$

with $v_j \in V$ and $U_j \in \mathrm{End}^0(U)$ $(j = 1, \ldots, r)$. It follows from remarks in §1.11

that $\psi_T \in \mathcal{A}(\sigma) \otimes V^0$.

Lemma 5.2.1.3. The linear mapping $T \mapsto \psi_T$ is a surjection of $\mathcal{T}[\mathcal{T}(P_2|P_1)]$ on $\mathcal{A}(\sigma, \tau_M^0)$ $[\mathcal{A}(\sigma, \tau^0_{P_1|P_2})$ for every $P_1, P_2 \in \mathcal{P}(A)]$. The mapping $T \mapsto \psi_T$ is bijective for $\mathcal{T} \to \mathcal{A}(\sigma, \tau_M^0)$ $[\mathcal{T}(P) \to \mathcal{A}(\sigma, \tau_P^0)$ for some $P \in \mathcal{P}(A)]$ if and only if σ is irreducible.

Proof. Given $\psi \in \mathcal{A}(\sigma, \tau_M^0) \subset \mathcal{A}(\sigma) \otimes V^0$, we have $\psi(k_1 : m : k_2) = \sum_{i=1}^{r} \mathrm{tr}(\sigma(m)U_i)v_i(k_2 : k_1)$ with $U_i \in \mathrm{End}^0(U)$ and $v_i \in V^0$ $(i = 1, \ldots, r)$. Set

$$\kappa_T(k_2 : k_1) = \int_{K_M \times K_M} \sum_{i=1}^{r} \sigma_0(m_2)U_i\sigma_0(m_1)v_i(m_2^{-1}k_2 : k_1m_1^{-1})dm_1dm_2 .$$

Since $\psi \in \mathcal{A}(\sigma, \tau_M^0)$, it follows that $\psi = \psi_T$. If $\psi \in \mathcal{A}(\sigma, \tau^0_{P_1|P_2})$, then $\psi = \psi_T = F_{P_1}(\tau^0)\psi_T F_{P_2}(\tau^0) = \psi_{F_{P_2} T F_{P_1}}$, so $\mathcal{T}(P_2|P_1) \to \mathcal{A}(\sigma, \tau^0_{P_1|P_2})$ is surjective too.

Next let us show that, if σ is irreducible, then $T \mapsto \psi_T$ is injective as a mapping from $\mathcal{T}$ to $\mathcal{A}(\sigma, \tau_M^0)$, consequently also as a mapping from $\mathcal{T}(P_2|P_1)$ to $\mathcal{A}(\sigma, \tau^0_{P_1|P_2})$ for every P_1, P_2. If $\psi_T = 0$, then $0 = \mathrm{tr}(\sigma(m)\kappa_T(k_2 : k_1))$ for all $m \in M$ and $(k_1, k_2) \in K \times K$. Therefore, $0 = \mathrm{tr}(\sigma(\alpha)\kappa_T(k_2 : k_1))$ for all $\alpha \in C_c^\infty(M)$ and all (k_1, k_2). If σ is irreducible, then $\sigma(C_c^\infty(M)) = \mathrm{End}^0(U)$. Since tr is a nondegenerate bilinear form on $\mathrm{End}^0(U)$, it follows that $\kappa_T = 0$ or $T = 0$.

Conversely, if σ is not irreducible, then, according to Lemma 1.11.1, there exists a nonzero $U_0 \in \mathrm{End}^0(U)$ such that $f_{U_0}(m) = \mathrm{tr}(\sigma(m)U_0) = 0$ for all $m \in M$. Let $\sigma_0(m_0)U_0\sigma_0(m_2) = U_0$ for all $(m_1, m_2) \in M_j^2$, let $K_j \subset K$ satisfy

$K_j \cap M = M_j$, and pick $v \in V^0$ such that $v(k_2 : k_1) = 1$, $(k_1, k_2) \in K_j^2$, and 0, otherwise. Set $\kappa_T(k_2 : k_1) = \int_{M_j^2} \sigma_0(m_2) U_0 \sigma_0(m_1) v(m_2^{-1} k_2 : k_1 m_1^{-1}) dm_1 dm_2$. Then $\kappa_T(1 : 1) \neq 0$, so $T \neq 0$. On the other hand,

$$\psi_T(k_1 : m : k_2) = \operatorname{tr}(\sigma(m) \kappa_T(k_2 : k_1))$$

$$= \operatorname{tr}(\sigma(m) U_0) \int_{M_j^2} v(m_2^{-1} k_2 : k_1 m_1^{-1}) dm_1 dm_2$$

$$= 0.$$

Thus, $T \mapsto \psi_T$ is not injective from $\mathcal{T}$ to $\mathcal{A}(\sigma, \tau_M^0)$.

The same argument, with obvious modifications, proves that, if σ is not irreducible, then the mapping is not injective from $\mathcal{T}(P)$ to $\mathcal{A}(\sigma, \tau_P^0)$ for any $P \in \mathcal{P}(A)$.

In the following lemma $\underset{\sim}{\alpha}^P \underset{M}{*} \psi_T$ signifies convolution on M composed with the product operation in V^0.

<u>Lemma</u> 5.2.1.4. Let $T \in \mathcal{T}$, $P \in \mathcal{P}(A)$, and $\alpha \in C_c^\infty(G)$. Then $\psi_{\alpha \underset{P}{*} T} = \underset{\sim}{\alpha}^P \underset{M}{*} \psi_T$ and $\psi_{T \underset{P}{*} \alpha} = \psi_T \underset{M}{*} \underset{\sim}{\alpha}^P$.

<u>Proof</u>. We check only the first relation, as the proof of the second is similar.

Since

$$\psi_{\alpha \underset{P}{*} T}(k_1 : m : k_2) = \operatorname{tr}(\sigma(m) \kappa_{T\pi_P(\alpha')}(k_2 : k_1)),$$

it suffices to show that

$$\operatorname{tr}(\kappa_{T\pi_P(\alpha')}(k_2 : k_1)\sigma(m)) = \int_{M \times K} \underset{\sim}{\alpha}^P (k_1 : x : \ell) \psi_T(\ell^{-1} : x^{-1} m : k_2) d\ell\, dx.$$

Proceeding, we have, for $h \in \mathcal{H}$,

$$\int_K \kappa_{T\pi_P(\alpha')}(k_2 : k_1)h(k_1^{-1})dk_1 = (T\pi_P(\alpha')h)(k_2)$$

$$= \int_K \kappa_T(k_2 : k)(\pi_P(\alpha')h)(k^{-1})dk$$

$$= \int_K \kappa_T(k_2 : k)\int_G \alpha'(x)(F_P h)(k^{-1}x)dxdk$$

$$= \int_K \kappa_T(k_2 : k)\int_G \alpha'(kx)(F_P h)(x)dxdk$$

$$= \int_K \kappa_T(k_2 : k)\int_{MNK} \alpha'(km_1 n_1 k_1)\delta_P^{\frac{1}{2}}(m_1)\sigma(m_1)(F_P h)(k_1)\, dm_1 dn_1 dk_1 dk$$

$$= \int_{K\times K\times M} \kappa_T(k_2 : k)\underset{\sim}{\alpha}'^P(k : m_1 : k_1)\sigma(m_1)h(k_1)dm_1 dk_1 dk$$

$$= \int_{K\times K\times M} \kappa_T(k_2 : k)\underset{\sim}{\alpha}^P(k_1^{-1} : m_1^{-1} : k^{-1})\sigma(m_1)h(k_1)dm_1 dk_1 dk$$

$$= \int_{M\times K\times K} \kappa_T(k_2 : k)\underset{\sim}{\alpha}^P(k_1 : m_1 : k^{-1})\sigma(m_1^{-1})h(k_1^{-1})dm_1 dk dk_1.$$

We have shown that

$$\kappa_{T\pi_P(\alpha')}(k_2 : k_1) = \int_M\int_K \kappa_T(k_2 : k)\underset{\sim}{\alpha}^P(k_1 : m_1 : k^{-1})\sigma(m_1^{-1})dm_1 dk.$$

It follows that

$$\mathrm{tr}(\kappa_{T\pi_P(\alpha')}(k_2 : k_1)\sigma(m)) = \int_M\int_K \underset{\sim}{\alpha}^P(k_1 : m_1 : k^{-1})\mathrm{tr}(\kappa_T(k_2 : k)\sigma(m_1^{-1}m))dm_1 dk,$$

as required.

Now we return to the case in which (V, τ) is arbitrary. Define a mapping of $\mathcal{C} \otimes V$ to $\mathcal{A}(M) \otimes V$ by sending $T \otimes v \mapsto \psi_T(1 : m : 1)v \overset{(\mathrm{def'n})}{=} \psi_{T\otimes v}(m)$. Let $\mathcal{U}$ denote the set of all $u \in \mathcal{C} \otimes V$ such that $\pi_0(k)u = u\tau(k)$ and $u\pi_0(k) = \tau(k)u$ for all $k \in K$ (i.e., $(\pi_0(k) \otimes 1)u = u(1 \otimes \tau(k)$, etc.). For P_1 and $P_2 \in \mathcal{P}(A)$ let $\mathcal{U}(P_2|P_1)$ denote the set of all $u \in \mathcal{U}$ such that u may be represented as $\Sigma T_i \otimes v_i$ with $T_i \in \mathcal{T}(P_2|P_1)$ and $v_i \in V(P_1|P_2)$. Write $\mathcal{U}(P) = \mathcal{U}(P|P)$ $(P \in \mathcal{P}(A))$.

Lemma 5.2.1.5. The mapping $u \mapsto \psi_u$ is a surjection of $\mathcal{U}[\mathcal{U}(P_1|P_2)]$ on $\mathcal{A}(\sigma, \tau_M)$ $[\mathcal{A}(\sigma, \tau_{P_2|P_1})$ for any $P_1, P_2 \in \mathcal{P}(A)]$. If σ is irreducible, then the mapping is bijective.

Proof. We divide the proof into three steps.

1) If $u \in \mathcal{U}[\mathcal{U}(P_1|P_2)]$, then $\psi_u \in \mathcal{A}(\sigma, \tau_M)[\mathcal{A}(\sigma, \tau_{P_2|P_1})]$.

Proof. If $u = \Sigma T_i \otimes v_i \in \mathcal{T} \otimes V$, then $\psi_u(m) = \Sigma\psi_{T_i}(1 : m : 1)v_i$

$= \Sigma \mathrm{tr}(\sigma(m)\kappa_{T_i}(1 : 1))v_i = \Sigma \mathrm{tr}(\sigma(m)S_i)v_i$, $S_i \in \mathrm{End}^0(U)$ and $v_i \in V$, so $\psi_u \in \mathcal{A}(\sigma) \otimes V$.

Moreover, if $u \in \mathcal{U}$ and $(m_1, m_2) \in K_M^2$, then

$$\begin{aligned}
\psi_u(m_1 m m_2) &= \Sigma \mathrm{tr}(\sigma(m_1 m m_2)\kappa_{T_i}(1 : 1))v_i \\
&= \Sigma \mathrm{tr}(\sigma(m)\kappa_{T_i}(m_2 : m_1))v_i \\
&= \Sigma \mathrm{tr}(\sigma(m)\kappa_{\pi_0(m_2)T_i\pi_0(m_1)}(1 : 1))v_i \\
&= \Sigma \mathrm{tr}(\sigma(m)\kappa_{T_i}(1 : 1))\,\tau(m_1)v_i\,\tau(m_2) \\
&= \tau_M(m_1)\psi_u(m)\,\tau_M(m_2) \quad (m \in M).
\end{aligned}$$

Finally, if $u \in \mathcal{U}(P_1|P_2)$, then

$$\begin{aligned}
F_{P_2}(\tau)\psi_u F_{P_1}(\tau) &= \Sigma\psi_{T_i} F_{P_2}(\tau)v_i F_{P_1}(\tau) \\
&= \Sigma\psi_{F_{P_1} T_i F_{P_2}} F_{P_2}(\tau)v_i F_{P_1}(\tau) \\
&= \Sigma\psi_{T_i'} v_i' ,
\end{aligned}$$

$\psi_u \in \mathcal{A}(\sigma, \tau_{P_2|P_1})$.

The mapping $u \mapsto \psi_u$ is a surjection of $\mathcal{U}[\mathcal{U}(P_1|P_2)]$ on $\mathcal{A}(\sigma, \tau_M)$ $[\mathcal{A}(\sigma, \tau_{P_2|P_1})]$.

Proof. Fix $\psi \in \mathcal{A}(\sigma, \tau_M)$ $[\mathcal{A}(\sigma, \tau_{P_2|P_1})]$ and let W be the (finite-dimensional) subspace of V spanned by $\tau(k_1)\psi(m)\tau(k_2)$ $(k_1, k_2 \in K;\ m \in M)$. We may assume that τ_W, the double representation τ in the space W, is unitary. Let $v_1, \ldots, v_r$ be an orthonormal base for W and set

$$\psi_i(k_1 : m : k_2) = (v_i, \tau_W(k_1)\psi(m)\tau_W(k_2))_W.$$

Then $\psi_i \in \mathcal{A}(\sigma, \tau_M^0)$. Since $T \mapsto \psi_T$ is a surjection (Lemma 5.2.1.3) we may choose $T_i \in \mathcal{T}$ such that $\psi_i = \psi_{T_i}$ $(i = 1, \ldots, r)$. Then

$$\tau(k_1)\psi(m)\tau(k_2) = \sum_i \psi_{T_i}(k_1 : m : k_2)v_i$$

$$= \sum \psi_{\pi_0(k_2)T_i\pi_0(k_1)}(1 : m : 1)v_i,$$

so $\psi(m) = \sum_i \psi_{\pi_0(k_2)T_i\pi_0(k_1)}(1 : m : 1)\ \tau(k_1)^{-1}v_i\ \tau(k_2)^{-1}$. Take $u = \int_{K\times K}\sum_i \pi_0(k_2)T_i\pi_0(k_1) \otimes \tau(k_1^{-1})v_i\,\tau(k_2^{-1})dk_1dk_2$. Then $u \in \mathcal{U}$ and $\psi = \psi_u$, so the first part is proved. If $\psi \in \mathcal{A}(\sigma, \tau_{P_2|P_1})$, then $F_{P_2}(\tau)\psi F_{P_1}(\tau) = \psi$. If $\psi = \sum \psi_{T_i'} v_i'$ with $\sum T_i' \otimes v_i' \in \mathcal{U}$, then we have $\psi = \sum \psi_{T_i'} F_{P_2}(\tau)v_i' F_{P_1}(\tau) = \sum \psi_{F_{P_1}T_i'F_{P_2}} F_{P_2}(\tau)v_i' F_{P_1}(\tau)$, so we may choose $u' = \sum T_i'' \otimes v_i'' \in \mathcal{U}$ with both $T_i'' \in \mathcal{T}(P_1|P_2)$ and $v_i'' \in V(P_2|P_1)$ for all i and $\psi = \psi_{u'}$.

) If σ is irreducible, then $u \mapsto \psi_u$ is an injection of $\mathcal{U}$ into $\mathcal{A}(\sigma, \tau_M^0)$.

Proof. Let $u = \Sigma T_i \otimes v_i \in \mathcal{U}$ with $v_1, \ldots, v_r$ linearly independent. If $\psi_u = 0$, then $\psi_{T_i}(1 : m : 1) = 0$ for all $i = 1, \ldots, r$ and $m \in M$. Moreover, for any $(k_1, k_2) \in K^2$,

$$\begin{aligned}\sum_i \psi_{T_i}(k_1 : m : k_2)v_i &= \sum_i \psi_{\pi_0(k_2)T_i\pi_0(k_1)}(1 : m : 1)v_i \\ &= \sum_i \psi_{T_i}(1 : m : 1)\tau(k_1)v_i\tau(k_2) \\ &= 0,\end{aligned}$$

so $\psi_{T_i} = 0$, $i = 1, \ldots, r$. However, by Lemma 5.2.1.3, we know that $T \mapsto \psi_T$ is bijective when σ is irreducible, so $T_i = 0$, $i = 1, \ldots, r$, i.e., $u = 0$.

§5.2.2. Definition and Elementary Properties of the Eisenstein Integral.

Let (V, τ) be any smooth unitary double representation of K with involution which satisfies associativity conditions (cf. §1.12). Let $\psi \in C(M, \tau_M)$ and $P \in \mathcal{P}(A)$. Extend $F_P(\tau)\psi$ to a function defined on G by setting $F_P(\tau)\psi(kmn) = \tau(k)F_P(\tau)\psi(m)$; this obviously makes sense. Define

$$E(P : \psi : x) = \int_K (F_P(\tau)\psi)(xk)\tau(k^{-1})\delta_P(xk)^{-\frac{1}{2}}dk.$$

Then, clearly, $E(P : \psi) = E(P : F_P(\tau)\psi) = E(P : \psi F_P(\tau)) \in C(G, \tau)$, i.e., $E(P) : C(M, \tau_M) \to C(M, \tau_P) \to C(G, \tau)$. We shall soon show that, if $\psi \in \mathcal{A}(M, \tau_M)$ then $E(P : \psi) \in \mathcal{A}(G, \tau)$.

Lemma 5.2.2.1. Let $\alpha \in C_c(G, \tau)$ and $\psi \in C(M, \tau_M)$. Then $\alpha * E(P : \psi) = E(P : \alpha^P \underset{M}{*} \psi)$.

Proof. Put $\Phi(x) = (F_P(\tau)\psi)(x)\delta_P(x)^{-\frac{1}{2}}$ and $\Phi' = \alpha * \Phi$. Then $E(P:\psi) = \int_K \Phi(xk)\tau(k^{-1})dk$ and $\alpha * E(P:\psi) = \int_K \Phi'(xk)\tau(k^{-1})dk$. We have

$$\begin{aligned}
\Phi'(m_0) &= \int_G \alpha(m_0x)\Phi(x^{-1})dx \\
&= \int_{MNK} \alpha(m_0mnk)\Phi(k^{-1}n^{-1}m^{-1})dmdndk \\
&= \int_{MN} \alpha(m_0mn)\Phi(m^{-1})dmdn \\
&= \int_M \delta_P(m_0m)^{-\frac{1}{2}}\alpha^P(m_0m)\Phi(m^{-1})dm \\
&= \delta_P(m_0)^{-\frac{1}{2}}(\alpha^P \underset{M}{*} F_P(\tau)\psi)(m_0) \\
&= \delta_P(m_0)^{-\frac{1}{2}}(\alpha^P \underset{M}{*} \psi)(m_0).
\end{aligned}$$

The lemma follows immediately.

Next we want to show that $E(P:\psi) * \alpha = E(P:\psi * \alpha^P)$ $(\alpha \in C_c(G,\tau))$. The argument makes use of the pre-hilbert structure for V. Given $\alpha, \beta \in C(G,\tau)$, we set $(\alpha,\beta)_G = \int_G (\alpha(x),\beta(x))_V dx$, provided the integral converges. In the following we assume Haar measures are normalized relative to K.

Lemma 5.2.2.2. Let $\alpha \in C_c(G,\tau)$ and $\psi \in C(M,\tau_M)$. Then $(\alpha, E(P:\psi))_G = (\alpha^P,\psi)_M$.

Proof.
$$\begin{aligned}
(\alpha, E(P:\psi))_G &= \int_{G\times K} (\alpha(x), F_P(\tau)\psi(xk)\tau(k^{-1}))_V \delta_P(xk)^{-\frac{1}{2}}dkdx \\
&= \int_G (\alpha(x), F_P(\tau)\psi(x))_V \delta_P(x)^{-\frac{1}{2}}dx \\
&= \int_{MN} (\alpha(mn),\psi(m))_V \delta_P(m)^{\frac{1}{2}}dmdn \\
&= \int_M (\alpha^P(m),\psi(m))_V dm \\
&= (\alpha^P,\psi)_M.
\end{aligned}$$

Recall that $\beta^{\sim}(x) = \beta^{*}(x^{-1})$, $\beta \in C_c(G, \tau)$. Recall also (Lemma 2.7.9) tha $(\beta^{\sim})^P = (\beta^P)^{\sim}$.

Lemma 5.2.2.3. Let $\psi \in C(M, \tau_M)$ and $\beta \in C_c(G, \tau)$. Then $E(P : \psi) * \beta = E(P : \psi \underset{M}{*} \beta^P)$.

Proof. Take $\alpha \in C_c(G, \tau)$. Then:

$$
\begin{aligned}
(\alpha, E(P : \psi) * \beta)_G &= (\alpha * \beta^{\sim}, E(P : \psi))_G \\
&= ((\alpha * \beta^{\sim})^P, \psi)_M \quad \text{(Lemma 5.2.2.2)} \\
&= (\alpha^P \underset{M}{*} (\beta^{\sim})^P, \psi)_M \quad \text{(Lemma 2.7.5)} \\
&= (\alpha^P \underset{M}{*} (\beta^P)^{\sim}, \psi)_M \\
&= (\alpha^P, \psi \underset{M}{*} \beta^P)_M \\
&= (\alpha, E(P : \psi * \beta^P))_G.
\end{aligned}
$$

Now let $T \in \mathcal{T}$, $\psi_T \in \mathcal{A}(\sigma, \tau_M^0)$. Then $E(P : \psi_T) \in C(G, \tau^0)$ and, recalling the bijection $j : C^{\infty}(G) \to C(G, \tau^0)$ (cf. §1.12), we set $E(P : T) = j^{-1}(E(P : \psi_T))$, so $E(P : T : x) = E(P : \psi_T : 1 : x : 1)$. We know that $E(P : \psi_T) = E(P : \psi_{F_P T F_P})$ (Lemma 5.2.1.5 and definition of $E(P : \psi_T)$), so $E(P : T) = E(P{:}F_P T$

Theorem 5.2.2.4. $E(P : T : x) = \mathrm{tr}(T\pi_P(x))$ $(T \in \mathcal{T})$.

Proof. We have $E(P : T : 1) = E(P : \psi_T : 1 : 1 : 1)$

$$
= \int_K (F_P(\tau^0)\psi_T)(1 : k : 1)\, \tau^0(k^{-1})dk
$$

$$= \int_K \psi_{F_P T F_P}(k : 1 : k^{-1})dk$$

$$= \int_K \mathrm{tr}(\kappa_{F_P T F_P}(k^{-1} : k)dk$$

$$= \mathrm{tr}(F_P T F_P) \quad \text{(Lemma 5.2.1.1).}$$

For the general case (x arbitrary) take $\alpha \in C_c^\infty(G)$. Then

$$j(\alpha * E(P : T)) = \underset{\sim}{\alpha} * E(P : \psi_T) \quad \text{(cf. §1.12)}$$

$$= E(P : \underset{\sim}{\alpha}^P_M * \psi_T) \quad \text{(Lemma 5.2.2.1)}$$

$$= E(P : \psi_{\alpha \underset{P}{*} T}) \quad \text{(Lemma 5.2.1.4)}$$

$$= E(P : \psi_{T\pi_P(\alpha')}) \quad \text{(definition of } \alpha \underset{P}{*} T\text{).}$$

Thus, $\alpha * E(P : T) = E(P : T\pi_P(\alpha'))$. Evaluating at $x = 1$, we have

$$\alpha * E(P : T : 1) = \mathrm{tr}(T\pi_P(\alpha'))$$

$$= \int_G \alpha(x^{-1})\mathrm{tr}(T\pi_P(x))dx.$$

Since $\int_G \alpha(x^{-1})[E(P : T : x) - \mathrm{tr}(T\pi_P(x)]dx = 0$ for all $\alpha \in C_c^\infty(G)$, the theorem is true.

Corollary 5.2.2.5. For any $T \in \mathcal{T}$, $E(P : T) \in \mathcal{A}(\pi_P)$ and $E(P : \psi_T) \in \mathcal{A}(\pi_P, \tau^0)$.

Proof. Since $j : \mathcal{A}(\pi_P) \to \mathcal{A}(\pi_P, \tau^0)$ (cf. §1.12) it is enough to know (§1.11) that $\mathrm{tr}(T\pi_P(x)) = E(P : T : x) \in \mathcal{A}(\pi_P)$.

Returning to the general case in which (V, τ) is arbitrary, we recall the space $\mathcal{U}$ (cf. §5.2.1). In particular, if $u \in \mathcal{U}$, then $\psi_u \in \mathcal{A}(\sigma, \tau_M)$, so $E(P : \psi_u) \in C(G, \tau)$.

<u>Lemma</u> 5.2.2.6. Let $u = \sum_{i=1}^{r} T_i \otimes v_i \in \mathcal{U}$. Then $E(P : \psi_u) = \sum_{i=1}^{r} E(P : T_i)v_i$. In particular, $E(P : \psi_u) \in \mathcal{A}(\pi_P, \tau)$.

<u>Proof.</u> $E(P : \psi_u : x) = \sum_{i=1}^{r} \int_K \psi_{T_i}(1 : xk : 1)v_i \tau(k^{-1})\delta_P(xk)^{-\frac{1}{2}}dk$

$$= \sum_i \int_K \psi_{\pi(k^{-1})T_i}(1 : xk : 1)v_i \delta_P(xk)^{-\frac{1}{2}}dk$$

$$= \sum_i \int_K \psi_{T_i}(1 : xk : k^{-1})v_i \delta_P(xk)^{-\frac{1}{2}}dk$$

$$= \sum_i E(T_i : x)v_i.$$

According to Corollary 5.2.2.5, $E(T_i : x) \in \mathcal{A}(\pi_P)$, so $E(P : \psi_u) \in (\mathcal{A}(\pi_P) \otimes V) \cap C(G$

Note that $E(P) : \mathcal{A}(\sigma, \tau_M) \to \mathcal{A}(\pi_P, \tau)$ factors through $\mathcal{A}(\sigma, \tau_P)$.

<u>Theorem</u> 5.2.2.7. For every smooth unitary double representation (V, τ) of K the mapping $\psi \mapsto E(P : \psi)$ of $\mathcal{A}(\sigma, \tau_P)$ to $\mathcal{A}(\pi_P, \tau)$ is surjective. The mapping is bijective for every (V, τ) if and only if π_P is irreducible.

<u>Proof.</u> Every (V, τ) can be regarded as a subrepresentation of a double representation of K which has an involution and satisfies associativity conditions, so it is enough to consider this case. Given $T \in \mathcal{T}$, write $f_T(x) = \mathrm{tr}(T\pi_P(x))$ $(x \in G)$. By Lemma 1.11.1 we may write $f(x) = \sum_{i=1}^{r} f_{T_i}(x)v_i$ $(x \in G)$, for any $f \in \mathcal{A}(\pi_P) \otimes V$, where $T_i \in \mathcal{T}$ $(1 \leq i \leq r)$ and $v_1, \ldots, v_r$ are linearly independent elements of V. We shall show that, if $f \in \mathcal{A}(\pi_P, \tau)$, then we may write $f = \sum_{i=1}^{r'} f_{T_i'}v_i''$ with $\sum_{i=1}^{r'} T_i' \otimes v_i' \in \mathcal{U}$. Using Lemmas 5.2.1.5 and 5.2.2.6, combined

with the fact that $E(P)$ factors through $\mathcal{A}(\sigma, \tau_P)$, we may then conclude that $= E(P : \psi)$ with $\psi \in \mathcal{A}(\sigma, \tau_P)$.

Since $f(x) = \tau(k_1^{-1}) f(k_1 x k_2) \tau(k_2^{-1})$, we have

$$f(x) = \sum_{i=1}^{r} f_{T_i}(k_1 x k_2) \tau(k_1^{-1}) v_i \tau(k_2^{-1})$$

$$= \sum_{i=1}^{r} f_{\pi_0(k_2) T_i \pi_0(k_1)}(x) \tau(k_1^{-1}) v_i \tau(k_2^{-1})$$

for all $(k_1, k_2) \in K \times K$ and $x \in G$. Let

$$u = \sum_{i=1}^{r} \int_{K \times K} \pi_0(k_2) T_i \pi_0(k_1) \otimes \tau(k_1^{-1}) v_i \tau(k_2^{-1}) dk_k dk_2.$$

Then, obviously, $u = \sum_{i=1}^{r'} T_i' \otimes v_i' \in \mathcal{U}$ and

$$f(x) = \Sigma f_{T_i'}(x) v_i'$$

$$= \Sigma E(T_i' : x) v_i'$$

$$= E(P : \psi_u : x).$$

This proves surjectivity.

To see that irreducibility implies injectivity, assume that $E(P : \psi) = 0$ for some nonzero $\psi \in \mathcal{A}(\sigma, \tau_P)$. We know that π_P irreducible implies σ irreducible, so $\mathcal{A}(\sigma, \tau_P)$ corresponds bijectively with $\mathcal{U}(P)$ (Lemma 5.2.1.5). There is a unique nonzero $u = \sum_{i=1}^{r} T_i \otimes v_i \in \mathcal{U}(P)$ such that $\psi = \psi_u$. We have $0 = E(P : \psi_u) = \sum_{i=1}^{r} f_{T_i} v_i$. Assuming, as we may, that $v_1, \ldots, v_r$ are linearly independent, we conclude that $f_{T_i} = 0$ $(i = 1, \ldots, r)$. Since $u \neq 0$, this contradicts Lemma 1.11.1.

Conversely, if $u \mapsto E(P : \psi_u)$ defines, for every (V, τ), an injection of $\mathcal{U}(P)$ into $\mathcal{A}(\pi_P, \tau_P)$, then, in particular, this is so for (V^0, τ^0). It is easy to see that this implies the injectivity of the mapping $T \mapsto f_T$, $\mathcal{T}(P) \to \mathcal{A}(\pi_P)$ (use Theorem 5.2.2.4 and the fact that $j(E(P : T)) = E(P : \psi_T))$. Again applying Theorem 1.11.1, we conclude that π_P is irreducible.

Let $(P, A) = (P = MN)$ and (P', A') $(P' = M'N')$ be p-pairs of G such that $(P', A') \succ (P, A)$. Set $(P \cap M', A) = ({}^*P, A) \prec (M', A')$. In the following extend $\delta_{{}^*P}$ to a function defined on G by setting $\delta_{{}^*P}(km'n') = \delta_{{}^*P}(m')$ $(k \in K,\ m' \in M',\ n' \in N')$. The functions $\kappa : G \to K$, $\mu : G \to M$ were defined in §0.6; κ', μ' and κ^*, μ^* are obvious variants.

<u>Lemma</u> 5.2.2.8. Let $\psi \in C(M, \tau_M)$. Then $E(P' : E({}^*P : \psi)) = E(P : \psi)$.

<u>Proof.</u> $E(P' : E({}^*P : \psi) : x) = \int_K \tau(\kappa'(xk))(F_{P'}(\tau)E)({}^*P : \psi : \mu'(xk))\tau(k^{-1})\delta_{P'}^{-\frac{1}{2}}(xk)dk$

$$= \int_K \tau(\kappa'(xk))F_{P'}(\tau)\int_{K_{M'}} \tau(\kappa^*(\mu'(xk)k_1))F_{{}^*P}(\tau)\psi(\mu^*(\mu'(xk)k_1))$$
$$\tau(k_1^{-1})\tau(k^{-1})\delta_{{}^*P}^{-\frac{1}{2}}(xkk_1)\delta_{P'}^{-\frac{1}{2}}(xk)dk_1dk$$

$$= \int_{K\times K_{M'}} \tau(\kappa'(xk))\tau(\kappa^*(\mu'(xk)k_1))F_P(\tau)\psi(\mu^*(\mu'(xk)k_1))$$
$$\tau(k_1^{-1}k^{-1})\delta_{{}^*P}^{-\frac{1}{2}}(xkk_1)\delta_{P'}^{-\frac{1}{2}}(xk)dk_1dk.$$

Noting that $xkk_1 \in \kappa'(xk)\kappa^*(\mu'(xk)k_1)\mu^*(\mu'(xk)k_1)N$, we may set $\kappa(xkk_1) = \kappa'(xk)\kappa^*(\mu'(xk)k_1)$ and $\mu(xkk_1) = \mu^*(\mu'(xk)k_1)$. Thus,

$$E(P' : E({}^*P : \psi) : x) = \int_{K\times K_{M'}} \tau(\kappa(xkk_1))(F_P(\tau)\psi)(\mu(xkk_1))\tau(kk_1)^{-1}$$
$$\delta_{{}^*P}^{-\frac{1}{2}}(xkk_1)\delta_{P'}^{-\frac{1}{2}}(xk)dk_1dk$$

$$= \int_{K \times K_{M'}} \tau(\kappa(xk))(F_P(\tau)\psi)(\mu(xk))\tau(k^{-1})$$

$$\delta_{*_P}^{-\frac{1}{2}}(xk)\delta_{P'}^{-\frac{1}{2}}(xkk_1^{-1})dk_1 dk.$$

Since $\delta_{P'}(xkk_1^{-1}) = \delta_{P'}(xk)$ $(k_1 \in K_{M'})$ and $\delta_{*_P}(xk)\delta_{P'}(xk) = \delta_P(xk)$, we may conclude that $E(P' : E({}^*P : \psi)) = E(P : \psi)$.

5.2.3. Eisenstein Integrals as Matrices for Induced Representations.

The purpose of this subsection is to bridge the gap between the Eisenstein integral defined as an integral and the Eisenstein integral seen as a matrix-valued function on G. We shall not apply the ideas developed in this section until 5.3 (see, however, the discussion of idempotents below).

Let W be a finite-dimensional complex vector space of dimension r. Besides the space End(W) of all endomorphisms of W we consider $\text{End}^2(W) = \text{End}(\text{End}(W)) \cong \text{End}(W) \otimes (\text{End}(W))'$, where $(\text{End}(W))'$ (according to our standard notation) denotes the dual vector space of End(W). The isomorphism "$\cong$" is the usual canonical mapping between $X \otimes X'$ and End(X) for an f.d.v.s. X. Using the trace form on End(W) we define a bijection $i : \text{End}(W) \to (\text{End}(W))'$ by setting $i(S) = S'$, where $\langle S', T\rangle = \text{tr}(ST)$ for all $T \in \text{End}(W)$. We may, thus, in a natural way, identify $\text{End}^2(W)$ and $\text{End}(W) \otimes \text{End}(W)$.

Fix a base $w_1, \ldots, w_r$ of W. Define $e_{ij} \in \text{End}(W)$ by setting $e_{ij}(w_\ell) = \delta_{j\ell} w_i$ $(\delta_{j\ell} = 1,\ j = \ell;\ 0, j \neq \ell)$, $1 \leq i, j, \ell \leq r$. The set of functions $\{e_{ij}\}_{1 \leq i,\ j \leq r}$ constitutes a base for End(W). The multiplicative properties of the basis elements are obvious and lead to the usual matrix multiplication, when elements $T \in \text{End}(W)$ are expressed in the form $T = \Sigma t_{ij} e_{ij}$, $1 \leq i,\ j \leq r$.

Write $\{e'_{ij}\}_{1 \leq i, j \leq r}$ for the dual basis of $(\mathrm{End}(W))'$. Then $I \in \mathrm{End}^2(W)$ has the expression $I = \Sigma e_{ij} \otimes e'_{ij} \in \mathrm{End}(W) \otimes (\mathrm{End}(W))'$. The identification given above between $(\mathrm{End}(W))'$ and $\mathrm{End}(W)$ yields $I = \Sigma e_{ij} \otimes e_{ji} \in \mathrm{End}(W) \otimes \mathrm{End}(W)$.

Now let $\mathcal{E}(K)$, as usual, be the set of classes of irreducible unitary representations of K. For $\underline{d} \in \mathcal{E}(K)$ let $\xi_{\underline{d}}$ denote the character and $d(\underline{d})$ the degree of $\underline{d}$. Write $\underline{d}^*$ for the element of $\mathcal{E}(K)$ whose character $\xi_{\underline{d}^*} = \overline{\xi}_{\underline{d}}$. If the Haar measure on K is normalized, then the function $\alpha_{\underline{d}} = d(\underline{d})\xi_{\underline{d}}$ is an idempotent in $C^\infty(K)$. For F a finite subset of $\mathcal{E}(K)$ write $\alpha_F = \sum_{\underline{d} \in F} \alpha_{\underline{d}}$. For convenience we generally assume that α_F is real-valued, i.e., that $\underline{d} \in F$ if and only if $\underline{d}^* \in F$.

Let π_0 be an admissible representation of K in a vector space $\mathcal{H}$. Fixing $F \subset \mathcal{E}(K)$, set $E = E_F = \int_K \alpha_F(k)\pi_0(k)dk$. Then E is a projection operator on $\mathcal{H}$. We write $\mathcal{H}_F = E\mathcal{H}$. The finite-dimensional vector space $\mathcal{H}_F = \bigoplus_{\underline{d} \in F} \mathcal{H}_{\underline{d}}$ where $\mathcal{H}_{\underline{d}}$ is the isotypic K-submodule of $\mathcal{H}$ corresponding to $\underline{d}$. Let $\tau = \tau_F$ be the double representation of K on $V = \mathrm{End}(\mathcal{H}_F) = E_F \mathrm{End}^0(\mathcal{H})E_F$. We may write $\mathrm{End}^2(\mathcal{H}_F) = V \otimes V \subset \mathrm{End}^0(\mathcal{H}) \otimes V$. As in §5.2.2, we let $\mathcal{U}$ denote the space of all elements $u = \Sigma T_i \otimes v_i \in \mathrm{End}^0(\mathcal{H}) \otimes V$ which satisfy the conditions $u\pi_0(k) = \tau(k)u$ and $\pi_0(k)u = u\tau(k)$ for all $k \in K$.

Lemma 5.2.3.1. As above, regard $\mathrm{End}^2(\mathcal{H}_F)$ as a subspace of $\mathrm{End}^0(\mathcal{H}) \otimes V$. Then the identity element $I \in \mathrm{End}^2(\mathcal{H}_F)$ is an element of $\mathcal{U}$.

Proof. We must show that $(\pi_0(k) \otimes 1)I = I(1 \otimes \tau(k))$ and $(1 \otimes \tau(k))I = I(\pi_0(k) \otimes 1)$.

We check only the first relation. We specialize the space W introduced above to be $\mathcal{H}_F$. Let $\pi_0(k) \otimes 1 = \Sigma k_{\mu\nu} e_{\mu\nu} \otimes 1$. Then

$$\Sigma k_{\mu\nu} e_{\mu\nu} \otimes 1 \cdot \Sigma e_{ij} \otimes e_{ji} = \Sigma k_{\mu i} e_{\mu j} \otimes e_{ji}$$

$$= \Sigma e_{\mu j} \otimes k_{\mu i} e_{ji}$$

$$= \Sigma e_{\mu j} \otimes e_{j\mu} \cdot \Sigma 1 \otimes k_{\mu i} e_{\mu i},$$

so $(\pi_0(k) \otimes 1)I = I(1 \otimes \tau(k))$.

Finally, let $P \in \mathcal{P}(A)$ $(P = MN)$. Let σ be an admissible representation of M and let $\pi = \mathrm{Ind}_P^G(\delta_P^{\frac{1}{2}}\sigma)$. Let $u = \Sigma T_i \otimes v_i \in \mathrm{End}^0(\mathcal{H}) \otimes \mathrm{End}(\mathcal{H}_F)$ $(\mathcal{H} = \mathcal{H}(P))$. Assume that $u \in \mathcal{U}$. Then, according to Lemma 5.2.2.6,

$$E(P : \psi_u : x) = \Sigma f_{T_i}(x) v_i$$

$$= \Sigma \mathrm{tr}(T_i \pi(x)) v_i \quad (x \in G).$$

In particular, letting $\mathcal{H}_F = W$ and taking $u = I = \sum_{1 \leq i,\, j \leq r} e_{ij} \otimes e_{ji}$, we have

$$E(P : \psi_I : x) = \sum_{1 \leq i, j \leq r} \mathrm{tr}(\pi(x) e_{ij})\, e_{ji}$$

$$= E_F \pi(x) E_F .$$

In other words, given the present context, we have, in effect, expressed the functions $\phi(x)$ of §5.1 as Eisenstein integrals.

§5.2.4. Functional Equations for the Eisenstein Integral.

Assume that $\sigma \in \omega \in \mathcal{E}_2(M)$. With respect to a fixed Haar measure dm^* on M/A, the formal degree $d(\omega)$ of ω can be defined so that

$$d(\omega)^{-1} \langle \tilde{u}_1, u_2 \rangle \langle \tilde{u}_2, u_1 \rangle = \int_{M/A} \langle \tilde{u}_1, \sigma(x) u_1 \rangle \langle \tilde{u}_2, \sigma(x^{-1}) u_2 \rangle dx^*$$

for any $u_i \in U$, $u_j^\sim \in U^\sim$, $1 \leq i, j \leq 2$. Recall that $\psi_T^\sim(k_1 : m : k_2) = \bar{\psi}_T(k_2^{-1} : m^{-1} : k_1^{-1})$ ($k_1, k_2 \in K$; $m \in M$; and $T \in \mathcal{T}$). Define

$$\|\psi_T\|^2 = \int_{M/A} \|\psi_T(x)\|^2_{V_0} dx^*.$$

<u>Lemma</u> 5.2.4.1. Let $S, T \in \mathcal{T}$. Then:

1) $\psi_T^\sim = \psi_{T^*}$

2) $\psi_S \underset{M/A}{*} \psi_T = d(\omega)^{-1}\psi_{TS}$

3) $E(P : \psi_T)^\sim = E(P : \psi_T^\sim)$

4) $\|\psi_T\|^2 = d(\omega)^{-1} tr(T^* T)$.

<u>Proof.</u>

1) $$\begin{aligned} \psi_{T^*}(k_1 : m : k_2) &= tr(\kappa_{T^*}(k_2 : k_1)\sigma(m)) \\ &= tr(\kappa_T(k_1^{-1} : k_2^{-1})^* \sigma^*(m^{-1})) \\ &= \overline{tr}(\sigma(m^{-1})\kappa_T(k_1^{-1} : k_2^{-1})) \\ &= \bar{\psi}_T(k_2^{-1} : m^{-1} : k_1^{-1}) \\ &= \psi_T^\sim(k_1 : m : k_2). \end{aligned}$$

2) It is sufficient to consider the case $T = h_1 \otimes h_1^\sim$ and $S = h_2 \otimes h_2^\sim$ ($h_i \in \mathcal{H}$, $h_i^\sim \in \mathcal{H}^\sim$, $i = 1, 2$).

$$\begin{aligned} \psi_S \underset{M/A}{*} \psi_T(k_1 : m : k_2) &= \int_K \int_{M/A} \psi_S(k_1 : x : \ell)\psi_T(\ell^{-1} : x^{-1}m : k_2)dx^* d\ell \\ &= \int_K \int_{M/A} tr(\sigma(x)\kappa_S(\ell : k_1))tr(\sigma(x^{-1}m)\kappa_T(k_2 : \ell^{-1}))dx^* d\ell \\ &= \int_K \int_{M/A} tr(\sigma(x)h_2(\ell) \otimes h_2^\sim(k_1^{-1}))tr(\sigma(x^{-1}m)h_1(k_2) \otimes h_1^\sim(\ell)) \\ &\qquad dx^* d\ell \\ &= \int_K \int_{M/A} \langle h_2^\sim(k_1^{-1}), \sigma(x)h_2(\ell)\rangle\langle h_1^\sim(\ell), \sigma(x^{-1}m)h_1(k_2)\rangle dx^* d\ell \end{aligned}$$

$$= d(\omega)^{-1}\int_K \langle h_1^{\sim}(\ell), h_2(\ell)\rangle d\ell \langle h_2^{\sim}(k_1^{-1}), \sigma(m)h_1(k_2)\rangle.$$

On the other hand,

$$\kappa_{T\cdot S}(k_2 : k_1) = \int_K \kappa_T(k_2 : \ell^{-1})\kappa_S(\ell : k_1)d\ell$$

$$= \int_K h_1(k_2) \otimes h_1^{\sim}(\ell) \cdot h_2(\ell) \otimes h_2^{\sim}(k_1^{-1})d\ell$$

$$= \int_K \langle h_1^{\sim}(\ell), h_2(\ell)\rangle d\ell \cdot h_1(k_2) \otimes h_2^{\sim}(k_1^{-1}),$$

so $d(\omega)^{-1}(\psi_S * \psi_T)(k_1 : m : k_2) = \mathrm{tr}(\sigma(m)\kappa_{T\cdot S}(k_2 : k_1))$, as required.

) $E(P : \psi_T)^{\sim} = E(P : \psi_{F_P T F_P})^{\sim} \xrightarrow{j^{-1}} E(P : F_P T F_P)^{\sim} = \mathrm{tr}(T\pi_P)^{\sim}.$

$$E(P : \psi_T^{\sim}) = E(P : \psi_{F_P T^* F_P}{}^{\sim}) \xrightarrow{j^{-1}} E(P : F_P T^* F_P) = \mathrm{tr}(T^*\pi_P).$$

We know that $\mathrm{tr}(T\pi_P)^{\sim} = \mathrm{tr}(T^*\pi_P)$.

) $\|\psi_T\|^2 = \int_{M/A}\int_{K\times K}\psi_T(k_1 : m : k_2)\overline{\psi}_T(k_1 : m : k_2)dk_1 dk_2 dm^*$

$$= \int_{M/A}\int_{K\times K}\psi_T(k_1 : m : k_2)\psi_T^{\sim}(k_2^{-1} : m^{-1} : k_1^{-1})dk_1 dk_2 dm^*$$

$$= \int_K \psi_T * \psi_T^{\sim}(k_1 : 1 : k_1^{-1})dk_1$$

$$= \int_K d(\omega)^{-1}\mathrm{tr}(\kappa_{T^*T}(k_1^{-1} : k_1))dk_1$$

$$= d(\omega)^{-1}\mathrm{tr}(T^*T) \quad \text{(Lemma 5.2.1.1)}.$$

Now assume that $\omega \in \mathcal{E}_2(M)$ is simple (cf. §§3.4 and 4.6). In this case, if $\psi \in \mathcal{A}(\omega, \tau_M)$, then $E(P : \psi) \in \mathcal{A}(C(\omega), \tau)$ and

$${}_vE_{P_2}(P_1 : \psi) \in \bigoplus_{s\in W(G/A)} \mathcal{A}(\omega^s, \tau_M)$$

$P_1, P_2 \in \mathcal{P}(A)$). We define linear mappings $c_{P_2|P_1}(s : \omega) : \mathcal{A}(\omega, \tau_M) \to \mathcal{A}(\omega^s, \tau_M)$ by setting

$$ {}_w E_{P_2}(P_1 : \psi) = \sum_{s \in W(G/A)} c_{P_2|P_1}(s : \omega)\psi. $$

Write $L(\omega, P) = \mathcal{A}(\omega, \tau_P)$ and $\mathcal{L}(\omega, P) = \mathcal{A}(\omega, \tau_{P|\bar{P}})$. Since $E(P_1 : \psi) = E(P_1 : F_{P_1}(\tau)\psi F_{P_1}(\tau))$ and $(\lambda(n_1)\rho(\bar{n}_2)E(P_1 : \psi))_{P_2} = E_{P_2}(P_1 : \psi)$ $(n_1 \in N_2,\ \bar{n}_2 \in \bar{N}_2;\ P_2 = MN_2,\ \bar{P}_2 = M\bar{N}_2)$, we may regard $c_{P_2|P_1}(s : \omega)$ as mapping $L(\omega, P_1)$ to $\mathcal{L}(\omega^s, P_2)$. That $c_{P_2|P_1}(s : \omega)$, regarded as such, is <u>injective</u> follows from Corollary 3.5.7 (Theorem 4.6.3) and Theorem 5.2.2.7 combined with Theorem 2.5.9.

In order to prove that $c_{P_2|P_1}(s : \omega) : L(\omega, P_1) \to \mathcal{L}(\omega^s, P_2)$ is <u>surjective</u> (i.e., bijective) we define the operator of the following theorem and study its properties. Notation: $c_{P|P} = c_{P|P}(1 : \omega)$.

<u>Theorem</u> 5.2.4.2. Fix $P \in \mathcal{P}(A)$. Then there exists a unique endomorphism c_P of $\mathcal{H}$ such that:

1) $c_P = F_{\bar{P}} c_P F_P$ (so $c_P : \mathcal{H}(P) \to \mathcal{H}(\bar{P})$) and

2) $c_{P|P}\psi_T = \psi_{c_P T F_P}$ for all $T \in \mathcal{T}$.

The operator c_P possesses the following additional properties:

3) The adjoint operator $c_P^* \in \mathrm{End}(\mathcal{H})$ exists and $c_{\bar{P}|P}\psi_T = \psi_{F_P T c_P^*}$ $(T \in \mathcal{T})$.

4) $c_{\bar{P}} = c_P^*$.

5) $c_P \pi_P(x) = \pi_{\bar{P}}(x) c_P$ $(x \in G)$.

6) c_P commutes with $\pi_0(k)$ for all $k \in K$.

7) Let $h_1, h_2 \in \mathcal{H}$ and set $h_{2,P} = F_P h_2 \in \mathcal{H}(P)$. Define $h_{2,P}(nmk) = \delta_P(m)^{\frac{1}{2}}\sigma(m)h_{2,P}(k)$ $(n \in N,\ m \in M,\ k \in K)$. Then $(h_1, c_P h_2) =$

$\lim_{\epsilon \to 0} \gamma^{-1} \int_{\bar{N}} \int_K q^{-\epsilon\langle\rho, H_P(\bar{n})\rangle} (h_1(k), h_{2,P}(\bar{n}k)) dk d\bar{n}$. (Assume measures normalized; cf. §5.3.4.)

Proof. Define a linear endomorphism $\mathcal{C}_P : \mathcal{T} \to \mathcal{T}$ by setting $c_{P|P}\psi_T = \psi_{\mathcal{C}_P(T)}$ ($T \in \mathcal{T}$). Since $c_{P|P}$ is injective and $T \mapsto \psi_T$ bijective, $\mathcal{T} \to \mathcal{A}(\sigma, \tau_M^0)$, $\mathcal{C}_P$ is well defined. Moreover, it is obvious that $F_{\bar{P}} \mathcal{C}_P(T) F_P = \mathcal{C}_P(T)$, so $\mathcal{C}_P(\mathcal{T}) \subset \mathcal{T}(\bar{P}|P)$. Recall also that $\mathcal{C}_P(T) = \mathcal{C}_P(F_P T F_P)$.

Let $\alpha, \beta \in C_c^\infty(G)$ and $T \in \mathcal{T}$. Then

$$(\underset{\sim}{\alpha} * E(P : \psi_T) * \underset{\sim}{\beta})_P = \underset{\sim}{\alpha}^P \underset{M}{*} E_P(P : \psi_T) \underset{M}{*} \underset{\sim}{\beta}^{\bar{P}} \quad \text{(Lemma 2.7.6)}$$

$$= \sum_{\chi \in \mathfrak{X}_{\pi_P}(P, A)} \underset{\sim}{\alpha}^P \underset{M}{*} E_{P,\chi}(P : \psi_T) \underset{M}{*} \underset{\sim}{\beta}^{\bar{P}} \quad \text{(cf. §3.1)},$$

so

$$(\underset{\sim}{\alpha} * E(P : \psi_T) * \underset{\sim}{\beta})_{P,1} = \underset{\sim}{\alpha}^P * c_{P|P}\psi_T * \beta^{\bar{P}}$$

$$= \underset{\sim}{\alpha}^P * \psi_{\mathcal{C}_P(T)} * \underset{\sim}{\beta}^{\bar{P}}$$

$$= \psi_{\pi_{\bar{P}}(\beta')\mathcal{C}_P(T)\pi_P(\alpha')} \quad \text{(Lemma 5.2.1.4)}.$$

On the other hand,

$$\underset{\sim}{\alpha} * E(P : \psi_T) * \underset{\sim}{\beta} = E(P : \underset{\sim}{\alpha}^P * \psi_T * \underset{\sim}{\beta}^P) \quad \text{(Lemmas 5.2.2.1-3)}$$

$$= E(P : \psi_{\pi_P(\beta')T\pi_P(\alpha')}) \quad \text{(Lemma 5.2.1.4)},$$

so

$$(\underset{\sim}{\alpha} * E(P : \psi_T) * \underset{\sim}{\beta})_{P,1} = \psi_{\mathcal{C}_P(\pi_P(\beta')T\pi_P(\alpha'))}.$$

Thus,

$$\mathcal{C}_P(\pi_P(\beta')T\pi_P(\alpha')) = \pi_{\bar{P}}(\beta')\mathcal{C}_P(T)\pi_P(\alpha').$$

We shall show that $\mathcal{C}_P(T) = c_P T F_P$ ($T \in \mathcal{T}$) where $c_P : \mathcal{H}(P) \to \mathcal{H}(\bar{P})$.

For this let F be any finite subset of $\mathcal{E}(K)$, let $\alpha_F \in C^\infty(K)$ be the corresponding idempotent, $K_F = \bigcap_{\underline{d} \in F} \ker \underline{d}$, and let $E_F = \pi_0(\alpha_F)$ (cf. §5.2.3). Write $\mathcal{T}_F = E_F \mathcal{T} E_F$, a finite dimensional simple algebra, and observe that $E_F F_P = F_P E_F$. Write $\mathcal{T}_F(P) = \mathcal{T}(P) \cap \mathcal{T}_F$. If $T \in \mathcal{T}_F$, then $\mathcal{C}_P(T) = \mathcal{C}_P(E_F T E_F) = \mathcal{C}_P(\pi_P(\alpha_F) T \pi_P(\alpha_F)) = \pi_{\bar{P}}(\alpha_F) \mathcal{C}_P(T) \pi_P(\alpha_F)$, so $\mathcal{C}_P : \mathcal{T}_F \to \mathcal{T}_F$. It is a standard fact in the theory of simple algebras that we may choose A and B in $\mathcal{T}_F$ such that $\mathcal{C}_P(T) = ATB$ for all $T \in \mathcal{T}_F$. Since $\mathcal{C}_P(T\pi_P(\alpha)) = \mathcal{C}_P(T)\pi_P(\alpha) \in \mathcal{T}_F$ for any $\alpha \in C_c(G/\!/K_F)$ and since $\pi_P(C_c(G/\!/K_F)) = \mathcal{T}_F(P)$, it follows that B is a constant multiple of $F_P E_F$. Thus, we may write $\mathcal{C}_P(T) = c_{P,F} T F_P$ $(T \in \mathcal{T}_F)$ with $c_{P,F} \in \mathcal{T}_F(\bar{P}|P) = \mathcal{T}_F \cap \mathcal{T}(\bar{P}|P)$. For any $F \subset \mathcal{E}(K)$ we have $c_{P,F} F_P = c_{P,F} = c_{P,F} E_F = F_{\bar{P}} c_{P,F}$. Since $\mathcal{C}_P(T) = c_{P,F'} T$ for any $T \in \mathcal{T}_F$ and $F' \supset F$, we may define $c_P : \mathcal{H}(P) \to \mathcal{H}(\bar{P})$ by setting $c_P | E_F \mathcal{H}(P) = c_{P,F}$.

The linear operator c_P clearly satisfies properties 1) and 2). The uniqueness of c_P, assuming 1) and 2), follows from the fact that, since c_P intertwines π_P and $\pi_{\bar{P}}$, c_P is determined up to a constant--the constant is in turn fixed by the relation $c_{P|P}\,{}^{\psi}T = {}^{\psi}c_P T$ for any nonzero $T \in \mathcal{T}(P)$. (Another uniqueness proof: If $c_{P|P}\,{}^{\psi}T = {}^{\psi}c'_P T F_P = {}^{\psi}c_P T F_P$ for all $T \in \mathcal{T}$, then $c'_P\, T - c_P T = 0$ for all $T \in \mathcal{T}(P)$, so $(c'_P - c_P) | \mathcal{H}(P) = 0$--however, by 1), this means that $c'_P = c_P$.)

3) The existence of c_P^* follows easily from the fact that $c_{P,F}^*$ exists for any $F \subset \mathcal{E}(K)$, since $c_{P,F} \in \mathcal{T}$. To see that $c_{\bar{P}|P}\,{}^{\psi}T = {}^{\psi}F_P T c_P^*$ let $T \in \mathcal{T}$. Then $c_{\bar{P}|P}\,{}^{\psi}T = (c_{P|P}({}^{\psi}\tilde{T}))^{\sim}$ (Corollary 3.2.7 and Lemma 5.2.4.1, 3))

$$= (c_{P|P}\psi_{T^*})^{\sim} \quad \text{(Lemma 5.2.4.1, 1))}$$

$$= (\psi_{c_P T^* F_P})^{\sim}$$

$$= \psi_{F_P T c_P^*}.$$

4) For any $F \subset \mathcal{E}(K)$ the function $E(P : E_F) = tr(E_F \pi_P) = \theta_F$ is independent of $P \in \mathcal{P}(A)$, since the equivalence class of π_P (and hence its character) is independent of P. Thus, $E(P : \psi_{E_F}) = E(\overline{P} : \psi_{E_F})$ (definition, cf. §5.2.2), so $c_{P|P}\psi_{E_F} = c_{P|\overline{P}}\psi_{E_F}$. Using 3), we obtain $\psi_{c_P E_F} = c_{P|P}\psi_{E_F} = c_{P|\overline{P}}\psi_{E_F} = \psi_{F_{\overline{P}} E_F c_{\overline{P}}^*}$. Since $c_P E_F = F_{\overline{P}} E_F c_{\overline{P}}^*$ for all $F \subset \mathcal{E}(K)$, we may obviously conclude that $c_P = c_{\overline{P}}^*$.

5) and 6) have essentially already been observed (and used). To be explicit: Let $\alpha \in C_c(G/\!/K_F)$. Then $\pi_{\overline{P}}(\alpha) c_{P,F'} = c_{P,F'} \pi_P(\alpha)$ for all $F' \supset F$. Therefore, $\pi_{\overline{P}}(x) c_P = c_P \pi_P(x)$ for all $x \in G$. For 6) take $k \in K$. Then:
$c_P \pi_0(k) = c_P F_P \pi_0(k) = c_P \pi_P(k) = \pi_{\overline{P}}(k) c_P = \pi_0(k) c_P$.

7) follows from the integral formula for $c_{P|P}(1 : \omega)$ (Theorem 5.3.4.1). An explicit proof is omitted.

Corollary 5.2.4.3. For any smooth double representation (V, τ) of K the mapping $c_{P|P}(1 : \omega) : L(\omega, P) \to \mathcal{L}(\omega, P)$ is bijective.

Proof. First, let $(V, \tau) = (V^0, \tau^0)$. Recall that, canonically, $\mathcal{T}(P) = \mathcal{H}(P) \otimes \mathcal{H}(P)$ and $\mathcal{T}(\overline{P}|P) = \mathcal{H}(\overline{P}) \otimes \mathcal{H}(P)$. For every finite $F \subset \mathcal{E}(K)$ $\dim \mathcal{T}_F(P) = \dim \mathcal{T}_F(\overline{P}|P)$ (cf. proof of Lemma 5.2.4.2 for notation). Since

$c_{P,F} : \mathcal{T}_F(P) \to \mathcal{T}_F(\bar{P}|P)$ is injective for every F, c_P is bijective. Combining the definition of c_P with Lemma 5.2.1.3 we conclude that $c_{P|P}(1 : \omega)$ is bijective in the present case.

Now let (V, τ) be arbitrary. We know that $c_{P|P}(1 : \omega)$ is injective. Given $\varphi \in \mathcal{L}(\omega, P)$ we want to exhibit $\psi \in L(\omega, P)$ such that $c_{P|P}(1 : \omega)\psi = \varphi$. Write $\tau(k_1)\varphi(m)\tau(k_2) = \sum_{i=1}^{r} \varphi_i(k_1 : m : k_2)v_i$, where $v_1, \ldots, v_r$ is a finite linearly independent subset of V. Notice that $\varphi_i \in \mathcal{L}(\omega, \tau_P^0)$. Therefore, we have $\psi_1, \ldots, \psi_r \in L(\omega, P)$ such that $c_{P|P}(1 : \omega)\psi_i = \varphi_i$ and, hence, $\tau(k_1)\varphi(m)\tau(k_2) = \sum_{i=1}^{r} (c_{P|P}(1 : \omega)\psi_i)(k_1 : m : k_2)v_i$ for all $(k_1, k_2) \in K \times K$. Setting $\psi = \Sigma\psi_i(1 : - : 1)v_i$, one sees both that $c_{P|P}(1 : \omega)\psi = \varphi$ and that $\psi \in L(\omega, P)$.

Theorem 5.2.4.4. There is a positive constant $\mu(\omega) = \mu$ such that, if $P_1, P_2 \in \mathcal{P}(A)$, $s, t \in W(G/A)$, and $\psi \in L(\omega^t, P_1)$, then $\mu\|c_{P_2|P_1}(s : \omega^t)\psi\|^2 = \|\psi\|^2$.

Proof. Using Theorem 5.2.4.2, 3) we have $c_{\bar{P}}c_P\pi_P(x) = \pi_P(x)c_{\bar{P}}c_P$ for all $x \in G$. Since $c_{\bar{P}}c_P = c_P^*c_P$ is a nonzero hermitian positive semi-definite operator which commutes with π_P on $\mathcal{H}(P)$ and satisfies $F_Pc_P^*c_PF_P = c_P^*c_P$, we may conclude that $c_P^*c_P = \mu_P^{-1}(\omega)F_P$, where $\mu_P(\omega)$ is a positive constant.

To check that $\mu = \mu_P(\omega^t)$ for all $P \in \mathcal{P}(A)$ and $t \in W(G/A)$ consider the function $f(x) = \theta_F(x) = E(P : E_F : x)$ $(x \in G)$, defined for finite $F \subset \mathcal{E}(K)$ (cf. the proof of 2), Theorem 5.2.4.2). The value of this Eisenstein integral is independent of P and t. Furthermore, $\|f_{P,1}\|^2 = \|f_{P_2,s}\|^2$ for all $P_2 \in \mathcal{P}(A)$ and $s \in W(G/A)$, by the Maass-Selberg relations. Explicitly, we have $\|f_{P,1}\|^2 = \|\psi_{c_PE_F}\|^2 = d(\omega)^{-1}tr(c_PE_Fc_P^*)$ (Lemma 5.2.4.1, 4)) $= d(\omega)^{-1}\mu^{-1}tr(F_PE_F)$ (the race of $F_P E_F$ being independent of P). Thus, μ is independent of P and t.

In addition, we have proved that $\mu \| c_{P|P}(1 : \omega)\psi_{E_F} \|^2 = \|\psi_{E_F F_P}\|^2 =$ $\| c_{P_2|P_1}(s : \omega^t)\psi_{E_F} \|^2$. The theory leading to the Maass-Selberg relations implies that the same relation holds if we replace E_F by any $T \in \mathcal{T}$.

Given a simple class $\omega \in \mathcal{E}_2(M)$, $s \in W(G/A)$, and $P_1, P_2 \in \mathcal{P}(A)$, we define

$$ {}^0c_{P_2|P_1}(s : \omega) = c_{P_2|P_2}(1 : \omega^s)^{-1} c_{P_2|P_1}(s : \omega) \text{ and} $$

$$ c^0_{P_2|P_1}(s : \omega) = c_{P_2|P_1}(s : \omega) c_{P_1|P_1}(1 : \omega)^{-1}. $$

We may regard ${}^0c_{P_2|P_1}(s : \omega)$ either as a mapping of $L(\omega, P_1)$ to $L(\omega^s, P_2)$ or as a mapping from $\mathcal{A}(\omega, \tau_M)$ to $L(\omega^s, P_2)$. The mapping $c^0_{P_2|P_1}(s : \omega)$ may similarly be regarded either as a mapping from $\mathcal{L}(\omega, P_1)$ to $\mathcal{L}(\omega^s, P_2)$ or as a mapping with domain $\mathcal{A}(\omega, \tau_M)$, provided, in the second case, we extend $c_{P_1|P_1}(1 : \omega)^{-1}$ to $\mathcal{A}(\omega, \tau_M)$ by composing with the projection of $\mathcal{A}(\omega, \tau_M)$ on $\mathcal{L}(\omega, P_1) : \psi \mapsto F_{P_1}(\tau)\psi F_{\overline{P}_1}(\tau) \mapsto c_{P_1|P_1}(1 : \omega)^{-1}(F_{P_1}(\tau)\psi F_{\overline{P}_1}(\tau))$. Sometimes we rely on the context to determine which version of these mappings we have in mind.

The following is just a reformulation of Theorem 5.2.4.4:

<u>Corollary</u> 5.2.4.5. ${}^0c_{P_2|P_1}(s : \omega)$ defines a unitary bijection of $L(\omega, P_1)$ onto $L(\omega^s, P_2)$. $c^0_{P_2|P_1}(s : \omega)$ defines a unitary bijection of $\mathcal{L}(\omega, P_1)$ onto $\mathcal{L}(\omega^s, P_2)$.

Theorem 5.2.4.6. Let $P_1, P_2, P_3 \in \mathcal{P}(A)$, $s, t \in W(G/A)$, and $\psi \in \mathcal{A}(\omega, \tau_M)$. Define

$$E^0(P_1 : \psi) = E(P_1 : c_{P_1|P_1}(1 : \omega)^{-1}\psi).$$

The following functional equations hold:

(1) $E^0(P_2 : c^0_{P_2|P_1}(s : \omega)\psi) = E^0(P_1 : \psi)$

and $c^0_{P_3|P_1}(st : \omega) = c^0_{P_3|P_2}(s : {}^t\omega)c^0_{P_2|P_1}(t : \omega)$.

(2) $E(P_2 : {}^0c_{P_2|P_1}(s : \omega)\psi) = E(P_1 : \psi)$

and ${}^0c_{P_3|P_1}(st : \omega) = {}^0c_{P_3|P_2}(s : {}^t\omega)\,{}^0c_{P_2|P_1}(t : \omega)$.

Proof. We verify only (2), as the proof of (1) is similar. Both Eisenstein integrals define elements of $\mathcal{A}(C(\omega), \tau)$. To verify that these elements are equal it suffices, according to the Maass-Selberg relations, to check that

$$E_{P_2, 1}(P_2 : {}^0c_{P_2|P_1}(s : \omega)\psi) = E_{P_2, s}(P_1 : \psi).$$

This is immediate from the definition of the operators c and ${}^0c$. The law of composition for the operators ${}^0c$ is of course clear.

Next define the following operation of $W = W(G/A)$ on $\mathcal{A}(M, \tau_M)$. Given $s \in W$ and $y = y(s) \in K$ a representative for s, set $(s\varphi)(m) = \tau(y)\varphi(y^{-1}my)\tau(y)$ $(m \in M)$ for any $\varphi \in \mathcal{A}(M, \tau_M)$. If $P \in \mathcal{P}(A)$, then $P^s = yPy^{-1}$.

Lemma 5.2.4.7. Let $P_1, P_2 \in \mathcal{P}(A)$ and $s, t \in W(G/A)$. Then

$$sc_{P_2|P_1}(t : \omega) = c_{P_2^s|P_1}(st : \omega)$$

and

$$c_{P_2|P_1}(t : \omega)s^{-1} = c_{P_2|P_1^s}(ts^{-1} : \omega^s).$$

The same relations hold with c replaced by either ${}^0c$ or c^0.

Proof. Noting that $\delta_{P_1}(x) = \delta_{P_1^s}(x^y)$ for all $x \in G$, one checks that $E(P_1^s : s\psi) = E(P_1 : \psi)$ for $\psi \in \mathcal{A}(\omega, \tau_M)$ (so $s\psi \in \mathcal{A}(\omega^s, \tau_M)$). It follows that

$\sum_{t \in W(A)} c_{P_2|P_1^s}(t : \omega^s)s\psi = \sum_{t \in W(A)} c_{P_2|P_1}(t : \omega)\psi$, so $c_{P_2|P_1^s}(ts^{-1} : \omega^s) = c_{P_2|P_1}(t : \omega)s^{-1}$. In the same way the first assertion of the lemma follows from the fact that $f_{P^s} = sf_P$ for any $f \in \mathcal{A}(G, \tau)$ (Lemma 2.7.3).

Since $c^0_{P_2|P_1}(t : \omega)c_{P_1|P_1}(1 : \omega) = c_{P_2|P_1}(t : \omega)$, we have $c^0_{P_2|P_1}(t : \omega)s^{-1} \cdot sc_{P_1|P_1}(1 : \omega)s^{-1} = c_{P_2|P_1}(t : \omega)s^{-1}$. Using what has already been proved this immediately yields the desired expression for $c^0_{P_2|P_1}(t : \omega)s^{-1}$.

On the other hand, $c^0_{P_3|P_1}(ut : \omega) = c^0_{P_3|P_2}(u : \omega^t)s^{-1} \cdot sc^0_{P_2|P_1}(t : \omega) = c^0_{P_3|P_2^s}(us^{-1} : \omega^{st}) \cdot sc^0_{P_2|P_1}(t : \omega)$, so $sc^0_{P_2|P_1}(t : \omega) = c^0_{P_2^s|P_1}(st : \omega)$.

Since $E(P_2^s : s\,{}^0c_{P_2|P_1}(t : \omega)\psi) = E(P_2 : {}^0c_{P_2|P_1}(t : \omega)\psi) = E(P_1 : \psi) = E(P_2^s : {}^0c_{P_2^s|P_1}(st : \omega)\psi)$, we may conclude that $s\,{}^0c_{P_2|P_1}(t : \omega) = {}^0c_{P_2^s|P_1}(st : \omega)$. The other relation follows via the law of composition for ${}^0c$ (Theorem 5.2.4.6, 2)).

Let A' be a standard torus of G and let $M' = Z_G(A')$. We may choose representatives for $W(A|A')$ in $W(G/A_0)$ and, for all elements $s \in W(G/A_0)$, choose representatives $y(s) \in K$. For every $s \in W(A|A')$ the inverse image of

A is a standard torus $A^{s^{-1}} = A_s$ with centralizer $M^{s^{-1}} = M_s = Z_G(A_s)$. For any $\gamma \in W(A|A')$ set $W_\gamma = W(G/A_\gamma)$ and ${}^*W_\gamma = W(M'/A_\gamma) = \{s' \in W(G/A_\gamma) | s'|A' = 1\}$. We may write $W(A|A') = \coprod_{\gamma \in \Gamma} y(\gamma)W_\gamma / {}^*W_\gamma$, where $\{A_\gamma | \gamma \in \Gamma\}$ is a complete set of M'-conjugacy class representatives for the set of all standard tori which are G-conjugate to A.

Corresponding to the class $\omega \in \mathcal{E}_2(M)$ and to any $s \in W(G/A_0)$, we have a class $\omega^s \in \mathcal{E}_2(M^s)$. For any $\psi \in \mathcal{A}(\omega, \tau_M)$ there is an element $s\psi \in \mathcal{A}(\omega^s, \tau_{M^s})$ defined by the relation $s\psi(m^y) = \tau(y)\psi(m)\tau(y^{-1})$. It is easy to see that, for any $P \in \mathcal{P}(A)$ with $P^s \in \mathcal{P}(A^s)$, we have $E(P : \psi) = E(P^s : s\psi)$. Thus, if $s \in W(G/A)$, then ${}^0c_{P^s|P}(1 : \omega)\psi = s\psi$. As we shall see later, it is also true that $c^0_{P^s|P}(1 : \omega)\psi = s\psi$.

Let $P' = M'N' \in \mathcal{P}(A')$. Let $P_i \in \mathcal{P}(A_\gamma)$ $(i = 1, 2)$ and assume that $(P', A') \succ (P_i, A_\gamma)$. Set $(P_i \cap M', A_\gamma) = ({}^*P_i, A_\gamma) \prec (M', A')$. Observe that both $L(\omega_\gamma, P_i) \subset L(\omega_\gamma, {}^*P_i)$ and $\mathcal{L}(\omega, P_i) \subset \mathcal{L}(\omega, {}^*P_i)$.

<u>Theorem</u> 5.2.4.8. Let $\psi \in \mathcal{A}(\omega, \tau_M)$ and $s \in {}^*W_\gamma$ $(\gamma \in \Gamma)$. Assume that $P \in \mathcal{P}(A)$ and that $P^{\gamma^{-1}} = P_1 \in \mathcal{P}(A_\gamma)$. Then:

(1) $E(P_2 : {}^0c_{{}^*P_2|{}^*P_1}(s : \omega_\gamma)\gamma^{-1}\psi) = E(P : \psi)$; in particular, ${}^0c_{P_2|P_1}(s : \omega_\gamma) = {}^0c_{{}^*P_2|{}^*P_1}(s : \omega_\gamma)|_{L(\omega_\gamma, P_1)}$.

(2) The subspace $\sum_{\gamma \in \Gamma} \sum_{t \in {}^*W_\gamma \backslash W_\gamma} \mathcal{A}(C^{M'}_{M_\gamma}(\omega^t_\gamma), \tau_{M'})$ of $\mathcal{A}(M', \tau_{M'})$ is a direct sum and contains ${}_wE^0_{P'}(P : \psi)$. The component of ${}_wE^0_{P'}(P : \psi) \in \mathcal{A}(C^{M'}_{M_\gamma}(\omega_\gamma), \tau_{M'})$ is $\sum_{t \in {}^*W_\gamma \backslash W_\gamma} E^0({}^*P_2 : c^0_{{}^*P_2|{}^*P_1}(t, \omega_\gamma)\gamma^{-1}\psi)$; in particular, $c^0_{P_2|P_1}(s : \omega_\gamma) =$

$${}^0c_{{}^*P_2|{}^*P_1}(s:\omega_\gamma)|\mathcal{L}(\omega_\gamma, P_1)\cdot$$

Proof. (1) Theorem 5.2.4.6,2) implies that $E({}^*P_2 : {}^0c_{{}^*P_2|{}^*P_1}(s:\omega_\gamma)\gamma^{-1}\psi) =$ $E({}^*P_1 : \gamma^{-1}\psi)$; both Eisenstein integrals define elements of $\mathcal{A}(M', \tau_{M'})$. Applying Lemma 5.2.2.8, we conclude that $E(P_2 : {}^0c_{{}^*P_2|{}^*P_1}(s:\omega_\gamma)\gamma^{-1}\psi) =$ $E(P' : E({}^*P_2 : {}^0c_{{}^*P_2|{}^*P_1}(s:\omega_\gamma)\gamma^{-1}\psi) = E(P_1 : \gamma^{-1}\psi) = E(P:\psi)$. Since the Eisenstein integral is a bijection from $L(\omega_\gamma^s, P_2)$ to $\mathcal{A}(C(\omega), \tau)$, it follows that

$${}^0c_{{}^*P_2|{}^*P_1}(s:\omega)|_{L(\omega, P_1)} = {}^0c_{P_2|P_1}(s:\omega).$$

(2) By Theorem 3.5.1 (analogue for $\omega \in \mathcal{E}_2(M)$), ${}_wE^0_{P'}(P:\psi) \in$ $\bigoplus_{\gamma\in\Gamma} \mathcal{A}(C^{M'}_{M_\gamma}(\omega^t_\gamma), \tau_{M'})$. We need check only that, for any fixed t, the component $t \in {}^*W_\gamma \setminus W_\gamma$ of $E^0_{P'}(P:\psi)$ in $\mathcal{A}(C^{M'}_{M_\gamma}(\omega^t_\gamma), \tau_{M'})$ equals $E^0({}^*P_2 : {}^0c_{P_2|P_1}(t:\omega_\gamma)\gamma^{-1}\psi)$. For this, we compute the constant terms and check that they coincide; the equality we want then follows from the Maas-Selberg relations.

First, use an obvious generalization of Lemma 5.2.4.7 to check that $E^0(P:\psi) = E^0(P_1 : \gamma^{-1}\psi)$. (The point is that we may choose $\gamma \in W(G/A_0)$ arbitrary, not necessarily in $W(G/A)$.) Here are the details:

$$\begin{aligned} E^0(P:\psi) &= E(P : c_{P|P}(1:\omega)^{-1}\psi) \\ &= E(P^{\gamma^{-1}} : \gamma^{-1}c_{P|P}(1:\omega)^{-1}\gamma\cdot\gamma^{-1}\psi) \\ &= E(P^{\gamma^{-1}} : c_{P^{\gamma^{-1}}|P^{\gamma^{-1}}}(1:\omega_\gamma)^{-1}\cdot\gamma^{-1}\psi) \end{aligned}$$

$$= E(P_1 : c_{P_1|P_1}(1 : \omega_\gamma)^{-1} \cdot \gamma^{-1}\psi)$$

$$= E^0(P_1 : \gamma^{-1}\psi).$$

We now see that

$$_wE^0_{P_2}(P : \psi) = {}_wE^0_{P_2}(P_1 : \gamma^{-1}\psi)$$

$$= \sum_{t \in W_\gamma} c^0_{P_2|P_1}(t : \omega_\gamma)\gamma^{-1}\psi,$$

also that $\sum_{t \in {}^*W_\gamma \backslash W_\gamma} {}_wE^0_{{}^*P_2}({}^*P_2 : c^0_{P_2|P_1}(t : \omega_\gamma)\gamma^{-1}\psi) =$

$= \sum_{\substack{t \in {}^*W_\gamma \backslash W_\gamma \\ u \in {}^*W_\gamma}} c^0_{{}^*P_2|{}^*P_2}(u : \omega^t_\gamma)c^0_{P_2|P_1}(t : \omega_\gamma)\gamma^{-1}\psi$. For $u = 1$, we know that

$$c^0_{{}^*P_2|{}^*P_2}(u : \omega^t_\gamma) = 1_{\mathcal{L}(\omega^t_\gamma, {}^*P_2)}.$$

Thus, the terms which correspond to the representatives $t \in {}^*W_\gamma \backslash W_\gamma$ in both ${}_wE^0_{P_2}(P_1 : \gamma^{-1}\psi)$ and ${}_wE^0_{{}^*P_2}({}^*P_2 : c^0_{P_2|P_1}(t : \omega_\gamma)\gamma^{-1}\psi)$ agree. The equality in (2), as one now sees, follows immediately from the Maass-Selberg relations.

To check that $c^0_{P_2|P_1}(s : \omega_\gamma) = c^0_{{}^*P_2|{}^*P_1}(s : \omega_\gamma)|_{\mathcal{L}(\omega_\gamma, P_1)}$ for $s \in {}^*W_\gamma$, observe as above that

$$E^0({}^*P_2 : c^0_{P_2|P_1}(s : \omega_\gamma)\gamma^{-1}\psi) = E^0({}^*P_2 : c^0_{{}^*P_2|{}^*P_1}(s : \omega_\gamma)\gamma^{-1}\psi)$$

for $\psi \in \mathcal{L}(\omega, P)$.

. 3. The Eisenstein Integral and Its Functional Equations: The Analytic Theory.

. 3.1. The Complex Structure on $\mathcal{E}_{\mathbb{C}}(M)$.

Let A be a special torus of G and let $M = Z_G(A)$. Let $H : M \to \mathfrak{a} =$
om$(X(M), \mathbb{R})$ be the mapping which satisfies $|\chi(m)| = q^{\langle \chi, H(m) \rangle}$ for all $m \in M$,
$\in X(M)$ (cf. §0.4). Recall that $\mathrm{Hom}(X(A), \mathbb{R}) = \mathrm{Hom}(X(M), \mathbb{R})$. Set
$^* = X(A) \underset{\mathbb{Z}}{\otimes} \mathbb{R}$, $\mathfrak{a}^*_{\mathbb{C}} = X(A) \underset{\mathbb{Z}}{\otimes} \mathbb{C}$. For $\nu \in \mathfrak{a}^*_{\mathbb{C}}$ let χ_ν denote the quasicharacter
M defined by setting $\chi_\nu(m) = q^{\sqrt{-1}\langle \nu, H_M(m) \rangle}$ $(m \in M)$.

Let $\sigma \in \omega \in \mathcal{E}_{\mathbb{C}}(M)$ and write $\chi_\sigma = \chi_\omega = \chi$ for the central exponent of
For any $\nu \in \mathfrak{a}^*_{\mathbb{C}}$ set $\sigma_\nu(m) = \sigma(m)\chi_\nu(m)$ $(m \in M)$. Then σ_ν is also irreduc-
le and admissible; we write ω_ν for its class and $\chi_{\omega,\nu} = \chi_{\omega_\nu}$ for its central ex-
onent. Let $\mathcal{O}_{\mathbb{C}} = \mathcal{O}_{\mathbb{C}}(\omega) = \{\omega' \in \mathcal{E}_{\mathbb{C}}(M) \mid \omega' = \omega_\nu$ for some $\nu \in \mathfrak{a}^*_{\mathbb{C}}\}$. The
ubset $\mathcal{O}_{\mathbb{C}}(\omega) \subset \mathcal{E}_{\mathbb{C}}(M)$ is called the (complex) orbit associated to ω.

The map $H : M \to \mathfrak{a}$ has kernel ${}^0M = \bigcap_{\chi \in X(M)} \ker|\chi|$. If $\ell = \dim A$,
en $M/{}^0M \cong \mathbb{Z}^\ell$; the subgroup $H(M) = L$ is a lattice in $\mathfrak{a}$. Define $L^* \subset \mathfrak{a}^*$ by
tting $\lambda \in L^*$ provided $\langle \lambda, L \rangle \subset \frac{2\pi}{\log q}\mathbb{Z}$. Then L^* is a lattice in $\mathfrak{a}^*$ and the
apping $\nu \mapsto \chi_\nu \in \mathfrak{X}(M)$ depends only on the coset of ν mod L^*. Consequently,
e class $\omega_\nu \in \mathcal{O}_{\mathbb{C}}(\omega)$ depends only upon ω and $\nu \in \mathfrak{a}^*_{\mathbb{C}}/L^*$. We also have
$_{\nu_1})_{\nu_2} = \omega_{\nu_1+\nu_2}$, i.e., the group $\mathfrak{a}^*_{\mathbb{C}}$ operates on $\mathcal{E}_{\mathbb{C}}(M)$. The stabilizer of any
ass ω is a lattice $L^*(\omega)$ in $\mathfrak{a}^*$ such that $L^*(\omega) \supset L^*$. Thus, every orbit has
natural structure of complex manifold. The set $\mathcal{E}_{\mathbb{C}}(M)$ itself, being a disjoint
nion of orbits, is a complex manifold.

Let $\mathcal{O}_{\mathbb{C}} \subset \mathcal{E}_{\mathbb{C}}(M)$ be an orbit and let $\sigma \in \omega \in \mathcal{O}_{\mathbb{C}}$. The restriction
${}_{{}^0M}$ is not necessarily irreducible; however, up to equivalence, $\sigma_{{}^0M}$ does not

depend upon the choice of $\omega \in \mathcal{O}_{\mathbb{C}}$. We observe that $\mathcal{O}_{\mathbb{C}} \subset {}^0\mathcal{E}_{\mathbb{C}}(M)$ if and only if the matrix coefficients of the admissible representation $\sigma_{{}^0M}$ are compactly supported functions on ${}^0M$. We write $\mathcal{E}_{2,\mathbb{C}}(M)$ for the union of all orbits $\mathcal{O}_{\mathbb{C}}$ such that $\sigma_{{}^0M}$ has square-integrable matrix coefficients for $\sigma \in \omega \in \mathcal{O}_{\mathbb{C}}$. Obviously, ${}^0\mathcal{E}_{\mathbb{C}}(M) \subset \mathcal{E}_{2,\mathbb{C}}(M)$.

Lemma 5.3.1.1. If $\mathcal{O}_{\mathbb{C}} \subset \mathcal{E}_{2,\mathbb{C}}(M)[{}^0\mathcal{E}_{\mathbb{C}}(M)]$, then $\mathcal{O}_{\mathbb{C}} \cap \mathcal{E}_2(M) \neq \phi$ $[\mathcal{O}_{\mathbb{C}} \cap {}^0\mathcal{E}(M) \neq \phi]$.

Proof. To prove both statements it obviously suffices to prove that, given $\omega \in \mathcal{E}_{2,\mathbb{C}}(M)$, there exists $\nu \in \mathfrak{a}_{\mathbb{C}}^*$ such that $\omega_\nu \in \mathcal{E}_2(M)$. If $\chi_\omega \in \hat{A}$, then, for any $(v, \tilde{v}) \in U \times \tilde{U}$ ($\sigma \in \omega$ acts in U), we have

$$\int_{M/A} |\langle \tilde{v}, \sigma(m)v \rangle|^2 dm^*$$

$$= \sum_{x \in M/{}^0MA} \int_{{}^0M} |\langle \tilde{\sigma}(x^{-1})\tilde{v}, \sigma(m)v \rangle|^2 d^0m < \infty, \text{ since } [M : {}^0MA] < \infty \text{ (cf. §0.4)}$$

and $\omega \in \mathcal{E}_{2,\mathbb{C}}(M)$. On the other hand, for any $\chi \in \mathfrak{X}(A)$, we have $|\chi(a)| = q^{\langle \nu, H(a) \rangle}$ for some $\nu \in \mathfrak{a}^*$, so $\chi \cdot \chi_{\sqrt{-1}\nu} \in \hat{A}$.

Let $\omega \in \mathcal{E}_2(M)$. Sometimes we write $\mathcal{O}$ or $\mathcal{O}(\omega)$ for $\mathcal{O}_{\mathbb{C}}(\omega) \cap \mathcal{E}_2(M)$. Clearly, $\mathcal{O}(\omega) = \mathfrak{a}^* \cdot \omega = (\mathfrak{a}^*/L^*(\omega))\omega$.

Let W be a subset of a complex manifold X. We call W h-dense in X if every holomorphic function on X which vanishes on W is identically zero on X. For example, if W is an open subset of $\mathfrak{a}^*$, then W is h-dense in $\mathfrak{a}_{\mathbb{C}}^*$.

Consider our fixed maximal split torus A_0 in G. Let $\mathfrak{a}_0 = \mathrm{Hom}(X(A_0), \mathbb{R})$ and let Q be a positive definite quadratic form on $\mathfrak{a}_0$ which is invariant under $W_0 = W(G/A_0)$. Let $B = B_{\mathfrak{a}_0}$ denote the bilinear form associated to Q. For any $\gamma \in W(A_0|A)$ set $B_{\mathfrak{a}}(H_1, H_2) = B_{\mathfrak{a}_0}(\gamma H_1, \gamma H_2)$ $(H_1, H_2 \in \mathfrak{a})$. Then $B_{\mathfrak{a}}$ does not depend upon the choice of γ. We may use $B_{\mathfrak{a}}$ to identify $\mathfrak{a}$ and $\mathfrak{a}^*$.

Given $P \in \mathcal{P}(A)$ we form $\bar{\omega}_P = \prod \alpha (\alpha \in \Sigma_r(P, A))$ and regard this expression as a polynomial function on $\mathfrak{a}$. Since we shall be interested only in the zeros of $\bar{\omega}_P$, the product defining $\bar{\omega}_P$ could be taken over all positive roots, all roots, etc., in particular the zero set of $\bar{\omega}_P$ does not depend upon the choice of $P \in \mathcal{P}(A)$. Using our identification via $B_{\mathfrak{a}}$ we may regard $\bar{\omega}_P$ as a function on $\mathfrak{a}^*$.

A class $\omega \in \mathcal{E}_{\mathbb{C}}(M)$ is said to be in <u>general position</u> if its central exponent χ_ω is unramified, i.e., if $\chi_\omega \circ s \neq \chi_\omega$ for all $s \in W(G/A)$. Clearly, if ω is in general position, then ω is unramified. The converse is in general false.

<u>Lemma</u> 5.3.1.2. Let $\omega \in \mathcal{E}_{\mathbb{C}}(M)$. Then, for all sufficiently small ν such that $\bar{\omega}_P(\nu) = \prod_{\alpha > 0} \langle \alpha, \nu \rangle \neq 0$, the class ω_ν is in general position.

<u>Proof.</u> We make an orbit $\mathcal{O}_{\mathbb{C}}(\omega)$ out of the central characters of elements of $\mathcal{O}_{\mathbb{C}}(\omega)$. If no element of $W(G/A)$ stabilizes $\mathcal{O}_{\mathbb{C}}(\omega)$, then every point of $\mathcal{O}_{\mathbb{C}}(\omega)$ is in general position. We may assume that $\chi_{\omega^s} = \chi_{\omega_{\nu(s)}}$ for certain elements $s \in W(G/A)$. For all $\nu(s) \neq 0$ pick a neighborhood N of zero in $\mathfrak{a}_{\mathbb{C}}^*/L^*(\omega)$ such that $N \cap (N^s + \nu(s)) = \phi$ for all $\nu(s) \neq 0$. If $\chi_{\omega^s} \neq \chi_\omega$ for all $s \in N(A)$, then $\nu \in N$ suffices. Assume $\chi_{\omega^s} = \chi_\omega$. Then, to conclude the proof, observe that the condition $\prod_{\alpha > 0} \langle \alpha, \nu \rangle \neq 0$ insures that $s\nu \neq \nu$ for all $s \in W(G/A)$.

Corollary 5.3.1.3. The subset of any orbit $\mathcal{O}_{\mathbb{C}} \subset \mathcal{E}_{\mathbb{C}}(M)$ which consists of classes in general position comprises an h-dense subset of $\mathcal{O}_{\mathbb{C}}$. More particularly, if $\omega \in \mathcal{E}(M)$ and $\mathcal{O}_{\mathbb{C}} = \mathcal{O}_{\mathbb{C}}(\omega)$, then $\mathcal{O} = \mathcal{O}_{\mathbb{C}} \cap \omega \cdot \mathfrak{a}^* = \mathcal{O}_{\mathbb{C}} \cap \mathcal{E}(M)$ contains an h-dense subset of $\mathcal{O}_{\mathbb{C}}$. Finally, any $\omega \in \mathcal{E}(M)$ is a limit point of a sequence $\{\omega_n\}_{n=1}^{\infty} \subset \mathcal{O}(\omega)$, where ω_n is in general position.

Let $\sigma \in \omega \in \mathcal{E}_{\mathbb{C}}(M)$ and let (V, τ) be a smooth unitary double representation of K. For $\psi \in \mathcal{A}(\omega, \tau_M)$ and $P \in \mathcal{P}(A)$ we write $E(P : \psi : \nu) = E(P : \psi_\nu) \in \mathcal{A}(\pi_{P,\omega_\nu}, \tau)$, where $\psi_\nu = \psi\chi_\nu \in \mathcal{A}(\omega_\nu, \tau_M)$ and $\pi_{P,\omega_\nu} = \mathrm{Ind}_P^G(\delta_P^{\frac{1}{2}}\sigma_\nu)$. It is clear that $E(P : \psi : \nu : x)$ is an entire function of $\nu \in \mathfrak{a}_{\mathbb{C}}^*$ for fixed $x \in G$ and a uniformly locally constant function in x for $\nu \in \mathfrak{a}_{\mathbb{C}}^*$. Thus, $E(P : \psi : \nu)$ is continuous on $\mathfrak{a}_{\mathbb{C}}^* \times G$.

Now let $\omega \in \mathcal{E}_2(M)$ and let F be a finite subset of $\mathcal{E}(K)$. Recall (Theorem 4.6.1) that, for all $\nu \in \mathfrak{a}^*$ such that ω_1 is unramified, $C(\pi_{P,(\omega_\nu)^s}) = C_M^G(\omega_\nu)$ is a class independent of $P \in \mathcal{P}(A)$ and $s \in W(G/A)$. It follows that, if $\psi_{s,F} = \psi_{E_F} \in \mathcal{A}(\omega^s, \tau_M^0)$ (cf. §5.2.4), then $E(P_2 : \psi_{s,F} : s\nu) = E(P_1 : \psi_{1,F} : \nu)$ for all $P_1, P_2 \in \mathcal{P}(A)$, $s \in W(G/A)$, and $\nu \in \mathfrak{a}_{\mathbb{C}}^*$. Setting $\theta(F : P : \omega^s : s\nu) = j^{-1}(E(P : \psi_s : s\nu))$, we conclude that the character of the admissible representation $\pi_{P,(\omega_\nu)^s}$ is, for all $\nu \in \mathfrak{a}_{\mathbb{C}}^*$, independent of $P \in \mathcal{P}(A)$ and $s \in W(G/A)$. This implies that $\pi_{P,(\omega_\nu)^s}$ is irreducible for one pair $(P, s) \in \mathcal{P}(A) \times W(G/A)$ if and only if the same is true for all such pairs; indeed, the composition series (cf. §5.4.1) of the admissible representation $\pi_{P,(\omega_\nu)^s}$ does not depend upon the choice of P and s.

We give some notation which will be needed later. Let $\omega \in \mathcal{E}(M)$. Write $\mathfrak{F} = \mathfrak{a}^*/L^*$ and $\mathfrak{F}(\omega) = \mathfrak{a}^*/L^*(\omega)$, $\mathfrak{F}_{\mathbb{C}} = \mathfrak{a}^*_{\mathbb{C}}/L^*$ and $\mathfrak{F}_{\mathbb{C}}(\omega) = \mathfrak{a}^*_{\mathbb{C}}/L^*(\omega)$. Write $\mathfrak{F}'_{\mathbb{C}}$ or $\mathfrak{F}'_{\mathbb{C}}(\omega)$ $[\mathfrak{F}'$ or $\mathfrak{F}'(\omega)]$ for the set of all $\nu \in \mathfrak{F}_{\mathbb{C}}$ or $\mathfrak{F}_{\mathbb{C}}(\omega)$ $[\mathfrak{F}$ or $\mathfrak{F}(\omega)]$ such that ω_ν is in general position. More generally, if A_1 is also a special torus of G, we write $\mathfrak{F}'_{\mathbb{C}}(A|A_1)$ or $\mathfrak{F}'_{\mathbb{C}}(\omega, A|A_1)$ $[\mathfrak{F}'(A|A_1)$ or $\mathfrak{F}'(\omega, A|A_1)]$ for the set of all $\nu \in \mathfrak{F}_{\mathbb{C}}$ or $\mathfrak{F}_{\mathbb{C}}(\omega)$ $[\mathfrak{F}$ or $\mathfrak{F}(\omega)]$ such that $\nu \circ s \neq \nu \circ t$ for all $s \neq t \in W(A|A_1)$.

We conclude this section with a result which we shall need in §5.5.

Let $\sigma \in \omega \in \mathcal{E}_2(M)$ and assume that $\sigma_\nu \simeq \sigma\chi_\nu$ acts in a vector space U which is independent of $\nu \in \mathfrak{a}^*_{\mathbb{C}}$. Since $\sigma_\nu|{}^0M$ is finitely generated, unitary, and independent of $\nu \in \mathfrak{a}^*_{\mathbb{C}}$, we may write $\sigma_\nu|{}^0M = n(\sigma_1 \oplus \ldots \oplus \sigma_r)$, where $\sigma_1, \ldots, \sigma_r$ are inequivalent discrete series representations of ${}^0M$ and, since M conjugates the classes, each occurs with a common multiplicity n. We write $\sigma_{i,\nu}$ for the extension of σ_i to ${}^0MA$ such that $\sigma_\nu|{}^0MA = n(\sigma_{1,\nu} \oplus \ldots \oplus \sigma_{r,\nu})$.

Let ${}^1M = {}^1M(\omega) = {}^1M(\sigma)$ denote a maximal subgroup of M with the following properties: (1) For each $i = 1, \ldots, r$, the representation $\sigma_{i,\nu}$ extends to n inequivalent irreducible representations $\sigma_{i,j,\nu}$ $(j = 1, \ldots, n)$ such that $\sigma_\nu|{}^1M = \bigoplus_{i,j} \sigma_{i,j,\nu}$. (2) For any choice of i, j, $\sigma_\nu \sim \mathrm{Ind}_{{}^1M}^{M} \sigma_{i,j,\nu}$. It is easy to see that ${}^1M$ exists and is independent of $\nu \in \mathfrak{a}^*_{\mathbb{C}}$. We also write ${}^2M$ for the maximal subgroup of M to which the representation $\bigoplus_{j=1}^{n} \sigma_{i,j,\nu}$ extends for any i and ν; we write $\sigma_\nu|{}^2M = \sigma^1_\nu \oplus \ldots \oplus \sigma^r_\nu$, which is again a sum of inequivalent irreducible representations, each characterized by the fact that $\sigma^i_\nu|{}^0M = n\sigma_i$. Since $M/{}^0M$ is abelian, it is of course obvious that ${}^1M$ and ${}^2M$ are normal subgroups,

${}^0M \subset {}^1M \subset {}^2M.$

There is a one-one correspondence between lattices $X^* \subset \mathfrak{a}^*$ such that $L_A^* \supset X^* \supset L_M^*$ and normal subgroups X of M such that ${}^0MA \subset X \subset M$. Given X, set $X^* = \{\nu \in \mathfrak{a}^* \mid \langle \nu, H(X) \rangle \subset \frac{2\pi\mathbb{Z}}{\log q}\}$ and vice versa. Define $M(\omega) = M(\sigma)$ by the relation $M(\omega)^* = L^*(\omega)$.

Lemma 5.3.1.4. The group $M(\omega)$ is a subgroup of ${}^1M$. More precisely, $[{}^1M : M(\omega)] = [{}^2M : {}^1M]$. In particular, if $\sigma|{}^0M$ decomposes simply, then ${}^1M = {}^2M = M(\omega)$.

Proof. For any open compact subgroup K_0 of ${}^1M$ there is a function $f_I(\nu : x) = f_{I(K_0)}(\nu : x) = \operatorname{tr}(\sigma_{i,j,\nu}(x)I)$ $(I \in \operatorname{End}_0(U))$, which represents the character of $\sigma_{i,j,\nu}$ as a functional on $C_c({}^1M/\!/K_0)$. Since $\sigma_{i,j,\nu}|{}^0M$ is irreducible, $f_I(\nu : x) \neq 0$ for all x in some set of representatives for ${}^1M/{}^0M$; in fact, since I is independent of ν and since $f_I(\nu+\nu') = \chi_{\nu'} f_I(\nu)$, the set of representatives is independent of ν, too. Thus, $\sigma_{i,j,\nu} \sim \sigma_{i,j,\nu+\nu_1}$ if and only if $\nu_1 \in {}^1M^*$. On the other hand, $\sigma_\nu \sim \sigma_{\nu+\nu_1}$ if and only if $\sigma_{i,j,\nu} \sim \sigma_{i,j',\nu+\nu_1}$ for any fixed i, $1 \leq j, j' \leq n$. Thus, $[L^*(\omega) : {}^1M^*] = n = [{}^2M : {}^1M]$, which implies both that $M(\omega) \subset {}^1M$ and that $[{}^1M : M(\omega)] = [{}^2M : {}^1M]$. The final statement of the lemma is clear.

§5.3.2. On the Simplicity of Exponents I.

Let (P, A) $(P = MN)$ be a standard p-pair of G. Fix a class $\omega \in \mathcal{E}_2(M)$ and let $\chi = \chi_\omega$ be its central exponent. Let $\sigma \in \omega$ act in a vector pace U. For any $\nu \in \mathfrak{a}^*_{\mathbb{C}}$ we have $\sigma_\nu \in \omega_\nu \in \mathcal{E}_{2,\mathbb{C}}(M)$, where $\sigma_\nu(m) =$ (m)$\chi_\nu(m)$ (cf. §5.3.1) may also be regarded, for all $m \in M$, as a linear trans- ormation of U. The central exponent $\chi_{\omega,\nu}$ of ω_ν satisfies $\chi_{\omega,\nu}(a) = \chi(a)\chi_\nu(a)$ or all $a \in A$. The class $\omega_\nu \in \mathcal{E}_2(M)$ if and only if $\nu \in \mathfrak{a}^*$.

Set $\pi_\nu = \mathrm{Ind}_P^G(\delta_P^{\frac{1}{2}}\sigma_\nu)$. We may identify the representation space $\mathcal{H}_\nu$ of ν with the subspace $\mathcal{H} \subset C^\infty(K : U)$ which consists of all functions β such that (pk) $= \sigma(p)\beta(k)$ $(p \in P \cap K,\ k \in K)$. If $\nu \in \mathfrak{a}^*$, then π_ν is a unitary representa- on of G.

Let $F \subset \mathcal{E}(K)$ as in §5.2.3. Write $E_F = E$ and set $\Phi(\nu : x) =$ $\pi_\nu(x)E$. Write $\pi_0 = \pi_\nu | K$ (of course, independent of ν). There is an obvious mooth double representation of K on $\mathrm{End}^0(\mathcal{H})$ naturally associated to π_0 (cf. 5.2.1). Write $\mathcal{H}_F = E\mathcal{H}$. Then $V = \mathrm{End}(\mathcal{H}_F)$ is a K-stable subspace of $\mathrm{End}^0(\mathcal{H})$ nd we let $\tau = \tau_F$ denote the double representation of K on V.

Let I be the identity element of V regarded as an element of $\subset \mathrm{End}^0(\mathcal{H}) \otimes V$ (cf. §§5.2.2 and 5.2.3). Then $\psi_I \in \mathcal{A}(\sigma, \tau_P)$ and we know, by emma 5.2.3.1, that

$$\Phi(\nu : x) = E(P : \psi_I : \nu : x)$$

$$= \int_K \psi_I(xk)\tau(k^{-1}) q^{\langle \sqrt{-1}\nu - \rho_P,\, H_P(xk)\rangle} dk.$$

or $\nu \in \mathfrak{a}^*$ the weak constant term ${}_w\Phi_{P_1}$ (cf. §4.5) exists for all p-pairs (P_1, A_1) of G.

The main result of this subsection is the following theorem.

Theorem 5.3.2.1. There is an integer $d \geq 1$ such that, if (P_1, A_1) $(P_1 = M_1N_1)$ is semi-standard p-pair, then

$$\prod_{s \in W(A|A_1)} (\rho(\alpha) - \hat{\alpha}(\chi_{\omega,\nu} \circ s))^d \, {}_w\Phi_{P_1}(\nu) = 0$$

for all $\alpha \in C_c^\infty(A_1)$ and all $\nu \in \mathfrak{a}^*$. (Recall that $(\chi_{\omega,\nu} \circ s)(a_1) = \chi_{\omega,\nu}(a_1^s)$ and, if $\eta \in \mathfrak{X}(A_1)$, then $\hat{\alpha}(\eta) = \int_{A_1} \alpha(a)\eta(a)da$).

Proof. For $\nu \in \mathfrak{a}^*$ and $\pi \subset \pi_\nu$ write $\mathcal{H}_\pi$ for the isotypic subspace of $\mathcal{H}$ associated to π. Let $S_\nu = \{\pi \subset \pi_\nu \mid \mathcal{H}_\pi$ is not orthogonal to $\mathcal{H}_F\}$. Then S_ν is a finite set. According to Corollary 4.5.12, the set ${}_w\mathfrak{X}_\pi(P_1, A_1) \subset \{\eta \in \mathfrak{X}(A_1) \mid \eta = \mathfrak{X}_{\omega,\nu} \circ s$ for some $s \in W(A|A_1)\}$ for any $\pi \subset \pi_\nu$. Let $d_\pi(\eta)$ denote the multiplicity of η in ${}_w\mathfrak{X}_\pi(P_1, A_1)$. Let $n(s,\nu) = n(\Phi, s, \nu) = \max_{\pi \in S_\nu} d_\pi(\chi_{\omega,\nu} \circ s)$. We intend to show that $\sup_{\nu \in \mathfrak{a}^*} n(s,\nu) < \infty$.

We have

$$\prod_{s \in W(A|A_1)} (\rho(\alpha) - \hat{\alpha}(\chi_{\omega,\nu} \circ s))^{n(s,\nu)} {}_w\Phi_{P_1}(\nu) = 0$$

for all $\alpha \in C_c^\infty(A_1)$. Let $K_F = \bigcap_{\underline{d} \in F} \ker \underline{d}$, an open normal subgroup of K. Choose an open subgroup K_0 of K_F such that $(\bar{N}_1 \cap K_0)(M_1 \cap K_0)(N_1 \cap K_0) = K_0$ (use Theorem 2.1.1). Set $E_0 = \int_{K_0} \pi(k)dk$ and $\mathcal{H}_0 = E_0\mathcal{H}$. Then $\mathcal{H}_0 \supset \mathcal{H}_F$ and $E_0E = EE_0 = E$. Set $\Phi_0(\nu) = E_0\pi_\nu E_0$. It is obviously sufficient to prove our theorem for Φ_0 in place of Φ.

Lemma 5.1.1, 6) implies that, given any $\nu \in \mathfrak{a}^*_{\mathbb{C}}$, we have $\Phi_{0,P_1}(ma) = \Phi_{0,P_1}(m)\Phi_{0,P_1}(a) = \Phi_{0,P_1}(a)\Phi_{0,P_1}(m)$ for all $m \in M_1$ and $a \in A_1$. Let $\dim \mathcal{H}_0 = d$. Using linear algebra, we see that, for $\nu \in \mathfrak{a}^*$, the relation

$$\prod_{\eta \in \bigcup_{\pi \in S_\nu} \mathfrak{X}_\pi(P_1, A_1)} (\rho(\alpha) - \hat{\alpha}(\eta))^d \Phi_{0,P_1}(\nu) = 0$$

holds, and this implies that

$$\prod_{s\in W(A|A_1)} (\rho(\alpha)-\hat{\alpha}(\chi_{\omega,\nu} \circ s))^d {}_w\Phi_{0,P_1}(\nu) = 0.$$

We conclude this section with some consequences of Theorem 5.3.2.1 for the case $\omega \in {}^0\mathcal{E}(M)$.

Corollary 5.3.2.2. Let $\omega \in {}^0\mathcal{E}(M)$. Then the conclusion of Theorem 5.3.2.1 is valid for all $\nu \in \mathfrak{a}^*_{\mathbb{C}}$. For any $\pi \prec \pi_\nu$ and $\nu \in \mathfrak{a}^*_{\mathbb{C}}$, $\mathfrak{X}_\pi(P_1, A_1) \subset \{\eta \in \mathfrak{X}(A_1) | \eta = \chi_{\omega,\nu} \circ s$ for some $s \in W(A|A_1)\}$. In particular, if $W(A|A_1) = \phi$, then $\Phi_{P_1}(\nu) = 0$.

Proof. Fix $\alpha \in C_c^\infty(A_1)$ and consider the function

$$\psi_\nu = \prod_{s\in W(A|A_1)} (\rho(\alpha)-\hat{\alpha}(\chi_{\omega,\nu} \circ s))^d \Phi_{P_1}(\nu).$$

This function is obviously defined for all $\nu \in \mathfrak{a}^*_{\mathbb{C}}$ and, as we know, identically zero for $\nu \in \mathfrak{a}^*$. To prove our corollary it is sufficient to show that ψ_ν is identically zero on all of $\mathfrak{a}^*_{\mathbb{C}}$. Since $\mathfrak{a}^*$ is h-dense in $\mathfrak{a}^*_{\mathbb{C}}$ and $\psi_\nu(m_1 a)$ is A_1-finite as a function on A_1 for arbitrary fixed $m_1 \in M_1$ and $\nu \in \mathfrak{a}^*_{\mathbb{C}}$, we need only show that $\psi_\nu(m_1 a)$ is holomorphic in $\mathfrak{a}^*_{\mathbb{C}}$ for any $m_1 \in M_1$, provided $a \in A_1^+(t)$ with t independent of ν.

Using Lemma 5.1.5, we obtain t independent of ν such that $\Phi_{0,P_1}(\nu : m_1 a) = \delta_{P_1}^{\frac{1}{2}}(m_1 a)\Phi_0(\nu : m_1 a)$ for all $a \in A_1^+(t)$. The same obviously holds with Φ in place of Φ_0. Let ω_1 be a compact subset of A_1 such that $\mathrm{Supp}\, \alpha \subset \omega_1$. Set $\omega_i = \omega_{i-1}\omega_1$ for $1 \leq i \leq d[W(A|A_1)] = db$. Choose a compact subset ω_0 of A_1 such that $\omega_0 \supset \bigcup_{i=1}^{db} \omega_i$. Taking t larger, we have $\Phi_{P_1}(\nu : m_1 a)$ holomorphic for all $a \in \omega_0 A_1^+(t)$. Then, clearly, $\psi_\nu(m_1 a)$, being a finite linear combination of holomorphic functions, is holomorphic for $a \in A_1^+(t)$. Using Lemma 2.6.2, we conclude that $\psi_\nu(m_1) = 0$, more generally that $\psi_\nu = 0$.

Corollary 5.3.2.3. Let π be an admissible irreducible representation of G and let (P, A) be a p-pair. If (P, A) is π-critical (cf. §2.4), then (P, A) is π-minimal (cf. ibid.).

Proof. Let $\eta \in \mathfrak{X}_\pi(P, A)$ be π-critical. Then there exists $\omega \in {}^0\mathcal{E}(M)$, $\nu \in \mathfrak{a}_{\mathbb{C}}^*$ and $\sigma_\nu \in \omega_\nu \in {}^0\mathcal{E}_{\mathbb{C}}(M)$ such that $\chi_{\omega,\nu} = \eta$ and $\pi \subset \operatorname{Ind}_{\overline{P}}^G(\delta_{\overline{P}}^{\frac{1}{2}}\sigma_\nu)$ (Corollary 2.4.2). If (P_1, A_1) is a p-pair with $(P_1, A_1) \lneqq (P, A)$, then $W(A|A_1) = \phi$ (since $\dim \mathfrak{a}_1 > \dim \mathfrak{a}$). Therefore, by Corollary 5.3.2.2, $\mathfrak{X}_\pi(P_1, A_1) = \phi$.

§5.3.3. On the Simplicity of Exponents II.

The assumptions of the preceding subsection remain in force. We shall show, in this subsection, that Theorem 5.3.2.1 is valid with $d = 1$. This will show that, for $\omega \in {}^0\mathcal{E}_{\mathbb{C}}(M)[\omega \in \mathcal{E}_2(M)]$ in general position, all exponents [all weak exponents] are simple. In §5.4.3 we shall show that the same is true when ω is merely unramified.

The idea behind the proof that $d = 1$ suffices in Theorem 5.3.2.1 is briefly as follows. Let $\Phi(\nu)$ be the function of the preceding section. If (P_1, A_1) is a p-pair and $\nu_0 \in \mathcal{F}'$, we may write ${}_w\Phi_{P_1}(\nu_0) = \sum_{s \in W(A|A_1)} \Phi_{P_1,s}(\nu_0)$, where $\Phi_{P_1,s}(\nu_0)$ denotes the component of $\Phi_{P_1}(\nu_0)$ which is annihilated by $(\rho(a) - \chi_{\omega,\nu_0}(a))^d$ for all $a \in A_1$. We first show that $\Phi_{P_1,s}$ has a continuation to a unction which is analytic in a neighborhood of $\nu_0 \in \mathfrak{F}'_{\mathbb{C}}$. It follows that ${}_w\Phi_{P_1}(\nu)$ s defined as an analytic function in a neighborhood of $\mathfrak{F}'$ in $\mathfrak{F}_{\mathbb{C}}$. We then show hat $d = 1$ on an h-dense subset of the domain of definition of ${}_w\Phi_{P_1}(\nu)$.

To facilitate the proof we first introduce a partial generalization of the receding ideas. Let $\omega \in {}^0\mathcal{E}(M)$ and $\nu_0 \in \mathfrak{F}_{\mathbb{C}}$. Given elements s and t f $W(A|A_1)$ we write $s \underset{\nu_0}{\sim} t$ if $\chi_{\omega,\nu_0} \circ s = \chi_{\omega,\nu_0} \circ t$. We then set $J(s) = J(s,\nu_0) =$ $\{t \in W(A|A_1) | t \underset{\nu_0}{\sim} s\}$. Let $\mathfrak{F}_{\mathbb{C}}(\nu_0, A|A_1) = \{\nu \in \mathfrak{F}_{\mathbb{C}} | \chi_{\omega,\nu} \circ s = \chi_{\omega,\nu} \circ t$ mplies $s \underset{\nu_0}{\sim} t$ $(s,t \in W(A|A_1))\}$. Given $\nu \in \mathfrak{F}_{\mathbb{C}}(\nu_0, A|A_1)$, we may write $\Phi_{P_1, J(s)}(\nu)$ for the component of $\Phi_{P_1}(\nu)$ which is annihilated by $\prod_{t \in J(s)} (\rho(\alpha) - \hat{\alpha}(\chi_{\omega,\nu} \circ t))^d$ for all $\alpha \in C_c^\infty(A_1)$. It then follows from Corollary 5.3.2.2 hat, for any $\nu \in \mathfrak{F}_{\mathbb{C}}(\nu_0, A|A_1)$, $\Phi_{P_1}(\nu) = \sum_{J \in \{J(s)\}} \Phi_{P_1, J}(\nu)$. In particular, if $\nu_0 \in \mathfrak{F}'_{\mathbb{C}}(A|A_1)$, then $J(s) = \{s\}$ for any $s \in W(A|A_1) = W_1$. If $\nu_0 \in \mathfrak{F}_{\mathbb{C}}$ and $\nu \in \mathfrak{F}'_{\mathbb{C}}(A|A_1)$, then $\Phi_{P_1, J(s)}(\nu) = \sum_{t \in J(s)} \Phi_{P_1, t}(\nu)$.

Lemma 5.3.3.1. Let L be a field, X and $\Lambda = (\Lambda_1, \ldots, \Lambda_r)$ indeterminates, and $\{n_1, \ldots, n_s\}$ a partition of the positive integer r. Let $m_0 = 0$, $m_i = \sum_{j \leq i} n_j$, and et $J_i = \{x \in \mathbb{Z} | m_{i-1} < x \leq m_i\}$ $(1 \leq i \leq s)$. Then there exist unique functions

$$h_i(X;\Lambda) = \sum_{j=0}^{n_i - 1} c_{ij}(\Lambda) X^j \in L(\Lambda)[X] \text{ such that } \prod_{k=1}^{r} (X - \Lambda_k)^{-1} = \sum_{i=1}^{s} h_i(X;\Lambda) \prod_{j \in J_i} (X - \Lambda_j)^{-1}.$$

Moreover, let $\lambda = (\lambda_1, \ldots, \lambda_r) \in L^r$ be any tuple which satisfies the following condition: If $\mu \in J_\alpha$ and $\nu \in J_\beta$ with $1 \leq \alpha < \beta \leq s$, then $\lambda_\mu \neq \lambda_\nu$. In this case all the coefficients $c_{ij}(\Lambda)$ are defined at $\Lambda = \lambda$.

Proof. To prove existence and uniqueness we shall show that there exist unique elements $c_{ij} \in L(\Lambda)$ such that the polynomial relation

$$1 = \sum_{i=1}^{s} h_i(X;\Lambda) \prod_{k \notin J_i} (X-\Lambda_k) = f(X)$$

is identically satisfied. Since $\deg f < r$, it is sufficient to show that there exist unique c_{ij}'s such that $f(\Lambda_\ell) = 1$ $(1 \leq \ell \leq r)$. To show this we solve the linear equations

$$\prod_{k \notin J_i} (\Lambda_\ell - \Lambda_k)^{-1} = \sum_{j=0}^{n_i - 1} c_{ij}(\Lambda)\Lambda_\ell^j \qquad (\ell \in J_i)$$

for $c_{ij}(\Lambda)$ $(1 \leq i \leq s;\ 0 \leq j < n_i)$. The unique solution is $c_{ij}(\Lambda) = V_j(\{\Lambda_k | k \in J_i\}) \cdot V^{-1}(\{\Lambda_k | k \in J_i\})$, where V is Van der Monde's determinant in the indicated variables and V_j is the modification required by Cramer's rule. By inspection it is clear that V divides V_j, so c_{ij} is actually defined whenever the numerator is defined. This proves the lemma.

Corollary 5. 3. 3. 2. If $L = \mathbb{C}$, the field of complex numbers, then there exist unique meromorphic functions $c_{ij}(\Lambda)$ such that the above relation is satisfied and such that c_{ij} is analytic at $\lambda \in \mathbb{C}^r$ provided the condition required above for λ is also satisfied.

Proof. The c_{ij}'s are rational functions, so they are analytic wherever they are defined; in fact, they are globally meromorphic.

Lemma 5.3. 3. 3. Let $\omega \in {}^0\mathcal{E}(M)$. Fix $s \in W_1$ and $\nu_0 \in \mathcal{F}_{\mathbb{C}}$. Let $J = J(s, \nu_0)$. Then, for any $m_1 \in M_1$, the function $\nu \mapsto \Phi_{P_1, J}(\nu : m_1)$ is analytic on

$\mathcal{F}_{\mathbb{C}}(\nu_0, A|A_1)$.

roof. For any $\nu_1 \in \mathcal{F}_{\mathbb{C}}(\nu_0, A|A_1)$ it follows easily from Lemma 1.4.4 that there xists $\alpha \in C_c^\infty(A_1)$ such that $\hat{\alpha}(\chi_{\omega,\nu_1} \circ s) \neq \hat{\alpha}(\chi_{\omega,\nu_1} \circ t)$ $(t \not\sim_{\nu_0} s)$. Write $_t(\nu) = \hat{\alpha}(\chi_{\omega,\nu} \circ t)$ for $t \in W(A|A_1)$ and $\nu \in \mathcal{F}_{\mathbb{C}}$. Then, for any t, the unction β_t is analytic on $\mathcal{F}_{\mathbb{C}}$, so there exists a neighborhood U of ν_1, uch that, if $\nu \in U$ and $t \not\sim_{\nu_0} s$, then $\beta_t(\nu) \neq \beta_s(\nu)$.

We now apply Lemma 5.3.3.1 and Corollary 5.3.3.2 in the following orm. Let $r = d[W(A|A_1)]$. Partition $\Lambda_1, \ldots, \Lambda_r$ into $[W(A|A_1)]$ subsets, each et consisting of d variables and each set indexed by an element of $W(A|A_1)$. We hus have a mapping $i \mapsto s(i)$ from $\{1, \ldots, r\} \to W(A|A_1)$. Now map $\{\Lambda_1, \ldots, \Lambda_r\} \to \{\mu_s | s \in W(A|A_1)\} \subset \mathbb{C}^{[W(A|A_1)]}$ by sending $\Lambda_i \mapsto \mu_{s(i)}$. We then ave

$$1 = \sum_{J \in \{J(s)\}} h_J(X : \{\mu_t | t \in W(A|A_1)\}) \prod_{t \notin J} (X - \mu_t)^d$$

rovided $\mu_s \neq \mu_t$, $t \not\sim_{\nu_0} s$. The functions $h_J(X : \{\mu_t\})$ are analytic functions of $(\mu_t)_{t \in W(A|A_1)}$ on their domains of definition. In this formula replace X by (α) and μ_t by $\beta_t(\nu)$ and apply both sides to $\Phi_{P_1}(\nu)$, $\nu \in U$. We obtain

$$\Phi_{P_1}(\nu) = \sum_{J \in \{J(s)\}} \Phi_{P_1, J}(\nu) = \sum_{J \in \{J(s)\}} \Psi_J(\nu), \text{ where}$$

$\Psi_J(\nu) = h_J(\rho(\alpha) : \{\beta_t(\nu) | t \in W(A|A_1)\}) \prod_{t \notin J} (\rho(\alpha) - \beta_t(\nu))^d \Phi_{P_1}(\nu)$. To check that $\Phi_{P_1, J}(\nu) = \Psi_J(\nu)$ for $\nu \in U$ it suffices to check that $\prod_{s \in J} (\rho(\alpha) - \beta_s(\nu))^d \Psi_J(\nu) = 0$. But this is formally clear from the definition of $\Psi_J(\nu)$.

To show that $\Phi_{P_1, J}(\nu : m_1)$ is holomorphic on $\mathcal{F}_{\mathbb{C}}(\nu_0, A|A_1)$ it suffices to show that $\Psi_J(\nu : m_1)$ is holomorphic on U. For this choose $t \geq 1$ such that, for $m \in m_1 A_1^+(t)$ and all $\nu \in \mathfrak{F}_{\mathbb{C}}$, $\Phi_{P_1}(\nu : m) = \Phi_{(P_1)}(\nu : m)$ (Lemma 5.1.5). Then, obviously, as a function of ν, $\Phi_{P_1}(\nu : m)$ is holomorphic. It follows that $\Phi_{P_1, J}(\nu : m) = \Psi_J(\nu : m)$ is holomorphic on U. Let T be an indeterminate and write

$$\prod_{s \in J} (1-\chi_{\omega,\nu}(a^s)^{-1}T)^d = 1 + b_1(\nu)T + \dots + b_{[J]d}(\nu)T^{[J]d}$$

$$= 1 - TF(\nu : T);$$

it is obvious that the functions $b_j(\nu)$ $(1 \leq j \leq [J]d)$ and, hence, $F(\nu : T)$ are holomorphic for $\nu \in \mathfrak{a}_{\mathbb{C}}^*$. We may write

$$(1-\chi_{\omega,\nu}(a^s)^{-1}\rho(a))^d = (1-\rho(a)F(\nu : \rho(a))).$$

Applying this to $\Phi_{P_1, J}(\nu : m)$ $(\nu \in U)$, we obtain

$0 = \Phi_{P_1, J}(\nu : m)-\rho(a)F(\nu : \rho(a))\Phi_{P_1, J}(\nu : m)$ or $\rho(a^{-1})\Phi_{P_1, J}(\nu : m) = F(\nu : \rho(a))\Phi_{P_1, J}(\nu : m)$. Evaluating the last equation at $m = m_1 a$ with $a \in A_1^+(t)$, we see that $\nu \mapsto \Phi_{P_1, J}(\nu : m_1)$ is analytic on U, hence, on all of $\mathfrak{F}_{\mathbb{C}}(\nu_0, A|A_1)$.

Corollary 5.3.3.4. Assume $\omega \in {}^0\mathcal{E}(M)$. Then $\Phi_{P_1}(\nu)$ is analytic on $\mathfrak{F}_{\mathbb{C}}$.

Proof. This is now obvious. Given $\nu_0 \in \mathfrak{F}_{\mathbb{C}}$, we may express Φ_{P_1} as a sum of functions, each of which is analytic at ν_0.

For the following two corollaries we revise the notation. Let the class ω (cf. §5.3.2) now be in ${}^0\mathcal{E}(M)$. Define $\Phi(\nu : x) = E_F \pi_\nu(x) E_F$ $(\nu \in \mathfrak{a}_{\mathbb{C}}^*;\ x \in G)$ as before. Let $(P_2, A_2) \succ (P, A)$ $(P_2 = M_2 N_2)$ and assume that $\sigma_2 \in \omega_2 \in \mathcal{E}_2(M_2)$. Also assume that $\sigma_2 \subset \mathrm{Ind}_{{}^*P}^{M_2}(\delta_{{}^*P}^{\frac{1}{2}} \sigma_{\nu_0})$ $(\nu_0 \in \mathfrak{a}_{\mathbb{C}}^*)$, where $({}^*P, A) =$

$(P \cap M_2, A) \prec (M_2, A_2)$. Regard $\mathfrak{a}^*_{2,\mathbb{C}}$ as a subspace of $\mathfrak{a}^*_{\mathbb{C}}$. It follows from Lemma 1.7.1 that $\sigma_{2,\nu} \subset \mathrm{Ind}^{M_2}_{{}^*P}(\delta^{\frac{1}{2}}_{{}^*P}\sigma_{\nu_0+\nu})$ for all $\nu \in \mathfrak{a}^*_{2,\mathbb{C}}$. Set $\pi_{2,\nu} = \mathrm{Ind}^G_{P_2}(\delta^{\frac{1}{2}}_{P_2}\sigma_{2,\nu})$ $(\nu \in \mathfrak{a}^*_{2,\mathbb{C}})$. Then $\pi_{2,\nu}$ is a subrepresentation of $\pi_{\nu_0+\nu}$, and it is easy to see that $\pi_{2,\nu}$ acts in a subspace $\mathcal{H}_2$ of $\mathcal{H}$ which does not depend upon ν. Define $\Phi_2(\nu : x) = E_F \pi_{2,\nu}(x) E_F$. Since $\mathcal{H}_2$ does not depend upon ν, we may also write $\Phi_2(\nu : x) = E\Phi(\nu_0+\nu : x)E$, where the projection operator E is constant.

Let (P_1, A_1) be, as before, a semi-standard p-pair of G.

Corollary 5.3.3.5. There is an open set $O \subset \mathfrak{a}^*_{2,\mathbb{C}}$ which contains $\mathfrak{a}^*_2$ and on which the weak constant term ${}_w\Phi_{2,P_1}(\nu)$ is holomorphic.

Proof. Set $J(\nu) = \{s \in W(A|A_1| \chi_{\omega,\nu} \circ s \in \hat{A}_1\}$, $\nu \in \mathfrak{a}^*_{\mathbb{C}}$. Then $J(\nu_0) = J(\nu_0+\nu)$ for all $\nu \in \mathfrak{a}^*_2$. It follows that $\Phi_{P_1, J(\nu_0)}(\nu_0+\nu)$ is analytic as a function of ν near $\mathfrak{a}^*_2$ in $\mathfrak{a}^*_{\mathbb{C}}$; therefore, the same is true of $\Phi_{2,P_1,J(\nu_0)}(\nu)$. On the other hand, ${}_w\Phi_{2,P_1}(\nu) = \Phi_{2,P_1,J(\nu_0+\nu)}(\nu) = \Phi_{2,P_1,J(\nu_0)}(\nu)$ for all $\nu \in \mathfrak{a}^*_2$. The corollary follows.

Corollary 5.3.3.6. Assume that $\omega_2 \in \mathcal{E}_2(M_2)$ is in general position. Then ${}_w\Phi_{2,P_2,s}(\nu)$ is analytic near $\nu = 0$ in $\mathfrak{a}^*_{2,\mathbb{C}}$ for any $s \in W(A_2)$.

Proof. Let χ be the element of $\hat{A}_2$ such that $\chi = \chi_{\omega_2,0} \circ s$. Let $J = \{t \in W(A|A_2)| \chi = \chi_{\omega_{\nu_0}} \circ t\}$. Then both $\Phi_{P_2,J}(\nu_0+\nu)$ and $\Phi_{2,P_2,J}(\nu)$ are holomorphic near $\nu = 0$ in $\mathfrak{a}^*_{\mathbb{C}}$. Since ω_2 is in general position, $\Phi_{2,P_2,J}(\nu) = {}_w\Phi_{2,P_2,s}(\nu)$ in some neighborhood of $\nu = 0$ in $\mathfrak{a}^*_{2,\mathbb{C}}$. This implies the corollary.

Now let $\omega_2 \in \mathcal{E}_2(M_2)$ and regard, as we may, ${}_w\Phi_{2,P_1}(\nu)$ as a function which is analytic on an open set O, $\mathfrak{a}_2^* \subset O \subset \mathfrak{a}_{2,\mathbb{C}}^*$. To show that Theorem 5.3.2.1 is valid with $d = 1$ we may show that there is an h-dense subset of O on which the function

$$\Psi(\nu) = \prod_{s \in W(A_2 | A_1)} (\rho(\alpha) - \hat{\alpha}(\chi_{\omega_2,\nu} \circ s))_w(\Phi_2)_{P_1}(\nu) = 0.$$

We may choose our h-dense subset in $\mathcal{F}'_{\mathbb{C}}(A_2 | A_1)$ and show that $(\rho(\alpha) - \hat{\alpha}(\chi_{\omega_2,\nu} \circ s))(\Phi_2)_{P_1,s}(\nu) = 0$ for all $s \in W(A_2 | A_1)$. In fact, it suffices to show this for weakly minimal exponents $\chi_{\omega_2,\nu} \circ s$, since otherwise we obtain weakly critical exponents and contradict Bruhat theory (§2.5). Finally, we observe that, in fact, it suffices to take $(P_1, A_1) = (P_2, A_2)$ and $s = 1$. This reduces the proof that $d = 1$ to the calculation of Theorem 5.3.4.1, where in fact we show that $d = 1$ in an h-dense subset of $\mathcal{F}'_{\mathbb{C}}(A | A_1)$.

§5.3.4. An Integral Formula for $E_{P,1}(P : \psi : \nu)$.

Let (V, τ) be a smooth unitary double representation of K. Let (P, A) $(P = MN)$ be a semi-standard p-pair of G. Let $\sigma \in \omega \in \mathcal{E}_2(M)$ and $\nu \in \mathfrak{a}_{\mathbb{C}}^*$. Let $\psi \in L(\omega, P)$ and extend ψ to G by setting $\psi(kmn) = \tau(k)\psi(m)$ $(k \in K$, $m \in M$, and $n \in N)$. Set $\psi_\nu(x) = \psi(x) q^{\langle \sqrt{-1}\nu, H_P(x) \rangle}$. Then the Eisenstein integral $E(P : \psi : \nu) \in \mathcal{A}(\pi_\nu, \tau)$, where $\pi_\nu = \operatorname{Ind}_P^G(\delta_P^{\frac{1}{2}} \sigma_\nu)$, σ_ν as in §5.3.1.

We know (§5.3.3) that there is an open set $O \subset \mathfrak{a}_{\mathbb{C}}^*$ which contains $\mathfrak{a}^*$ and on which ${}_wE'_P(P : \psi : \nu)$ is defined as an analytic function of ν. Moreover, if ω_ν is in general position, then ${}_wE_P(P : \psi : \nu) = \sum_{s \in W(A)} E_{P,s}(P : \psi : \nu)$, where $E_{P,s}$ denotes the $\chi_{\omega,\nu} \circ s$-projection of ${}_wE_P$. The left hand side and each of the terms on the right are analytic functions of ν near any $\nu_0 \in U$ such that ω_{ν_0} is in general position. In this subsection we shall derive an integral formula for $E_{P,1}$. It will be evident from this formula that $\chi_{\omega_\nu} \circ 1 = \chi_{\omega_\nu}$ is a simple exponent for π_ν whenever ω_ν is in general position and whenever the integral formula is valid. The formula will be valid on an open h-dense subset, denoted $\mathfrak{F}_{\mathbb{C}}(P)$ or $\mathfrak{F}_{\mathbb{C}}(\overline{P})$, of $\mathfrak{F}_{\mathbb{C}}$, which open set we shall now make explicit.

To define $\mathfrak{F}_{\mathbb{C}}(P)$ recall that $\mathfrak{a}$ is equipped with a positive definite quadratic form which permits us to identify $\mathfrak{a}$ and $\mathfrak{a}^*$. We may regard the set of roots $\Sigma(P,A)$ as a subset of $\mathfrak{a}$. We let $\mathfrak{a}^{*+} = \{\nu \in \mathfrak{a}^* | \langle\nu,\alpha\rangle > 0$ for all $\alpha \in \Sigma(P,A)\}$. Let $\mathfrak{F}_{\mathbb{C}}(P) \subset \mathfrak{F}_{\mathbb{C}}$ be the image of $\mathfrak{a}^* + \sqrt{-1}\,\mathfrak{a}^{*+}$ under the projection of $\mathfrak{a}^*_{\mathbb{C}}$ on $\mathfrak{F}_{\mathbb{C}}$.

In the following theorem we shall need the functions κ, μ, η introduced in §0.6. Since multiplication injects $\overline{N} \times P$ into a dense open subset of G, the product measure $d\overline{n}\,d_rp$ on $\overline{N} \times P$ obviously induces a Haar measure on G. Given $\varphi \in C_c^\infty(G)$, we have $\int_G \varphi(x)dx = \int_{\overline{N}\cdot P} \varphi(\overline{n}p)d\overline{n}d_rp =$
$\int_{\overline{N}\cdot P} \varphi(\kappa(\overline{n})\mu(\overline{n})\eta(\overline{n})p)\delta_P(p)d\overline{n}\,d_\ell p = \int_{\overline{N}\cdot P} \varphi(\kappa(\overline{n})p)\delta_P(\mu(\overline{n}))^{-1}d\overline{n}\,d_rp$. We also observe that, if $\int_K dk = \int_{\overline{N}\cap K} d\overline{n} = \int_{P\cap K} dp = 1$, then the Haar measures $dk\,d_rp$ and $\gamma^{-1}d\overline{n}d_rp$ on G are equal, where the constant $\gamma = \gamma(G/P) = \int_{\overline{N}} \delta_P(\overline{n})^{-1}d\overline{n}$ is defined with respect to the above normalization of the Haar measure on $\overline{N}$ (cf. §1.2.2).

<u>Theorem</u> 5.3.4.1. Let $\nu \in \mathfrak{F}_{\mathbb{C}}(\overline{P}) = -\mathfrak{F}_{\mathbb{C}}(P)$ and $\psi \in L(\omega, P)$. Then

$$E_{P,1}(P:\psi:\nu:m) = \gamma^{-1}\int_{\overline{N}} \psi_\nu(m\mu(\overline{n})^{-1})\tau(\kappa(\overline{n}))^{-1}\delta_P^{-\frac{1}{2}}(\overline{n})d\overline{n}.$$

Proof. The preliminary remarks imply that

$$E(P:\psi:\nu:x) = \gamma^{-1}\int_{\bar{N}} \psi_\nu(x\kappa(\bar{n}))\tau(\kappa(\bar{n}))^{-1}\delta_P^{-\frac{1}{2}}(x\kappa(\bar{n}))\delta_{\bar{P}}^{-1}(\bar{n})d\bar{n}.$$

Since $ma\kappa(\bar{n}) \in \bar{n}^{ma} m\mu(\bar{n})^{-1}aN$, it follows easily that $\gamma\chi_{\omega,\nu}(a^{-1})E_{(P)}(P:\psi:\nu:ma)$ $= \int_{\bar{N}} \psi_\nu(\bar{n}^{ma} m\mu(\bar{n})^{-1})\,\tau(\kappa(\bar{n}))^{-1}\delta_P^{-\frac{1}{2}}(\bar{n}^{ma})\delta_P^{-\frac{1}{2}}(\bar{n})d\bar{n}$.

It is obvious that we may choose an open subset $O_1 \subset O$ in which $\lim_{a \to_P \infty}(E_P - {}_wE_P)(P:\nu:ma)$ exists and is zero for any $m \in M$ and $\nu \in O_1$. Moreover, for any $\nu \in \mathcal{F}_{\mathbb{C}}(\bar{P}) \cap O_1$, $\nu = \nu_{\mathbb{R}} + \sqrt{-1}\nu_I$ ($\nu_{\mathbb{R}} \in \mathcal{F}$, $\nu_I \in \mathfrak{a}^{*+}$) there exists $\epsilon = \epsilon(\nu) > 0$ such that $\langle \nu_I, \alpha \rangle \geq \epsilon \langle \rho, \alpha \rangle$ for all $\alpha \in \Sigma(\bar{P}, A)$.

Given $\nu \in O_1$ and $a \in A^+(t)$, $t >> 1$, we have

$$\chi_{\omega,\nu}(a^{-1})E_{(P)}(P:\psi:\nu:ma) = \sum_{s \in W(A)} E_{P,s}(P:\psi:\nu:ma)\chi_{\omega,\nu}(a^{-1})$$
$$+ (E_P - {}_wE_P)(P:\psi:\nu:ma)\chi_{\omega,\nu}(a^{-1}).$$

Under our assumptions on O_1 the last term disappears as $a \to_P \infty$ and, for $s \neq 1$, $\chi_{\omega,\nu}^{-1}(a)\chi_{\omega,\nu}(a^s) \to 0$, $a \to_P \infty$. We may conclude that, if $\lim_{a \to_P \infty}\chi_{\omega,\nu}(a^{-1})E_P(P:\psi:\nu:ma)$ exists, then its value can only be $E_{P,1}(P:\psi:\nu:m)$ $(\nu \in O_1)$. On the other hand, since we may extend $E_{P,1}$ as an analytic function, once we show that an integral formula represents $E_{P,1}$ on $O_1 \cap \mathcal{F}_{\mathbb{C}}(\bar{P})$, we may conclude immediately that the extended function $E_{P,1}$ is analytic on the full domain of convergence of the integral formula, i.e., as we shall see, $\mathcal{F}_{\mathbb{C}}(\bar{P})$.

Using Theorem 4.4.4, we see that, for any $r > 0$, there exists $C_r > 0$ such that

$$\|\psi_\nu(\bar{n}^{ma} m\mu(\bar{n})^{-1})\tau(\kappa(\bar{n}))^{-1}\delta_P^{-\frac{1}{2}}(\bar{n}^{ma})\delta_P^{-\frac{1}{2}}(\bar{n})\|_V$$
$$\leq C_r\, \Xi_M(x)(1+\sigma_M^*(x))^{-r} q^{-(1-\epsilon)\langle\rho, H_P(\bar{n}^{ma})\rangle - (1+\epsilon)\langle\rho, H_P(\bar{n})\rangle},$$

here $x = \bar{n}^{ma} m\mu(\bar{n})^{-1}$ and $\nu \in \mathfrak{F}_{\mathbb{C}}(\bar{P})$. *It now obviously follows from Theorem* 3.16 *that*

$$\lim_{a \to \infty, P} \int_{\bar{N}} \psi_\nu(\bar{n}^{ma} m\mu(\bar{n})^{-1}) \tau(\kappa(\bar{n}))^{-1} \delta_P^{-\frac{1}{2}}(\bar{n}^{ma}) \delta_P^{-\frac{1}{2}}(\bar{n}) d\bar{n}$$

xists $(\nu \in \mathfrak{F}_{\mathbb{C}}(\bar{P}))$.

Let K_0 be an open compact subgroup of G such that $\psi_\nu \in C(G/\!\!/K_0 : V)$ d let $C_1(a) = \{\bar{n} \in \bar{N} \mid \bar{n}^{ma} \in K_0\}$. Then

$$\left\| \int_{\bar{N} - C_1(a)} \psi_\nu(\bar{n}^{ma} m\mu(\bar{n})^{-1}) \tau(\kappa(\bar{n}))^{-1} \delta_P^{-\frac{1}{2}}(\bar{n}^{ma}) \delta_P^{-\frac{1}{2}}(\bar{n}) d\bar{n} \right\|_V$$

$$\leq \int_{\bar{N} - C_1(a)} \frac{C_r \, \Xi_M(\bar{n}^{ma} m\mu(\bar{n})^{-1})}{(1+\sigma^*_M(\bar{n}^{ma} m\mu(\bar{n})^{-1}))^r} q^{-(1-\epsilon)\langle\rho, H_P(\bar{n}^{ma})\rangle - (1+\epsilon)\langle\rho, H_P(\bar{n})\rangle} d\bar{n}.$$

he right hand side is arbitrarily small provided t is chosen large and $a \in A^+(t)$. he theorem follows immediately.

3.5. Analytic Functional Equations for the Eisenstein Integral.

Fix $\omega \in \mathcal{E}_2(M)$. Then, for all ν in an open subset of $\mathfrak{F}'_{\mathbb{C}}(\omega)$, ω_ν is mple (§§5.3.2-4). Since on an h-dense subset of this open set the functions $_{P_2,s}(P_1 : \psi : \nu)$ $(P_1, P_2 \in \mathcal{P}(A), \psi \in \mathcal{A}(\omega, \tau_M))$ are analytic functions of ν, we ay interpret the formal functional equations of §5.2.4 as analytic relations be- een the analytic function $_w E_{P_2}$ and its components. In the case $\omega \in {}^0\mathcal{E}(M)$ already know, essentially, that these relations are analytic on $\mathfrak{F}'_{\mathbb{C}}$. In neral, we establish in the succeeding section (§5.4) that these relations are eromorphic on $\mathfrak{F}_{\mathbb{C}}$, though not, so far as we presently know, necessarily analytic on all of $\mathfrak{F}'_{\mathbb{C}}$ (for $\omega \in \mathcal{E}_2(M) - {}^0\mathcal{E}(M)$). We shall state the functional equations in as general a form as we know them in this subsection, leaving the complete proofs until later.

Let (V, τ) be a smooth unitary double representation of K. For $\psi \in \mathcal{A}(\omega, \tau_M)$ regarded as a function on G (cf. §5.2.2) we have

$$E(P_1 : \psi : \nu : x) = E(P_1 : \psi_\nu : x)$$

$$= \int_K (F_{P_1}(\tau)\psi)(xk)\tau(k^{-1}) q^{<\sqrt{-1}\nu - \rho_{P_1}, H_{P_1}(xk)>} dk.$$

We know that $E(P_1 : \psi : \nu) \in \mathcal{A}(\pi_\nu, \tau)$ for any $\nu \in \mathfrak{F}_{\mathbb{C}}$ $(\pi_\nu = \mathrm{Ind}_{P_1}^G(\delta_{P_1}^{\frac{1}{2}}\sigma_\nu)$, $\sigma_\nu \in \omega_\nu \in \mathcal{E}_{2,\mathbb{C}}(M))$.

Theorem 5.3.5.1. Let $P_1, P_2 \in \mathcal{P}(A)$. Then for any ν in an open subset of $\mathfrak{F}'_{\mathbb{C}}$ which contains $\mathfrak{F}'$ and $s \in W(G/A)$ there exists a unique linear mapping $c_{P_2|P_1}(s : \omega : \nu) : \mathcal{A}(\omega, \tau_M) \to \mathcal{A}(\omega^s, \tau_M)$ such that ${}_wE_{P_2}(P_1 : \psi : \nu : m) = \sum_{s \in W(G/A)} (c_{P_2|P_1}(s : \omega : \nu)\psi)(m) q^{<\sqrt{-1}s\nu, H_P(m)>}$ for any $\psi \in \mathcal{A}(\omega, \tau_M)$. The mapping $c_{P_2|P_1}(s : \omega : \nu)$ factors: $\mathcal{A}(\omega, \tau_M) \xrightarrow{\pi} L(\omega, P_1) \to \mathcal{L}(\omega^s, P_2) \xrightarrow{\iota} \mathcal{A}(\omega^s, \tau_M)$, with π and ι being, respectively, the natural projection and injection. As a function of ν, $c_{P_2|P_1}(s : \omega : \nu)$ is meromorphic on $\mathfrak{F}_{\mathbb{C}}$.

Proof. For ν in an open subset of $\mathfrak{F}'_{\mathbb{C}}$ we have ${}_wE_{P_2}(P_1 : \psi : \nu : m) =$

$$\sum_{s \in W(G/A)} E_{P_2, s}(P_1 : \psi : \nu : m) = \sum_{s \in W(G/A)} (c_{P_2|P_1}(s : \omega : \nu)\psi)(m) q^{<\sqrt{-1}s\nu, H(m)>}$$

$(m \in M)$. The projections to the components of $\oplus \mathcal{A}(\omega_\nu{}^s, \tau_M)$ are unique for

$\nu \in \mathcal{F}'_{\mathbb{C}}$ intersected with the domain of existence of ${}_wE_{P_2}(P_1 : \psi : \nu)$, so the mappings $c_{P_2|P_1}(s : \omega : \nu)$ are uniquely defined. We know that the mappings factor as indicated. The meromorphicity of the c-functions will be proved in §5.4, specifically Theorem 5.4.2.1 and Corollary 5.4.3.2.

Recall that, if $\psi \in \mathcal{A}(\omega, \tau_M)$, then $\|\psi\|^2 = \int_{M/A} \|\psi(m)\|_V^2 dm^*$. With respect to this metric $\mathcal{A}(\omega, \tau_M)$ is a pre-hilbert space. It is easy to see (i.e., for every finite-dimensional subspace of V) that the adjoint $(c_{P_2|P_1}(s : \omega : \nu))^*$ is defined for any ν in the domain of definition of $c_{P_2|P_1}(s : \omega : \nu)$.

<u>Theorem</u> 5.3.5.2. There exists a complex-valued meromorphic function $\nu \mapsto \mu(\omega : \nu)$ on $\mathcal{F}_{\mathbb{C}}(\omega)$ such that $\mu(\omega : \nu)(c_{P_2|P_1}(s : \omega : \bar{\nu}))^* c_{P_2|P_1}(s : \omega : \nu)\psi = F_{P_1}(\tau)\psi F_{P_1}(\tau)$ for every $P_1, P_2 \in \mathcal{P}(A)$, $s \in W(G/A)$, and $\psi \in \mathcal{A}(\omega, \tau_M)$. Moreover, $\mu(\omega^s : s\nu) = \mu(\omega : \nu)$ and $\mu(\omega : \nu)$ is holomorphic and nonnegative on $\mathcal{F}(\omega)$.

<u>Proof.</u> Given that the c-functions are meromorphic on $\mathcal{F}_{\mathbb{C}}$, one obtains this theorem by analytic continuation from the corresponding statement for $\nu \in \mathcal{F}'$, which is a modification of Theorem 5.2.4.4.

Regard $c_{P_2|P_1}(s : \omega : \nu)$ as a mapping from $L(\omega, P_1)$ to $\mathcal{L}(\omega^s, P_2)$.

Define $${}^0c_{P_2|P_1}(s : \omega : \nu) = c_{P_2|P_2}(1 : \omega^s : s\nu)^{-1} c_{P_2|P_1}(s : \omega : \nu)$$

and $$c^0_{P_2|P_1}(s : \omega : \nu) = c_{P_2|P_1}(s : \omega : \nu) c_{P_1|P_1}(1 : \omega : \nu)^{-1}$$

as meromorphic functions on $\mathcal{F}_{\mathbb{C}}$. Corollary 5.2.4.5 implies that these operators are holomorphic and unitary for $\nu \in \mathcal{F}'$. Set

$$E^0(P : \psi : \nu) = E(P : c_{P|P}(1 : \omega : \nu)^{-1}\psi : \nu)$$

($\psi \in \mathcal{L}(\omega, P)$, $P \in \mathcal{P}(A)$, ν in an open subset of $\mathfrak{F}'_{\mathbb{C}}$). Then $E^0(P : \psi : \nu)$ is holomorphic in ν on $\mathfrak{F}$. By composing with the obvious projections and injections we sometimes regard the functions ${}^0c_{P_2|P_1}(s : \omega : \nu)$ and $c^0_{P_2|P_1}(s : \omega : \nu)$ as mappings from $\mathcal{A}(\omega, \tau_M)$ to $\mathcal{A}(\omega^s, \tau_M)$.

All the functional equations proved in §5.2.4 have obvious analogues in the present context. For completeness we give the statements. The following functional equations hold for ν in an open subset of $\mathfrak{F}'_{\mathbb{C}}$ which contains $\mathfrak{F}'$ [all of $\mathfrak{F}'_{\mathbb{C}}$, provided $\omega \in {}^0\mathcal{E}(M)$]. They may be regarded as meromorphic relations for $\nu \in \mathfrak{F}_{\mathbb{C}}$.

<u>Theorem</u> 5.3.5.3. Let $s, t \in W(G/A)$, $\psi \in \mathcal{A}(\omega, \tau_M)$, and $P_1, P_2, P_3 \in \mathcal{P}(A)$. Then:

(1) $E(P_2 : {}^0c_{P_2|P_1}(s : \omega : \nu)\psi : s\nu) = E(P_1 : \psi : \nu)$ and

$${}^0c_{P_3|P_1}(st : \omega : \nu) = {}^0c_{P_3|P_2}(s : \omega^t : t\nu)\,{}^0c_{P_2|P_1}(t : \omega : \nu).$$

(2) $E^0(P_2 : c^0_{P_2|P_1}(s : \omega : \nu)\psi : s\nu) = E^0(P_1 : \psi : \nu)$ and

$$c^0_{P_3|P_1}(st : \omega : \nu) = c^0_{P_3|P_2}(s : \omega^t : t\nu)\,c^0_{P_2|P_1}(t : \omega : \nu).$$

(3) (For $\varphi \in \mathcal{A}(M, \tau_M)$ we have $(s\varphi)(m) = \tau(y)\,\varphi(y^{-1}my)\tau(y^{-1})$ $(y = y(s) \in K)$.)

$$sc_{P_2|P_1}(t : \omega : \nu) = c_{P_2^s|P_1}(st : \omega : \nu) \text{ and}$$

$$c_{P_2|P_1}(t : \omega : \nu)s^{-1} = c_{P_2|P_1^s}(ts^{-1} : \omega^s : s\nu).$$

The same relations hold for ${}^0c$ and c^0.

We recall some notation from §5.2.4. Let A' be a standard torus of G with $M' = Z_G(A')$. For every $s \in W(G/A_0)$ we assume representatives $y = y(s) \in K$. For any $s \in W(G/A_0)$ we write $A_s = A^{s^{-1}}$ and $M_s = Z_G(A_s)$. There is a mapping $\mathcal{A}(M, \tau_M) \to \mathcal{A}(M^t, \tau_{M^t})$ for any $t \in W(G/A_0)$ defined by setting $(t\psi)(m^{y(t)}) = \tau(y)\psi(m)\tau(y^{-1})$ for $\psi \in \mathcal{A}(\omega, \tau_M)$ $(m \in M)$.

We choose representatives for $W(A|A')$ in $W(G/A_0)$. For any $\gamma \in W(A|A')$ we set $W_\gamma = W(G/A_\gamma)$ and ${}^*W_\gamma = W(M'/A_\gamma)$. Then $W(A|A') = \coprod_{\gamma \in \Gamma} y(\gamma)W_\gamma/{}^*W_\gamma$, where Γ indexes M'-conjugacy class representatives for $A^G \cap M'$.

Let $P' = M'N' \in \mathcal{P}(A')$. Fix $\gamma \in \Gamma$. Assume that $(P', A') \succ (Q_i, A_\gamma)$, $i = 1, 2$, and set $(Q_i \cap M', A_\gamma) = ({}^*Q_i, A_\gamma)$. Let $P_1^{\gamma^{-1}} = Q_1$.

Continuing with Theorem 5.3.4.3:

(4) For any $u \in {}^*W_\gamma$, $E(P' : E({}^*Q_2 : {}^0c_{{}^*Q_2|{}^*Q_1}(u : \omega_\gamma : \gamma^{-1}\nu)\gamma^{-1}\psi : u\gamma^{-1}\nu) = E(P_1 : \psi : \nu)$; in particular, ${}^0c_{Q_2|Q_1}(u : \omega_\gamma : \gamma^{-1}\nu) = {}^0c_{{}^*Q_2|{}^*Q_1}(u : \omega_\gamma : \gamma^{-1}\nu)|L(\omega_\gamma, Q_1)$.

(5) On a dense subset of $\mathcal{F}_{\mathbb{C}}$, $E^0_{P'}(P_1 : \psi : \nu)$ is the sum of the functions $E^0({}^*Q_2 : c^0_{{}^*Q_2|{}^*Q_1}(t : \omega_\gamma : \gamma^{-1}\nu)\gamma^{-1}\psi : t\gamma^{-1}\nu)$ over $\gamma \in \Gamma$ and $t \in {}^*W_\gamma \backslash W_\gamma$. In particular, for all $t \in {}^*W_\gamma$,

$$c^0_{Q_2|Q_1}(t : \omega_\gamma : \gamma^{-1}\nu) = c^0_{{}^*Q_2|{}^*Q_1}(t : \omega_\gamma : \gamma^{-1}\nu)|\mathcal{L}(\omega_\gamma, Q_1).$$

We close this subsection with integral formulas for the c-functions. In view of Theorem 5.3.5.3 (3), it suffices to write down these formulas for $c_{P_2|P_1}(1 : \omega : \nu)$ $(P_1, P_2 \in \mathcal{P}(A))$.

Fix P_1 and P_2 and let $\Phi_i = \Sigma_r(P_i, A)$ $(i = 1,2)$. Set $\Phi_{1,2}^- = \Phi_1 \cap (-\Phi_2)$ and $\Phi_{1,2}^+ = \Phi_1 \cap \Phi_2$. Let $\bar{N}^- = \bar{N}_1 \cap N_2$ and $\bar{N}^+ = \bar{N}_1 \cap \bar{N}_2$. Then $\bar{N}^- = \prod_{\alpha \in \Phi_{1,2}^-} \bar{N}_\alpha$ and $\bar{N}^+ = \prod_{\alpha \in \Phi_{1,2}^+} \bar{N}_\alpha$.

Theorem 5.3.5.4. For $\psi \in \mathcal{A}(\omega, \tau_M)$ and $\nu \in \mathfrak{F}_{\mathbb{C}}(\bar{P}_2)$ set

$$(J_{P_2|P_1}(\nu)\psi)(m) = \int_{\bar{N}^-\bar{N}^+} \tau(\kappa_1(\bar{n}^-))\psi(\mu_1(\bar{n}^-)m\mu_1(\bar{n}^+)^{-1})\tau(\kappa_1(\bar{n}^+))^{-1} \cdot q^{\langle\sqrt{-1}\nu, H_{P_1}(\bar{n}^-)-H_{P_1}(\bar{n}^+)\rangle} \cdot q^{-\langle\rho_{P_1}, H_{P_1}(\bar{n}^-)+H_{P_1}(\bar{n}^+)\rangle} d\bar{n}^- d\bar{n}^+.$$

Then $J_{P_2|P_1}(\nu)\psi$ is uniformly convergent on compact subsets of $\mathfrak{F}_{\mathbb{C}}(\bar{P}_2)$ and

$$J_{P_2|P_1}(\nu)\psi = \begin{cases} \gamma(G/P_1)c_{P_2|P_1}(1 : \omega : \nu)\psi & (\psi \in L(\omega, P_1)) \\ \gamma(G/P_1)(c_{\bar{P}_2|\bar{P}_1}(1 : \omega : \bar{\nu}))^*\psi & (\psi \in \mathcal{L}(\omega, \bar{P}_2)). \end{cases}$$

Sketch of the proof. In the proof of Theorem 5.3.4.1 we have seen that, for $\psi \in L(\omega, P_1)$, we have $\gamma(G/P_1)\chi_{\omega,\nu}(a^{-1})E(P_1 : \psi : \nu : ma)$

$$= \delta_{P_1}^{-\frac{1}{2}}(ma)\int_{\bar{N}_1} \psi_\nu(\bar{n}^{ma} m\mu(\bar{n})^{-1})\tau(\kappa(\bar{n}))^{-1}\delta_{P_1}^{-\frac{1}{2}}(\bar{n}^{ma})\delta_{P_1}^{-\frac{1}{2}}(\bar{n})d\bar{n}.$$

Write $\bar{n} = \bar{n}_+\bar{n}_-$, where $\bar{n}_+ \in \bar{N}^+$ and $\bar{n}_- \in \bar{N}^-$. Notice that $\delta_{P_2}^{\frac{1}{2}}(ma)\delta_{P_1}^{-\frac{1}{2}}(ma)d\bar{n}_- = d(\bar{n}_-^{ma})$. Thus, since $\bar{n}_+^{ma} \to 1$, $a \xrightarrow[P_2]{} \infty$, $\gamma(G/P_1)\chi_{\omega,\nu}(a^{-1})\delta_{P_2}^{\frac{1}{2}}(ma)E(P_1 : \psi : \nu : ma) \to$

$$_{1}\tau(\kappa(\bar{n}_-)\psi(\mu(\bar{n}_-)m\mu(\bar{n}_+)^{-1})\tau(\kappa(\bar{n}_+))^{-1}$$

$$q^{<\sqrt{-1}\nu, H_{P_1}(\bar{n}_-)-H_{P_1}(\bar{n}_+)> - <\rho_1, H_{P_1}(\bar{n}_+)+H_{P_1}(\bar{n}_-)>} d\bar{n}\chi_\nu(m),$$

$$a_{P_2} \to \infty.$$

ne convergence arguments go as in the proof of Theorem 5. 3. 4. 1. We delay the oof of the formulas for the adjoints until §5. 4. 3.

5. 4. Applications and Further Development of the Analytic Theory.

Let A be an A_0-standard torus of G, let $M = Z_G(A)$, and $\mathcal{P}(A)$ the t of A-associated parabolic subgroups of G. Let $W = W(G/A)$. Given $\epsilon\ {}^0\mathcal{E}_{\mathbb{C}}(M)$ or $\mathcal{E}_2(M)$, write $\pi_{P,\omega} = \mathrm{Ind}_P^G(\delta_P^{\frac{1}{2}}\sigma)$, $\sigma \in \omega$. If X is an admissible - or M-module, we write ${}_wX$ for the maximal tempered submodule.

5. 4. 1. Some Ideas of Casselman.

Casselman proved Theorem 5. 4. 4. 1 in the case that A is a maximal plit torus of G (see [5b]). In doing so he developed essentially all the results to e proved in this subsection.

By a composition series for an admissible G-module we mean a composition series for the corresponding $C_c^\infty(G)$-module.

heorem 5. 4. 1. 1. Let $\omega \in {}^0\mathcal{E}_{\mathbb{C}}(M)[\mathcal{E}_2(M)]$. Let π_{ω, P_1} act in the vector space $\mathcal{H}$. Let $\bar{\pi}$ denote the representation of M on $\mathcal{H}/\mathcal{H}(P_2)[{}_w(\mathcal{H}/\mathcal{H}(P_2))]$. Then $\bar{\pi}$ admissible (§2. 3). *Moreover, $\bar{\pi}$ has a finite composition series of length* $[W]$

and the character of $\bar{\pi}$ is

$$\theta_{\bar{\pi}} = \delta_{P_2}^{+\frac{1}{2}} \sum_{s \in W} \theta_{\omega^s},$$

where θ_{ω^s} denotes the character of ω^s.

Proof. Assume first that ω is unitary and in general position. In this case $\pi = \pi_{\omega, P_1}$ is irreducible and unitary and $C(\pi) = C(\pi_{\omega^s, P_2})$ for any $s \in W(A)$ and $P_2 \in \mathcal{P}(A)$. Moreover, since ${}_W\mathfrak{X}_\pi(\bar{P}_2, A) = \{\omega^s \mid s \in W(A)\}$, it follows that only the above class exponents occur in the composition series for $\bar{\pi}$. Since π is irreducible, $\delta_{P_2}^{+\frac{1}{2}} \omega^s$ occurs as a quotient exactly once in $\bar{\pi}$. Since exponents are simple, $\delta_{P_2}^{+\frac{1}{2}} \omega^s$ occurs in the composition series exactly once. The theorem is proved for the case ω in general position and unitary.

To prove the general case it is sufficient to show that the character $\theta_{\bar{\pi}_\nu}$ is an analytic function of $\nu \in \mathfrak{a}_{\mathbb{C}}^*$ [ν in an open set containing $\mathfrak{a}^*$]. For $\nu \in \mathfrak{a}_{\mathbb{C}}^*$ [an open subset] we have $\pi_\nu = \mathrm{Ind}_{P_1}^G (\delta_{P_1}^{\frac{1}{2}} \sigma_\nu)$ and $\bar{\pi}_\nu$ the representation of M on $\mathcal{H}_\nu / \mathcal{H}_\nu(P_2)[{}_W(\mathcal{H}_\nu/\mathcal{H}_\nu(P_2))]$. Let $\{K_j\}_{j=1}^\infty$ be a sequence of open compact subgroups of G as in Theorem 2.1.1. Fix an index j_0 and $\alpha \in C_c(M /\!/ M_{j_0})$ $(M_{j_0} = K_{j_0} \cap M)$. By Corollary 5.1.10 there exists an index j such that, for all ν,

$$\theta_{\bar{\pi}_\nu}(\alpha) = \int_{M_{j_0}} \alpha(m) \delta_{P_2}(m)^{+\frac{1}{2}} \phi_{j, \bar{P}_2}(\nu : m) dm,$$

where $\phi_{j, \bar{P}_2}(\nu : m)$ [i.e., ${}_W\phi_{j, \bar{P}_2}(\nu : m)$] is a certain analytic function of ν (cf. §5. Moreover, the integral on the right-hand side is analytic. Therefore, the left-hand side is analytic and the theorem is proved.

We present two more lemmas of Casselman and an interesting consequence.

Lemma 5.4.1.2. Let $P = MN$ be a parabolic subgroup of G. If $0 \to V_1 \to V_2 \to V_3 \to 0$ is an exact sequence of P-modules or N-modules, then so are both $0 \to V_1(N) \to V_2(N) \xrightarrow{\varphi} V_3(N) \to 0$ and $0 \to V_1/V_1(N) \to V_2/V_2(N) \to V_3/V_3(N) \to 0$.

Proof. That $V_1(N) = V_1 \cap V_2(N)$ is immediate from Lemma 2.2.1; the surjectivity of φ follows easily from the definition of $V(N)$. The exactness of the second sequence is even easier, so the proof is left to the reader.

Corollary 5.4.1.3. Let $0 \to V_1 \to V_2 \to V_3 \to 0$ be an exact sequence of tempered admissible G-modules. Then

$$0 \to {}_w(V_1/V_1(N)) \to {}_w(V_2/V_2(N)) \to {}_w(V_3/V_3(N)) \to 0$$

is an exact sequence of tempered M-modules.

Proof. Write $V_i/V_i(N) = \bigoplus_{\chi \in \mathfrak{X}(A)} V_i/V_i(N)(\chi)$ and use the fact that $V_i/V_i(N)(\chi)$ is tempered if and only if $\chi \in \hat{A}$ (Corollary 4.5.4).

Lemma 5.4.1.4. Let $0 \to W \to V \xrightarrow{\varphi} U \to 0$ be an exact sequence of [tempered] admissible G-modules. Assume that there exists $\chi \in \mathfrak{X}(Z)[\hat{Z}]$ such that $\pi(z)v = \chi(z)v$ for all $z \in Z$ and $v \in V$, where π denotes the representation of G on V. Assume that either W or U is irreducible and supercuspidal [of class $\mathcal{E}_2(G)$]. Then the sequence splits.

Proof. We prove only the unbracketed statement, as the proofs are obviously the same. Let σ be an irreducible and supercuspidal representation of G in the space U. Note that $\mathcal{A}(\sigma)$ is a convolution algebra on G/Z. Let K_0 be an open compact subgroup of G and let $\theta = \theta_{K_0}$ be the corresponding idempotent in $\mathcal{A}(\sigma)$. Assume that $\int_{G/Z}\theta(x)\sigma(x^{-1}) \cdot \bar{v}dx^* = \bar{v}$ for some $\bar{v} \in U-(0)$. Let $v \in \varphi^{-1}(\bar{v}) \subset V$. Then $v_1 = \int_{G/Z}\theta(x)\pi(x^{-1})vdx^* \in \varphi^{-1}(\bar{v})$, too. Set $\psi(\sigma(g)\bar{v}) = \pi(g)v_1$. Observe that $\psi : U \to V$ is a G-module mapping and that $\varphi \cdot \psi = 1_U$. The proof in the case that W is irreducible and supercuspidal goes similarly.

Lemma 5.4.1.5. Let $\sigma \in \omega \in {}^0\mathcal{E}_{\mathbb{C}}(M)[\mathcal{E}_2(M)]$ and let $\pi = \mathrm{Ind}_P^G(\delta_P^{\frac{1}{2}}\sigma)$, $P \in \mathcal{P}(A)$. Then, if $\pi_1 \prec \pi$, $C(\pi_1) \notin {}^0\mathcal{E}_{\mathbb{C}}(G)[\mathcal{E}_2(G)]$, provided $A \neq Z$.

Proof. Let π act in $\mathcal{H}$. Assume that $\pi_1 \prec \pi$, where π_1 is irreducible and $C(\pi_1) \in {}^0\mathcal{E}_{\mathbb{C}}(G)[\mathcal{E}_2(G)]$. Without loss of generality assume that π_1 occurs as a quotient in the finitely generated submodule $\mathcal{H}_0$ of $\mathcal{H}$. It then follows from Lemma 5.4.1.4 that $\pi_1 \subset \pi$. Let $\mathcal{H}_1 \subset \mathcal{H}$ be the representation space of π. Then $\Delta|\mathcal{H}_1$ intertwines $\pi_1|P$ with $\delta_P^{\frac{1}{2}}\sigma$, $\Delta\beta = \beta(1)$ $(\beta \in \mathcal{H})$, so π_1 has nontrivial [tempered] class exponents, so π_1 is not supercuspidal [square-integrable].

We may also argue via the Eisenstein integral. Let $f \in \mathcal{A}(\pi_1)$. Then $\underset{\sim}{f} = E(P : \psi)$ for some $\psi \in L(\omega, P)$. For any $\alpha \in {}^0C_c^\infty(G)[{}^0\mathcal{C}(G)]$ we have

$$(\underset{\sim}{\alpha}, E(P : \psi))_G = (\underset{\sim}{\alpha}^P, \psi)_M$$

$$= 0 .$$

In particular, taking $\alpha = j^{-1}(E(P : \psi))$, where the support of α is restricted to G_T (cf. Chapter 4 for definition), we obtain a contradiction.

Remark. In particular, no matrix coefficient of π can be a supercusp form [cusp form].

The following theorem was first proved by Roger Howe ([8b]), with indefinite restrictions on G.

Theorem 5.4.1.6. Let π be an admissible representation of G in a vector space V. If V is a G-module of finite type, then V has a finite composition series.

Proof. It suffices to consider the case in which V is a cyclic G-module. We may also obviously assume that there is an element $\chi \in \mathfrak{X}(Z)$ and $\ell \in \mathbb{N}$ such that $\prod_{i=1}^{\ell}(\chi(z_i)-\pi(z_i))v = 0$ for every $v \in V$ and tuple $(z_1, \ldots, z_\ell) \in Z^\ell$. Let $G \cdot v_0 = V$ and $Z \cdot v_0 = W$. Let $W_{\ell-1} = \{w \in W \mid \prod_{i=1}^{\ell-1}(\chi(z_i)-\pi(z_i))w = 0$ for any $(z_1, \ldots, z_{\ell-1}) \in Z^{\ell-1}\}$. As an induction hypothesis we assume that $G \cdot W_{\ell-1}$ has a finite composition series. We shall show that $\bar{V} = V/G \cdot W_{\ell-1}$ has a finite composition series. Clearly, $(\pi(z)-\chi(z))\bar{v} = 0$ for any $z \in Z$ and $\bar{v} \in \bar{V}$.

If $\sigma \in \omega \in {}^0\mathcal{E}_{\mathbb{C}}(G)$ and $\sigma \prec \pi$, then, according to Lemma 5.4.1.4, $\sigma \subset \pi$ and there is a projection $\varphi : V \to W$ where σ acts in W. Since $v_0 \in V_{K_0}$ for some open compact subgroup K_0 of G and φ is a G-module morphism, $\varphi(v_0) \in W_{K_0}$. Since $\dim V_{K_0} < \infty$ there can be only finitely many supercuspidal factors in π. Splitting them off, we may assume that V contains no supercuspidal subquotients.

Let $P_1, \ldots, P_\ell$ be all the standard maximal parabolic subgroups of G (with respect to some fixed (P_0, A_0)). Map $V \to \oplus \Sigma V/V(P_i)$, where $V/V(P_i)$ is an M_i-module ($P_i = M_iN_i$, $i = 1, \ldots, \ell$). Each quotient $V/V(P_i)$ is admissible and finitely generated. By an obvious induction hypothesis each module $V/V(P_i)$ has a finite composition series. Since V contains no supercuspidal subquotients it follows from Lemma 5.4.1.2 that $\ell(V) < \Sigma \ell(V/V(P_i))$. This completes the proof of Theorem 5.4.1.6.

For completeness and because of the importance in applications of these ideas, we close this section with the following two properties of the "Jacquet functor". The proofs are easy and left to the reader.

Assume that V is an admissible G-module and (P, A) $(P = MN)$ a semistandard p-pair.

Lemma 5.4.1.7. (Universal Mapping Property). Any P-module morphism of V to a P-module on which N acts trivially factors (uniquely) through $V/V(P)$.

Lemma 5.4.1.8. (Transitivity). Let $(P', A') \succ (P, A)$ $(P' = M'N')$ and set $({}^*P, A) = (P \cap M', A)$. Then $(V/V(P'))/(V/V(P'))({}^*P)$ is isomorphic to $V/V(P)$.

§5.4.2. Some Results for the Case p-rank Equal One.

Assume that $\dim A/Z = 1$. In this case $\mathcal{P}(A) = \{P, \overline{P}\}$, where $P = MN$ and $\overline{P} = M\overline{N}$ are opposite parabolic subgroups of G. The group $W = W(G/A)$ consists of either one or two elements.

Fix $\omega \in \mathcal{E}_2(M)$ with central exponent χ_ω and orbit $\mathcal{O}_{\mathbb{C}} = \omega \cdot \mathfrak{a}^*_{\mathbb{C}}$. Choose an element $\alpha \in \mathfrak{a}$ such that $<\nu, \alpha> = 0$ $(\nu \in \mathfrak{a}^*)$ if and only if $\nu \in \mathfrak{z}^* + L^*$. Set $z = z(\nu) = q^{\sqrt{-1}<\nu, \alpha>}$. The mapping $\nu \mapsto z$ sends $\mathfrak{a}^*_{\mathbb{C}}$ onto $\mathbb{C} - \{0\} = \mathbb{C}^\times$. The c-functions and their adjoints are defined on $\mathcal{F}_{\mathbb{C}}$; hence they may be regarded as functions of z. The function $\mu(\omega : \nu)$ factors through the projection of $\mathfrak{a}^*_{\mathbb{C}}$ on $\mathfrak{a}^*_{\mathbb{C}}/(L^*(\omega) + \mathfrak{z}^*_{\mathbb{C}})$.

Theorem 5.4.2.1. For any $s \in W$ and $P_1, P_2 \in \mathcal{P}(A)$ the functions $c_{P_2|P_1}(s : \omega : \nu)$ and $c_{P_2|P_1}(s : \omega : \overline{\nu})^*$ are meromorphic on $\mathcal{F}_{\mathbb{C}}$, analytic on $\mathcal{F}'$. Regarded as nctions of $z = z(\nu) \in \mathbb{C}^{\times}$, the c-functions have at most simple poles at $z_0 = z(\nu_0)$, $\in \mathfrak{F}_{\mathbb{C}} - \mathfrak{F}'_{\mathbb{C}}$, a finite set of points $z \in \mathbb{C}$. If $\omega \in {}^0\mathcal{E}(M)$, these are the ly singularities. In general, there can be at most finitely many poles in $_{\mathbb{C}}^{*}/(L^* + \mathfrak{z}_{\mathbb{C}}^{*})$. The functions $c_{P_2|P_1}(s : \omega : \nu)$ and $c_{\overline{P}_2|P_1}(s : \omega : \overline{\nu})^*$ are holomorphic the open half space $\mathfrak{F}_{\mathbb{C}}(\overline{P}_2^{s^{-1}})$. The function $\mu(\omega : \nu)$ is meromorphic on $\mathfrak{F}_{\mathbb{C}}(\omega)$.

roof. Theorems 5.3.5.3(3) and 5.3.5.4 (the integral formulas, etc.) imply olomorphicity in the indicated half-space. Also applying Theorem 5.3.5.2 and orollary 5.3.3.6, we see that the only singularities of the c-functions e outside $\mathfrak{F}'_{\mathbb{C}}$. In the supercuspidal case we know there can be no other singu- rities (§5.3.3). In general, the singularities lie in a lower dimensional set. hat the singularity set for the c-functions on $\mathfrak{a}_{\mathbb{C}}^{*}/(L^* + \mathfrak{z}_{\mathbb{C}}^{*})$ is at most a finite set llows from the fact that singularities can occur only along with multiple exponents r the constant term (not just the weak constant term!) and the occurrence of ultiple exponents depends upon hyperplane conditions which admit only finitely any solutions.

Let us show that, at $\nu_0 \in \mathfrak{F}_{\mathbb{C}} - \mathfrak{F}'_{\mathbb{C}}$, the c-functions, regarded as functions z, can have at most simple poles. Assume ω is not in general position. Then $V] = 2$. Choose $a \in A$ such that $\chi_{\omega,\nu}(a) \neq \chi_{\omega,\nu}(a^s)$ for ν near ν_0, $s \neq 1 \in W$. ithout loss of generality we may assume that $z(\nu_0) = 1$. We may take $\chi_{\omega,\nu}(a) =$ $\nu)^{\ell}$ for some $\ell \neq 0$ and $\chi_{\omega,\nu}(a^s) = z(\nu)^{-\ell}$. Write $\beta_t(\nu) = \chi_{\omega,\nu}(a^t)$ for $t \in W$. or any $\psi \in L(\omega, P_1)$ and ν near ν_0 we may write

$$_wE_{P_2}(P_1 : \psi : \nu) = \left[\frac{\rho(a)-\beta_s(\nu)}{\beta_1(\nu)-\beta_s(\nu)}\right]_w E_{P_2}(P_1 : \psi : \nu) + \left[\frac{\rho(a)-\beta_1(\nu)}{\beta_s(\nu)-\beta_1(\nu)}\right]_w E_{P_2}(P_1 : \psi : \nu)$$

$$= E_{P_2,1}(P_1 : \psi : \nu) + E_{P_2,s}(P_1 : \psi : \nu)$$

$$= c_{P_2|P_1}(1 : \omega : \nu)\psi\chi_{\omega,\nu} + c_{P_2|P_1}(s : \omega : \nu)\psi\chi_{\omega,s\nu}.$$

Rewriting this in terms of $z = z(\nu)$, we have

$${}_wE_{P_2}(P_1 : \psi : z(\nu)) = [\frac{\rho(a)-z^{-\ell}}{z^{\ell}-z^{-\ell}}]_wE_{P_2}(P_1 : \psi : z) + [\frac{\rho(a)-z^{\ell}}{z^{-\ell}-z^{\ell}}]_wE_{P_2}(P_1 : \psi : z)$$

$$= [\frac{z^{\ell}\rho(a)-1}{z^{2\ell}-1}]_wE_{P_2}(P_1 : \psi : z) + [\frac{\rho(a)-z^{2\ell}}{1-z^{2\ell}}]_wE_{P_2}(P_1 : \psi : z).$$

Since ${}_wE_{P_2}(P_1 : \psi : z)$ is analytic at $z = 1$ and $z^{2\ell}-1$ has simple zeroes, the above expressions imply that the c-functions have at most simple poles at $z = 1$ or $\nu_0 = 0$. The same follows for the adjoints, by the Maass-Selberg relations.

To check that the remaining singularities are at most poles one uses the analyticity of $E_{P_2}(P_1 : \psi : \nu)$ and argues similarly (it is not clear that the poles are simple in this case!). We omit the details.

That $\mu(\omega : \nu)$ is meromorphic on $\mathfrak{F}_{\mathbb{C}}(\omega)$ follows from the fact that the c-functions are meromorphic.

Corollary 5.4.2.2. Let $\omega \in \mathcal{E}_2(M)$. Then:

1) If ω is unramified, then $c_{P_2|P_1}(s : \omega : \nu)$ is holomorphic at $\nu = 0$ for all $P_1, P_2 \in \mathcal{P}(A)$ and $s \in W$. Moreover, all elements of ${}_w\mathfrak{X}_{C_M^G(\omega)}(P_2, A)$ are simple

2) If ω is unramified $\mu(\omega : 0) > 0$.

Proof.

1) Let ω be unramified. To prove that in this case the c-functions are analytic at $\nu = 0$ it suffices to consider the case $[W] = 2$. We have ${}_wE_{P_2}(P_1 : \psi : \nu) =$

$c_{P_2|P_1}(1 : \omega : \nu)\psi\chi_\nu + c_{P_2|P_1}(s : \omega : \nu)\psi\chi_{s\nu}$ for any $\psi \in \mathcal{A}(\omega, \tau_M)$. Either both c-functions are analytic near $\nu = 0$ or, regarded as a function of z, each has a simple pole. In the second case, we have

$$= \lim_{\nu\to 0} \nu\, {}_w E_{P_2}(P_1 : \psi : \nu : m) =$$

$$\lim_{\nu\to 0} \nu c_{P_2|P_1}(1 : \omega : \nu)\psi(m)\chi_\nu(m) + \lim_{\nu\to 0} \nu c_{P_2|P_1}(s : \omega : \nu)\psi(m)\chi_{s\nu}(m) = \psi_1(m) + \psi_s(m),$$

where $\psi_1 \in \mathcal{L}(\omega, P_2)$ and $\psi_s \in \mathcal{L}(\omega^s, P_2)$ and neither is zero. This is impossible.

The simplicity of the exponents is a consequence of the analyticity of the c-functions: We know that $(\chi_{\omega,\nu}(a^s) - \rho(a))E_{P_2, s}(P_1 : \psi : \nu) = 0$ for all ν near $\nu = 0$. Since $E_{P_2, s}(P_1 : \psi : \nu)$ is analytic at $\nu = 0$ and

$$\chi_{\omega,\nu}(a^s)E_{P_2, s}(P_1 : \psi : \nu) = \rho(a)E_{P_2, s}(P_1 : \psi : \nu)$$

near $\nu = 0$, we may conclude by taking limits that this relation holds at $\nu = 0$.

2) We know that $\mu(\omega : 0) \geq 0$. If $\mu(\omega : 0) = 0$, then $c_{P_2|P_1}(s : \omega : \nu)$ must have a pole at $\nu = 0$, since $\lim_{\nu\to 0} \mu(\omega : \nu) \| c_{P_2|P_1}(s : \omega : \nu)\psi \|^2 = \|\psi\|^2$ for all $\psi \in L(\omega, P)$.

Corollary 5.4.2.3. Let ω be ramified. Then the following are equivalent:

1) $\pi_{P,\omega}$ is reducible for some $P \in \mathcal{P}(A)$.

2) $\pi_{P,\omega}$ is reducible for all $P \in \mathcal{P}(A)$.

3) $c_{P_2|P_1}(s : \omega : \nu)$ is analytic at $\nu = 0$ for some [every] $P_1, P_2 \in \mathcal{P}(A)$ and $s \in W$.

4) $\mu(\omega : 0) > 0$.

5) The elements of ${}_w\mathcal{X}_{\pi_{P,\omega}}(P, A)$ are simple for some [every] $P \in \mathcal{P}(A)$.

Proof. The equivalence of 1) and 2) follows, e.g., from the equality of the characters $\pi_{P,\omega}$ for all $P \in \mathcal{P}(A)$, the equivalence of 3) and 4) from Theorem 5.3.5.2. To prove that 3) and 4) imply 5) one can argue as in the proof of Corollary 5.4.2.2. The equivalence of 5) and 1)-2) follows easily from Theorem 5.4.1.1 and Theorem 1.7.10.

Let us show that 2) implies 3). It is sufficient to show that, for all $\psi \in L(\omega, P_1)$, $(\rho(a)-\chi_\omega(a))E_{P_2}(P_1 : \psi : 0) \neq 0$, if $c_{P_2|P_1}(1 : \omega : \nu)$ is singular at $\nu = 0$. Note that the "minus-one" coefficients in the Laurent expansions for $E_{P_2,1}$ and $E_{P_{2,s}}$ are negatives of one another. Therefore, since $E_{P_2}(P_1 : \psi : \nu)$ is holomorphic at $\nu = 0$,

$$(\rho(a)-\chi_\omega(a))E_{P_2}(P_1 : \psi : 0) = \lim_{\nu\to 0}(\rho(a)-\chi_\omega(a))E_{P_2}(P_1 : \psi : \nu)$$

$$= \lim_{\nu\to 0}((\chi_{\omega,\nu}(a^s)-\chi_\omega(a))E_{P_2,s}(P_1 : \psi : \nu) + (\chi_{\omega,\nu}(a)-\chi_\omega(a))E_{P_2,1}(P_1 : \psi : \nu))$$

$$= \lim_{\nu\to 0}((\chi_{\omega,\nu}(a^s)-\chi_\omega(a))\psi_{-1}\nu^{-1} + (\chi_{\omega,\nu}(a)-\chi_\omega(a))(-\psi_{-1}\nu^{-1}))$$

$$= \lim_{\nu\to 0}\left(\frac{\chi_{\omega,\nu}(a^{-1})-\chi_{\omega,\nu}(a)}{\nu}\right)\psi_{-1}$$

$\neq 0$, since $0 \neq \psi_{-1} \in \mathcal{L}(\omega, P_2)$.

Remarks.

1) If π is reducible, the components are inequivalent. This follows immediately either from Theorem 1.7.10 combined with Theorem 5.4.1.1 or from Theorem 2.5.9.

2) If $\omega \in {}^0\mathcal{E}(M)$, $\mu(\omega : \nu)$ has no zeros off $\mathfrak{a}^* + \mathfrak{z}^*_{\mathbb{C}}$ because the c-functions are analytic off $\mathfrak{a}^* + \mathfrak{z}^*_{\mathbb{C}}$.

For the next lemma we assume $\omega \in {}^0\mathcal{E}(M)$, since in general we do not know that the c-functions are analytic off $\mathfrak{a}^*$.

Lemma 5.4.2.4. Let $\omega \in {}^0\mathcal{E}(M)$ and let $\nu_0 \in \sqrt{-1}\,\mathfrak{a}^* - \sqrt{-1}\,\mathfrak{z}^*$. Then $\pi = \pi_{P,\omega_{\nu_0}}$ is irreducible if and only if $\mu(\omega : \nu)$ is analytic at $\nu = \nu_0$.

Proof. If $\mu(\omega : \nu)$ is analytic at $\nu = \nu_0$, then, by Theorems 5.3.5.2 and 5.4.2.1, $c_{P|P}(1 : \omega : \nu_0)$ is bijective, so the Eisenstein integral $E(P : \nu_0) : L(\omega, P) \to \mathcal{A}(\pi, \tau)$ is injective. Combined with Theorem 5.2.2.7, this implies that π is irreducible.

Suppose that $\mu(\omega : \nu)$ has a pole at $z(\nu) = z(\nu_0)$. Then either $c_{P|P}(1 : \omega : \nu_0)$ or $c_{P|P}(1 : \omega : \bar{\nu}_0)^*$ has a kernel. Assume that $c_{P|P}(1 : \omega : \nu_0)$ has a kernel. Arguing as in the first part of the proof of Theorem 5.2.4.2, one observes that the kernel of $c_{P|P}(1 : \omega : \nu_0)$ corresponds to a G-submodule of $\mathcal{A}(\pi)$ (i.e., we have $j : \mathcal{A}(\pi) \to \mathcal{A}(\pi, \tau^0)$. Given $\alpha, \beta \in C_c^\infty(G)$ and $\psi_T \in L(\omega, P)$, we obtain $(\underset{\sim}{\alpha} * E(P : \psi_T) * \underset{\sim}{\beta})_{P,1} = \psi_{\pi_{\bar{P}}(\beta')\mathcal{C}_P(T)\pi_P(\alpha')}$, etc.). It remains to show that this double G-submodule of $\mathcal{A}(\pi)$ is not all of $\mathcal{A}(\pi)$. This implies that $c_{P|P}(1 : \omega : \nu_0)$ and $c_{P|P}(1 : \omega : \bar{\nu}_0)^*$ must simultaneously have kernels, so our assumption above was not restrictive. To prove that $c_{P|P}(1 : \omega : \nu_0) \not\equiv 0$, we shall show that $\nu_0 \in \mathfrak{X}_\pi(P, A)$. By Theorem 3.2.4, this suffices, since

$$c_{P|P}(s : \omega : \nu_0)\psi\chi_{s\nu_0} \in \mathcal{L}((\omega_{\nu_0})^s, P), \quad \psi \in L(\omega, P).$$

Consider $\pi_{\bar{P},\omega_{\nu_0}}$ and let $\mathcal{H} \subset C^\infty(K : U)$ be the representation space. Defining $\Delta\beta = \beta(1)$ $(\beta \in \mathcal{H})$, we see that $\omega_{\nu_0} \in \mathfrak{X}_{\pi_{\bar{P},\omega_{\nu_0}}}(P, A)$. On the other hand,

$\pi_{\overline{P},\omega_{\nu_0}}$ and $\pi_{P,\omega_{\nu_0}}$ have the same class exponents, because they have the same characters and, consequently, the same irreducible subquotients. It follows from Lemma 5.4.1.2 that the class exponents are determined by the irreducible subquotients.

For simplicity, assume that $Z = 1$ in the next lemma.

Lemma 5.4.2.5. For any $\omega \in \mathcal{E}_2(M)$, the function $\mu(\omega : \nu)$ has at most finitely many poles for $\nu \in \mathcal{F}_{\mathbb{C}}$.

Proof. Since the c-functions and their adjoints have at most finitely many poles, we need consider only points ν_0 where both $c_{P|P}(1 : \omega : \nu_0)$ and $c_{P|P}(1 : \omega : \overline{\nu}_0)^*$ are analytic. In this case the function $\mu(\omega : \nu)$ has a pole at $\nu = \nu_0$ if and only if $c_{P|P}(1 : \omega : \nu_0)$ has a kernel (cf. Proof of Lemma 5.4.2.4). Let (V, τ) be a finite dimensional smooth double representation of K. Then both $L(\omega, P)$ and $\mathcal{L}(\omega, P)$ have the same finite dimension. We know that $c_{P|P}(1 : \omega : \nu)$ is a bijection for ν in an open subset of $\mathcal{F}_{\mathbb{C}}(\omega)$. We have an integral formula for $c_{P|P}(1 : \omega : \nu)$ convergent for $\nu \in \mathcal{F}_{\mathbb{C}}(\overline{P})$. We use this formula to express $c_{P|P}(1 : \omega : \nu)$ by a matrix power series. Indeed, identifying bases for $L(\omega, P)$ and $\mathcal{L}(\omega, P)$ we obtain a power series for the determinant of $c_{P|P}(1 : \omega : \nu) = \sum_{k=0}^{\infty} c_k q^{\sqrt{-1}k\nu} = \sum_{k=0}^{\infty} c_k z^k$. Since $c_{P|P}(1 : \omega : \nu) \neq 0$ on an open set, it follows that, as a function of z, there are only finitely many zeros for $|z| < 1$. To conclude, we observe that $\mu(\omega : \nu)$ has poles in the other half space only at the zeros of $c_{\overline{P}|\overline{P}}(1 : \omega : \nu)$.

.4.3. The Product Formula and Consequences.

Let (V, τ) be a smooth unitary double representation of K. Let A an A_0-standard torus, $M = Z_G(A)$, and $P, P' \in \mathcal{P}(A)$ $(P = MN,\ P' = MN')$. Let $\in \mathcal{E}_2(M)$ and set $\Phi = \Sigma_r(P, A)$, the set of reduced roots for (P, A). Given $\in \Phi$, let A_α be the maximal subtorus of A which lies in the kernel of the root aracter α and let $M_\alpha = Z_G(A_\alpha)$. The p-pair $({}^*P_\alpha, A) = (M_\alpha \cap P, A) \prec (M_\alpha, A_\alpha)$ a maximal p-pair of M_α and ${}^*P_\alpha = MN_\alpha$, where $N_\alpha = M_\alpha \cap N$, is its Levi composition. Let $\mu_\alpha(\omega : \nu)$ have the same meaning for (M_α, A) as $\mu(\omega : \nu)$ s for (G, A). We also write $\gamma(M_\alpha/{}^*P_\alpha)$ for the constant defined relative to $_\alpha$, $K_\alpha = K \cap M_\alpha$, and $\overline{N}_\alpha$ $({}^*\overline{P}_\alpha = M\overline{N}_\alpha = MN_{-\alpha})$ just as $\gamma(G/P)$ was defined relative to G, K, and $\overline{N}$.

The Lie algebra $\mathfrak{n}$ of $\underset{\sim}{N}$ may be written as $\mathfrak{n} = \bigoplus_{\alpha \in \Phi} \mathfrak{n}_\alpha$, where $\mathfrak{n}_\alpha$ the Lie algebra of $\underset{\sim}{N}_\alpha$. Under the adjoint representation of $\underset{\sim}{A}$ on $\mathfrak{n}$, the subace $\mathfrak{n}_\alpha$ is the sum of all the root subspaces corresponding to α and its multiples. Given any ordering of Φ, we may write $N = \prod_{\alpha \in \Phi} N_\alpha$ and $\overline{N} = \prod_{\alpha \in \Phi} \overline{N}_\alpha$.

A point $H \in \mathfrak{a}$ is called regular if $\alpha(H) \neq 0$ for every $\alpha \in \Phi$ and mi-regular if $\alpha(H) = 0$ for exactly one $\alpha \in \Phi$. Let $\mathfrak{C} = \mathfrak{C}_0$ be the positive amber associated to Φ : $\mathfrak{C} = \{H \in \mathfrak{a} \,|\, \alpha(H) > 0 \text{ for all } \alpha \in \Phi\}$. We may choose $_0 \in \mathfrak{C}$ and $H_{r+1} \in -\mathfrak{C}$ such that the line segment L joining them consists of only gular and semi-regular points, and such that L passes through the chamber $'$ corresponding to P'. List the semi-regular points $H_1, \ldots, H_r$ $(r = [\Phi])$ in e order in which they occur as we pass from H_0 to H_{r+1} along L. We obtain bijection between Φ and $\{H_1, \ldots, H_r\}$ and thus an ordering of Φ: $\alpha_i \in \Phi$ is the ot which changes sign as we pass the point H_i $(i = 1, \ldots, r)$ along L. The segment (H_i, H_{i+1}) lies in the chamber $\mathcal{L}_i$ which corresponds to the parabolic subgroup $Q_i = MN_i \in \mathcal{P}(A)$; we have $Q_0 = P$, $Q_r = \overline{P}$, and $Q_s = P'$, $0 \leq s \leq r$. To each group Q_i there corresponds an opposite group $\overline{Q}_i = M\overline{N}_i \in \mathcal{P}(A)$.

According to our ordering of Φ the roots $\{\alpha_1, \ldots, \alpha_j\} = \Phi \cap \Sigma_r(\overline{Q}_j, A)$ and $\{\alpha_{j+1}, \ldots, \alpha_r\} = \Phi \cap \Sigma_r(Q_j, A)$, $j = 0, \ldots, r$. We note that $\overline{N} \cap N_j = \prod_{i \leq j} \overline{N}_{\alpha_i}$ and $\overline{N} \cap \overline{N}_j = \prod_{i > j} \overline{N}_{\alpha_i}$. Of course M normalizes N_α and $\overline{N}_\alpha$ for all $\alpha \in \Phi$. For later use we observe that $\overline{N}_{\alpha_j}$ normalizes $\overline{N} \cap N_{j-1}$; N_{α_j} normalizes N_{j-1}.

Let $\psi \in \mathcal{A}(\omega, \tau_M)$ and let κ and μ be the locally constant mappings defined in §0.6. Set

$$(c^-_{Q_j|P}(\nu)\psi)(m) = \int_{\overline{N} \cap N_j} \tau(\kappa(\overline{n}))\psi(\mu(\overline{n})m) q^{\langle \sqrt{-1}\nu - \rho_P, H_P(\overline{n}) \rangle} d\overline{n}$$

and

$$(c^+_{Q_j|P}(\nu)\psi)(m) = \int_{\overline{N} \cap \overline{N}_j} \psi(m\mu(\overline{n})^{-1}) \, \tau(\kappa(\overline{n}))^{-1} q^{\langle \sqrt{-1}\nu - \rho_P, H_P(\overline{n}) \rangle} d\overline{n}.$$

These integrals converge absolutely and uniformly on compact subsets of $\mathfrak{F}_{\mathbb{C}}(\overline{Q}_j)$ (Theorem 5.3.5.4). The two operators obviously commute and, as we shall see, the analytic functions on $\mathfrak{F}_{\mathbb{C}}(\overline{Q}_j)$ extend to meromorphic functions on $\mathfrak{F}_{\mathbb{C}}$. According to Theorem 5.3.5.4,

$$\gamma(G/P)c_{P'|P}(1 : \omega : \nu)\psi = c^+_{P'|P}(\nu)c^-_{P'|P}(\nu)\psi \quad \text{for } \psi \in L(\omega, P).$$

Set $c^-_\alpha(\nu) = \gamma(M_\alpha/{}^*P_\alpha)c_{{}^*\overline{P}_\alpha|{}^*P_\alpha}(1 : \omega : \nu)$ and $c^+_\alpha(\nu) = \gamma(M_\alpha/{}^*P_\alpha)c_{{}^*P_\alpha|{}^*P_\alpha}(1 : \omega : \nu)$ for all $\nu \in \mathfrak{F}_{\mathbb{C}}$ such that these functions are defined.

heorem 5.4.3.1. For all $\nu \in \mathfrak{F}_{\mathbb{C}}$ where these functions are defined

$$c^+_{P'|P}(\nu)c^-_{P'|P}(\nu) = c^+_{\alpha_{s+1}}(\nu)\ldots c^+_{\alpha_r}(\nu)c^-_{\alpha_s}(\nu)\ldots c^-_{\alpha_1}(\nu)$$

$$= c^-_{\alpha_s}(\nu)\ldots c^-_{\alpha_1}(\nu)c^+_{\alpha_{s+1}}(\nu)\ldots c^+_{\alpha_r}(\nu).$$

roof. It is sufficient to establish the above relations for ν lying in an h-dense ubset of $\mathfrak{F}_{\mathbb{C}}$, then to use analytic continuation to justify the general asser- ions. We shall check only the first relation; the second of course follows from he commutativity of c^+ and c^-.

Let $\psi \in \mathcal{A}(\omega, \tau_M)$. It is sufficient to check that $c^-_{Q_j|Q_0}(\nu)\psi =$ $c^-_{\alpha_j}(\nu)c^-_{Q_{j-1}|Q_0}(\nu)\psi[c^+_{Q_{r-j}|Q_0}(\nu)c^-_{P'|P}(\nu)\psi = c^+_{\alpha_{r-j}}(\nu)c^+_{Q_{r-j+1}|Q_0}(\nu)c^-_{P'|P}(\nu)\psi]$. Let $\mathfrak{F}_{\mathbb{C}}(\alpha_j) = \{\nu \in \mathfrak{F}_{\mathbb{C}} | \langle\nu_I, \alpha_j\rangle > 0,\ \nu = \nu_{\mathbb{R}} + \sqrt{-1}\nu_I\}$. We shall check only the unbracketed tatement for $\nu \in \bigcap_{1 \le i \le j} \mathfrak{F}_{\mathbb{C}}(\alpha_i)$.

We have

$$c^-_{Q_j|P}(\nu)\psi(m) = \int_{\bar{N} \cap N_j} \tau(\kappa(\bar{n}))\psi(\mu(\bar{n})m)q^{\langle\sqrt{-1}\nu - \rho_P, H_P(\bar{n})\rangle} d\bar{n}$$

$$= \int_{\bar{N}_{\alpha_j}} \int_{\bar{N} \cap N_{j-1}} \tau(\kappa(\bar{n}\bar{n}_j))\psi(\mu(\bar{n}\bar{n}_j)m)q^{\langle\sqrt{-1}\nu - \rho_P, H_P(\bar{n}\bar{n}_j)\rangle} d\bar{n}d\bar{n}_j$$

ince $\bar{N}_{\alpha_j}$ normalizes $\bar{N} \cap N_{j-1}$, we may change the order of the variables (with- ut changing the measure!). Write $\bar{n}_j = \kappa_j\mu_j\eta_j$ with $\kappa_j \in K \cap M_{\alpha_j}$, $\mu_j \in M$, and $\eta_j \in N_{\alpha_j}$. Since η_j normalizes N_{j-1}, we may write $\bar{n}^{\eta_j} = \bar{n}'n'$ ($\bar{n}' \in \bar{N} \cap N_{j-1}$, $n' \in N \cap N_{j-1}$), "push" n' to the right, and again change variables, ridding the notation of both n' and η_j--which are seen to be negligible. We have $\kappa(\kappa_j\mu_j\bar{n}) \in \kappa_j\kappa(\bar{n}^{\mu_j}) P \cap K$ and $\mu(\kappa_j\mu_j\bar{n}) \in K_M\mu(\bar{n}^{\mu_j} \cdot \mu_j)$. Thus,

$$c_{Q_j}|P^{(\nu)}\psi(m) = \int_{\bar{N}_{\alpha_j}} \tau(\kappa_j)\int_{\bar{N}\cap N_{j-1}} \tau(\kappa(\bar{n}^{\mu_j}))\psi(\mu(\bar{n}^{\mu_j})\mu_j m)$$

$$q^{\langle\sqrt{-1}\nu - \rho_P, H_P(\bar{n}^{\mu_j}) + H(\mu_j)\rangle} d\bar{n}d\bar{n}_j$$

$$= \int_{\bar{N}_{\alpha_j}} \tau(\kappa_j)(c_{Q_{j-1}}|P^{(\nu)}\psi)(\mu_j m) q^{\langle 2\rho_P - \sum_{i=j}^{r} m_i\alpha_i, H(\mu_j)\rangle}$$

$$q^{\langle\sqrt{-1}\nu - \rho_P, H(\mu_j)\rangle} d\bar{n}_j,$$

where the change of variables $\bar{n}^{\mu_j} \to \bar{n}'$ accounts for the factor $q^{\langle 2\rho_P - \sum_{i=j}^{r} m_i\alpha_i, H(\mu_j)\rangle}$ introduced above (m_i is the multiplicity of α_i and its mult (counted with the right factors) in $\mathcal{M}_{\alpha_i}$).

To conclude that

$$c_{Q_j}|P^{(\nu)}\psi(m) = (c_{\alpha_j}(\nu)c_{Q_{j-1}}|P^{(\nu)}\psi)(m),$$

we must show that $\langle\sqrt{-1}\nu + \rho_P - \sum_{i=j}^{r} m_i\alpha_i, H_P(\mu_j)\rangle = \langle\sqrt{-1}\nu - \rho_{*P_{\alpha_j}}, H(\mu_j)\rangle$.

Since $\rho_P - \sum_{i=j}^{r} m_i\alpha_i = -\rho_{Q_{j-1}}$ and $\langle\rho_{Q_{j-1}}, H(\mu_j)\rangle = \langle\rho_{*P_{\alpha_j}}, H(\mu_j)\rangle$ (note that $\mu_j \in M_{\alpha_j}$ and that α_j is a <u>simple</u> root in $\Sigma(Q_{j-1}, A)!$), the desired equality holds.

Corollary 5.4.3.2. For any $P_1, P_2 \in \mathcal{P}(A)$ and $s \in W(G/A)$ the function $c_{P_2|P_1}(s : \omega : \nu)$ is meromorphic on $\mathfrak{a}^*_{\mathbb{C}}$.

Proof. Since $c_{P_2|P_1}(1 : \omega : \nu)$ is a product of meromorphic functions (Theorems 5.4.2.1 and 5.4.3.1), it is also meromorphic. Theorem 5.3.5.3 (3) implies that the same is true for $c_{P_2|P_1}(s : \omega : \nu)$ for any $s \in W(G/A)$.

Corollary 5.4.3.3. $\mu(\omega : \nu)$ is meromorphic on $\mathfrak{a}^*_{\mathbb{C}}$. $\gamma^{-2}(G/P)\mu(\omega : \nu) = \prod_{\alpha \in \Phi} \gamma^{-2}(M_\alpha/{}^*P_\alpha)\mu_\alpha(\omega : \nu)$, where $\Phi = \Sigma_r(P, A)$ for any $P \in \mathcal{P}(A)$, the remaining symbols as defined previously.

Proof. The right-hand side above is meromorphic, as it is a product of meromorphic functions. To prove the corollary it suffices to justify the product formula. Since $\gamma(G/P)c_{\overline{P}|P}(1 : \omega : \nu) = c^-_{\alpha_r}(\nu)\ldots c^-_{\alpha_1}(\nu)$ and $\gamma(G/P)(c_{\overline{P}|P}(1 : \omega : \overline{\nu}))^* = c^-_{\alpha_1}(\overline{\nu}))^*\ldots(c^-_{\alpha_r}(\overline{\nu}))^*$, the present corollary follows immediately via Theorem 5.3.5.2.

The following corollary justifies the notation $\gamma(G/M) = \gamma(G/P)$ for any $P \in \mathcal{P}(A)$.

Corollary 5.4.3.4. $\gamma(G/P)$ does not depend upon $P \in \mathcal{P}(A)$.

Proof. The preceding corollary implies that it is sufficient to consider P and $\overline{P}$ maximal and opposite. However, by conjugacy, it is clear that $\gamma(G/P_0) = \gamma(G/\overline{P}_0)$ when P_0 and $\overline{P}_0$ are minimal and opposite. Thus, $\gamma(G/P_0) = \gamma(G/P)\gamma(M/P_0 \cap M) = \gamma(G/\overline{P})\gamma(M/\overline{P}_0 \cap M)$, which implies the corollary.

Remarks.

1) It is not true that $\gamma(G/M) = \prod \gamma(M_\alpha/M)$.

2) This corollary may also be deduced from the Plancherel formula for G (§5.5.2), i.e., one can derive a Plancherel's formula "independent of $P \in \mathcal{P}(A)$" without assuming this result.

Corollary 5.4.3.5. The formula for the adjoints in Theorem 5.3.5.4:

$$\gamma(G/M)(c_{\bar{P}_2|\bar{P}_1}(1:\omega:\bar{\nu}))^*\psi = J_{P_2|P_1}(\nu)\psi \quad (\psi \in \mathcal{L}(\omega, \bar{P}_2)).$$

Proof. It follows from Theorem 5.4.3.1 that it is sufficient to check the case p-rank equal one. It is thus sufficient to check that the same integrals represent, respectively, $(c_{\bar{P}|P}(1:\omega:\bar{\nu}))^*$ and $c_{P|\bar{P}}(1:\omega:\nu)$, $(c_{\bar{P}|\bar{P}}(1:\omega:\bar{\nu}))^*$ and $c_{P|P}(1:\omega:\nu)$; indeed, we need check these relations only for $\nu \in \mathfrak{F}'$. Since $\chi_\nu c_{P_2|P_1}(1:\omega_0:\nu) = c_{P_2|P_1}(1:\omega_\nu:0)\chi_\nu$ and, consequently, a similar equality holds for the adjoints, it is sufficient to check that we have identical integrals representing $c_{\bar{P}|P}(1:\omega_0)^*$ and $c_{P|\bar{P}}(1:\omega_0)$, respectively, $c_{\bar{P}|\bar{P}}(1:\omega_0)^*$ and $c_{P|P}(1:\omega_0)$ for ω_0 in general position.

We apply the formal theory, specifically Theorem 5.2.4.2. We will consider the case of the standard double representation $(V^0, \tau^0) = (C^\infty(K \times K), \tau^0)$, and leave to the reader the problem of carrying the results over to an arbitrary smooth double representation--the proof of the general case follows a pattern exhibited in the proof of Corollary 5.2.4.3.

Write $c_{P_2|P_1} = c_{P_2|P_1}(1:\omega_0)$. According to Theorem 5.2.4.2, we have operators $c_P, F_P, c_P^* = c_{\bar{P}}, F_{\bar{P}}$ which act in a certain K-module $\mathcal{H}$

cf. §5.2.4) and satisfy the relations $c_{P|P}\psi_T = \psi_{c_P T F_P}$ and $c_{\bar{P}|P}\psi_T = \psi_{F_P T c_{\bar{P}}}$ for $T \in \mathcal{T} = \mathrm{End}^0(\mathcal{H})$. We first show that $c^*_{P|P}\,\psi_T = \psi_{c_{\bar{P}} T F_P}$ and $c^*_{\bar{P}|P}\,\psi_T = \psi_{F_P T c_P}$. For this recall that $(\psi_T, \psi_S) = d(\omega_0)^{-1}\mathrm{tr}(TS^*)$ $(T, S \in \mathcal{T})$. Therefore, $d(\omega_0)(c_{P|P}\psi_T, \psi_S) = \mathrm{tr}(c_P T F_P S^*) = \mathrm{tr}(T(c_{\bar{P}} S F_P)^*)$, so $c^*_{P|P}\,\psi_S = \psi_{c_{\bar{P}} S F_P}$. Similarly, $d(\omega_0)(c_{\bar{P}|P}\psi_T, \psi_S) = \mathrm{tr}(F_P T c_{\bar{P}} S^*) = \mathrm{tr}(T(F_P S c_P)^*)$, so $c^*_{\bar{P}|P}\,\psi_S = \psi_{F_P S c_P}$. Thus, $c^*_{\bar{P}|\bar{P}}\psi_S = \psi_{c_P S F_{\bar{P}}}$ and $c^*_{P|\bar{P}}\psi_S = \psi_{F_{\bar{P}} S c_{\bar{P}}}$.

We have integral (rather principal value integral) representations:

$$\gamma c_{P|P}\psi_T(m) = \lim_{\epsilon \to 0} \int_{\bar{N}} \psi_T(m\mu(\bar{n})^{-1})\tau^0(\kappa(\bar{n})^{-1})q^{-\langle(1+\epsilon)\rho,\, H_P(\bar{n})\rangle}\, d\bar{n}$$

$$\gamma c_{\bar{P}|P}\psi_T(m) = \lim_{\epsilon \to 0} \int_{\bar{N}} \tau^0(\kappa(\bar{n}))\psi_T(\mu(\bar{n})m)q^{-\langle(1+\epsilon)\rho,\, H_P(\bar{n})\rangle}\, d\bar{n}$$

for $T \in \mathcal{T}(P)$, $\psi_T \in L(\omega, P)$. In order to conclude that, for $T \in \mathcal{T}(P|\bar{P})$, $\psi_T \in \mathcal{L}(\omega, \bar{P})$, in the first case, and $T \in \mathcal{T}(\bar{P}|P)$, $\psi_T \in \mathcal{L}(\omega, P)$, in the second case, these same integrals represent, respectively, $c^*_{\bar{P}|\bar{P}}\psi_T$ and $c^*_{P|\bar{P}}\psi_T$ (both elements of $L(\omega, \bar{P})$), we have to observe that, if $F_P T = T$, in the first case, and $TF_P = T$, in the second, then the integrals give, respectively, $\psi_{c_P T}$ and $\psi_{T c_{\bar{P}}}$. However, Theorem 5.2.4.2 informs us that

c_P [hence, $c_{\bar{P}}$] is uniquely determined by two conditions ((1) $c_P = F_{\bar{P}} c_P F_P$ and (2) $c_{P|P}\psi_T = \psi_{c_P T F_P}$ [(1) $c_{\bar{P}} = F_P c_{\bar{P}} F_{\bar{P}}$ and (2) $c_{\bar{P}|P}\psi_T = \psi_{F_P T c_{\bar{P}}}$] for all $T \in \mathcal{T}$), both of which conditions are seen to be satisfied by our integral expressions. The general formulas now follow by analytic continuation.

Recall that the Lie algebra $\mathfrak{a}_0$ of A_0 is equipped with a $W(G/A_0)$-

invariant scalar product, which induces a $W(G/A)$-invariant scalar product on $\mathfrak{a}$. We can speak of the reflection s_α of $\mathfrak{a}$ associated to a reduced root $\alpha \in \Sigma_r(G, A)$. For any $\omega \in \mathcal{E}_2(M)$ the ramification group $W(\omega) = \{s \in W(G/A) | \omega^s = \omega\}$ is defined. Let $\mathcal{O}$ be an orbit of $\mathcal{E}_2(M)$. Let $\Sigma_r(\mathcal{O}) = \{\alpha \in \Sigma_r(G, A) | s_\alpha \in W(\omega)$ for some $\omega \in \mathcal{O}\}$. For every $\alpha \in \Sigma_r(\mathcal{O})$ choose $a_\alpha \in A$ such that $s_\alpha(a_\alpha) \neq a_\alpha$

Theorem 5.4.3.6. Fix $P_1, P_2 \in \mathcal{P}(A)$, $s \in W(G/A)$, and $\omega \in \mathcal{O}$. Then the function $c_{P_2|P_1}(s : \omega : \nu)$ is meromorphic on $\mathcal{F}_{\mathbb{C}}(\omega)$. If $\omega \in {}^0\mathcal{E}(M)$, it is holomorphic on $\mathcal{F}'_{\mathbb{C}}(\omega)$. If $\omega \in \mathcal{E}_2(M)$, it is holomorphic on $\mathcal{F}_{\mathbb{C}}(\bar{P}_2^{s^{-1}})$. Moreover, $\nu \mapsto \prod_{\alpha \in \Sigma_r(\mathcal{O})} (\chi_{\omega,\nu}(a_\alpha^{s_\alpha}) - \chi_{\omega,\nu}(a_\alpha)) c_{P_2|P_1}(s : \omega \cdot \nu)$ is holomorphic on $\mathfrak{a}^*$ [on $\mathfrak{a}^*_{\mathbb{C}}$, if $\omega \in {}^0\mathcal{E}(M)$].

Proof. Use Theorems 5.4.3.1 and 5.4.2.1.

Theorem 5.4.3.7. Let $\omega \in {}^0\mathcal{E}(M)$, $\nu_0 \in \mathfrak{a}^*_{\mathbb{C}}$, and assume that ω_{ν_0} is unramified. Then the representation $\pi = \mathrm{Ind}_P^G(\delta_P^{\frac{1}{2}}\sigma_{\nu_0})$ is irreducible ($P \in \mathcal{P}(A)$, $\sigma_{\nu_0} \in \omega_{\nu_0}$) if and only if $\mu(\omega : \nu)$ is analytic at $\nu = \nu_0$.

Proof. Suppose $\mu(\omega : \nu)$ is not analytic at $\nu = \nu_0$. Then one of the factors $\mu_\alpha(\omega : \nu)$ is also not analytic at ν_0. It follows from Lemma 5.4.2.4 that $\mathrm{Ind}_{{}^*P}^{M_\alpha}(\delta_{{}^*P}^{\frac{1}{2}} \sigma_{\nu_0})$ is reducible. Therefore, so is $\mathrm{Ind}_P^G(\delta_P^{\frac{1}{2}}\sigma_{\nu_0}) =$ $\mathrm{Ind}_{P_\alpha}^G(\delta_{P_\alpha}^{\frac{1}{2}} \mathrm{Ind}_{{}^*P}^{M_\alpha}(\delta_{{}^*P}^{\frac{1}{2}} \sigma_{\nu_0}))$ (${}^*P = M_\alpha \cap P$ --we are assuming without loss of generality that α is a simple root).

Conversely, if $\mu(\omega : \nu)$ is analytic at $\nu = \nu_0$, then, since ϕ_{ν_0} is un-
amified, the c-functions are defined at ν_0 (Theorem 5.4.2.1 and 5.4.3.1) and,
s follows from Theorem 5.3.5.2, they are injective $L(\omega, P) \to \mathcal{L}(\omega, P)$. Therefore,
he Eisenstein integral is injective from $L(\omega_{\nu_0}, P)$ to $\mathcal{A}(\pi_{\nu_0}, \tau)$, so π_{ν_0} is
rreducible (Theorem 5.2.2.7).

xercises. Let $\sigma \in \omega \in \mathcal{E}_2(M)$ and $P \in \mathcal{P}(A)$ $(P = MN)$.

If $\mu(\omega : 0) = 0$, then ω is ramified and there exists $\chi \in \mathfrak{X}_\pi(P, A)$,
$= \mathrm{Ind}_P^G(\delta_P^{\frac{1}{2}}\sigma)$, such that χ is not simple (cf. Corollaries 5.4.2.2 and 5.4.2.3).

) If $\mu(\omega : 0) > 0$ and ω is fixed by a reflection s_α with respect to a reduced
oot $\alpha \in \Sigma_r(G, A)$, then $\pi = \mathrm{Ind}_P^G(\delta_P^{\frac{1}{2}}\sigma)$ is reducible (cf. Corollary 5.4.2.3).

If ω is unramified, then $\mu(\omega : 0) > 0$ and all $\chi \in \mathfrak{X}_\pi(P, A)$ are simple
f. Corollary 5.4.2.2).

5.4.4. The Composition Series Theorem.

Let A be an A_0-standard torus, let $M = Z_G(A)$, and $W = W(G/A)$.
ssume that $\sigma \in \omega \in {}^0\mathcal{E}_{\mathbb{C}}(M)$ or $\mathcal{E}_2(M)$ and set $\pi_{P,\omega} = \mathrm{Ind}_P^G(\delta_P^{\frac{1}{2}}\sigma)$, $P \in \mathcal{P}(A)$.
he purpose of this subsection is to prove the following theorem, first proved
y Casselman in the case $A = A_0$.

heorem 5.4.4.1. (1) For any $P \in \mathcal{P}(A)$ the representation $\pi_{P,\omega}$ has a finite
omposition series of length no greater than $[W]$. If $P_1, P_2 \in \mathcal{P}(A)$, $s \in W$, then
$\pi_{P_1,\omega}$ and π_{P_2,ω^s} have isomorphic composition series.

(2) Let A' be a special torus of G and $M' = Z_G(A')$. If $\omega \in \mathcal{E}_2(M)$,
t $\omega' \in \mathcal{E}_2(M')$; if $\omega \in {}^0\mathcal{E}_{\mathbb{C}}(M)$, let $\omega' \in {}^0\mathcal{E}_{\mathbb{C}}(M')$. Then, if A' is not conjugate

to A, $\pi_{P,\omega}$ and $\pi_{P',\omega'}$ have no equivalent composition factors. Let $A = A'$. If $\omega' \neq \omega^s$ for any $s \in W(G/A)$, then, again, $\pi_{P',\omega'}$ and $\pi_{P,\omega}$ are disjoint.

Proof. First, consider the case $\omega \in \mathcal{E}_2(M)$. Theorem 2.5.8 implies a stronger statement than that the composition series is at most of length [W]; it implies that the commuting algebra has at most this dimension. Theorem 4.6.1 and analytic continuation, say, of the characters (cf. §5.3.1), imply the equivalence of $\pi_{P_1,\omega}$ and π_{P_2,ω^s}. This is also easily deduced from Frobenius reciprocity combined with the tempered version of Theorem 5.4.1.1. All of statement (2) follows from Theorem 2.5.8. Theorem 2.5.8 implies even more than we have formulated here (i.e., $\omega \in \mathcal{E}'(M)$ instead of $\mathcal{E}_2(M)$ suffices in the hypothesis).

Now let $\omega \in {}^0\mathcal{E}_{\mathbb{C}}(M)$. Since the characters of $\pi_{P_1,\omega}$ and π_{P_2,ω^s} are equal (cf. §5.3.1), the composition series of these representations are isomorphic; for this, it is sufficient to take $\omega \in \mathcal{E}_{2,\mathbb{C}}(M)$.

Let us next prove (2) for $\omega \in {}^0\mathcal{E}_{\mathbb{C}}(M)$. Let π_0 be a component of $\pi_{P,\omega}$, $P \in \mathcal{P}(A)$. By Lemma 5.4.1.5, π_0 is not supercuspidal. Therefore, by Corollary 2.4.2 and Theorem 3.3.1, there exists a semistandard p-pair (P', A') $(P' = M'N')$ and $\omega' \in {}^0\mathcal{E}_{\mathbb{C}}(M')$ such that π_0 is a quotient of $\pi_{P',\omega'}$. By Corollar 5.3.2.2, A' is conjugate to a subtorus of A; in fact, by the same argument, A' and A are necessarily conjugate, so we may assume $A = A'$ and $P' \in \mathcal{P}(A)$. Thus, $\omega' \in \mathfrak{X}_{\pi_0}(P', A) \subset \mathfrak{X}_{\pi_{P,\omega}}(P', A)$ for some $P' \in \mathcal{P}(A)$, by Theorem 3.3.1 and Lemma 5.4.1.2 (note also Theorem 3.2.4). Now Theorem 5.4.1.1 and Lemma 1.13.1 imply that $\omega' = \omega^s$ for some $s \in W$. Thus, (2) is proved.

To prove that, for $\omega \in {}^0\mathcal{E}_{\mathbb{C}}(M)$, the length of a composition series of $\pi_{P,\omega}$ $(P \in \mathcal{P}(A))$ is bounded by $[W]$ we shall argue by induction on $\dim A/Z$.

Lemma 5.4.4.2. Theorem 5.4.4.1 is true for the case $\dim A/Z = 1$.

Proof. As before, Lemma 5.4.1.5 and Corollary 5.3.2.2 imply that if $\pi_0 \prec \pi_{P,\omega}$, then either $\mathfrak{X}_{\pi_0}(P, A)$ or $\mathfrak{X}_{\pi_0}(\bar{P}, A)$ is nonempty. By using Theorem 5.4.1.1 and Lemma 5.4.1.5 we conclude that $\pi_{P,\omega}$ has a composition series of length at most two (if $[W] = 1$, this is immediately clear; if $[W] = 2$, it follows from the fact that P and $\bar{P}$ are conjugate).

It remains to check only that, in the case $[W] = 1$, $\pi_{P,\omega}$ is irreducible. For $\omega \in {}^0\mathcal{E}(M)$ this is contained in Theorem 2.5.9 (or Theorems 5.4.1.1 and 5.7.7). Replacing $\pi_{P,\omega}$ by $\pi_{\bar{P},\omega}$, if necessary, we need consider only the case $\omega = (\omega_0)_\nu$, where $\omega_0 \in {}^0\mathcal{E}(M)$ and $\operatorname{Im}\nu > 0$. If $\mathfrak{X}_{\pi_0}(P, A)$ and $\mathfrak{X}_{\pi_0}(\bar{P}, A)$ are both non-empty, then $\pi_0 = \pi_{P,\omega}$ is irreducible. If $\mathfrak{X}_{\pi_0}(\bar{P}, A) = \phi$, say, then Theorem 5.4.4 implies that $C(\pi_0) \in \mathcal{E}_{2,\mathbb{C}}(G)$, whence Corollary 3.2.6 implies $\mathfrak{X}_{\pi_0}(\bar{P}, A) \neq \phi$. Contradiction.

Corollary 5.4.4.3. Let $\dim A/Z = 1$. Then, if $0 \neq f \in \mathcal{A}(\pi_{P,\omega})$, $f_Q \neq 0$ for all $Q \in \mathcal{P}(A)$.

Proof. If $f_Q = 0$ for all $Q \in \mathcal{P}(A)$, then, by Corollary 5.3.2.2 and Lemma 2.6.3 $f_{Q'} = 0$ for all proper p-pairs (Q', A') of G, so f is a supercusp form. We know (Lemma 5.4.1.5) that this is impossible. Let $f_{Q_1} \neq 0$ $(Q_1 \in \mathcal{P}(A))$. If $[W(A)] = 1$, then $\pi_{P,\omega}$ is irreducible. Since $\mathfrak{X}_{\pi_{P,\omega}}(Q_1, A) = \mathfrak{X}_{\pi_{P,\omega}}(\bar{Q}_1, A)$ and both are nonempty, we conclude that $f_{\bar{Q}_1} \neq 0$ too. If $[W(A)] = 2$, then

$\underset{\sim}{f}_{Q_1}(m^y) = \tau(y)\underset{\sim}{f}_{\bar{Q}_1}(m)\tau(y^{-1})$ $(m \in M)$ for some $y \in K$, so both are nonzero.

Let π be an admissible representation of G in a vector space V. Given $C \in \mathcal{E}_{\mathbb{C}}(G)$, we say that π is pure of class C or C-pure if every irreducible subquotient of π is of class C. For $C \in \mathcal{E}_{\mathbb{C}}(G)$ fixed we write V_C for the (obviously unique) subspace of V which is pure of class C and contains every such subspace of V. Obviously, if $C_1 \neq C_2 \in \mathcal{E}_{\mathbb{C}}(G)$, then $V_{C_1} \cap V_{C_2} = (0)$. In general it is not true, however, that $V = \sum_{C \in \mathcal{E}_{\mathbb{C}}(G)} V_C$ (e.g., for $SL(2,\Omega)$ the representation $\mathrm{Ind}_{P_0}^G 1$ does not contain the special representation as a subrepresentation--the only pure subspace is one-dimensional.

Lemma 5.4.4.4. Let π be an admissible representation of G in V. Let $C \in {}^0\mathcal{E}_{\mathbb{C}}(G)$. Then there exists a G-submodule V_C of V which is pure of class C and C does not occur as a subquotient in V/V_C.

Proof. Since π is admissible, C can occur as a subquotient of π only finitely often. Therefore, without loss of generality, we may assume that V is a G-module of finite type. We may even assume that there exists $\chi \in \mathfrak{X}(Z)$ such that $V = V_\chi = \{v \in V \mid \prod_{i=1}^{\ell}(\chi(z_i) - \pi(z_i))v = 0$ for every $(z_1, \ldots, z_\ell) \in Z^\ell\}$. If $\ell = 1$, the lemma follows from Lemma 5.4.1.4. The general case follows via an obvious inducti

Corollary 5.4.4.5. If every irreducible subquotient of π is supercuspidal, then V is the direct sum of its C-pure components.

For $s \in W(G/A)$ we write $\mathcal{A}(M)_{\omega^s}$ for the (right) ω^s-pure subspace of $\mathcal{A}$

Corollary 5.4.4.6. Let $f \in \mathcal{A}(\pi_{P,\omega})$. Then $f_P \in \bigoplus_{s \in W/W(\omega)} \mathcal{A}(M)_{\omega^s}$. Moreover, if $\underset{\sim}{f}_Q = 0$ for some $Q \in \mathcal{P}(A)$, then $f = 0$.

Proof. Let $f \in \mathcal{A}(\pi_{P,\omega})$. Using Theorem 3.2.4 and Lemma 5.4.1.2, we see that, if $\omega' \in \mathfrak{X}_f(P, A)$, then $\omega' \in \mathfrak{X}_{\pi_0}(P, A)$ for some $\pi_0 < \pi_{P,\omega}$. Thus, $\mathfrak{X}_f(P, A) \subset \{\omega^s\}_{s \in W}$ (Theorem 5.4.1.1). By Corollary 5.4.4.5, $f_P \in \oplus \mathcal{A}(M)_{\omega^s}$.

To deduce that $\underset{\sim}{f}_Q = 0$ for some $Q \in \mathcal{P}(A)$ only if $f = 0$, we note first that $\underset{\sim}{f}_{Q_1} \neq 0$ for some $Q_1 \in \mathcal{P}(A)$ (Corollary 5.3.2.2). Let Q_1, Q_2 be adjacent elements of $\mathcal{P}(A)$. Let $(P', A') \succ (Q_i, A)$ $(i = 1, 2)$ and assume that $\dim A/A' = 1$. Let $P' = M'N'$ and set $(Q_i^*, A) = (Q_i \cap M', A)$. These are maximal p-pairs of (M', A'). The assumption $0 \neq \underset{\sim}{f}_{Q_1} = (\underset{\sim}{f}_{P'})_{*Q_1}$ implies that $(\underset{\sim}{f}_{P'})_{*Q_2} = \underset{\sim}{f}_{Q_2} \neq 0$ (Corollary 5.4.4.3). Hence $\underset{\sim}{f}_Q \neq 0$ for all $Q \in \mathcal{P}(A)$, if $f \neq 0$.

Let us conclude the proof of Theorem 5.4.4.1. By Corollary 5.4.4.6, $\mathfrak{X}_{\pi_0}(Q, A) \neq \phi$ for all components π_0 of $\pi_{P,\omega}$ and all $Q \in \mathcal{P}(A)$. Therefore, by the exactness of the Jacquet functor (Lemma 5.4.1.2), the length of a composition series for $\pi_{P,\omega}$ is no greater than the length of a composition series for the Jacquet module, hence, no greater than $[W]$ (Theorem 5.4.1.1).

5.4.5. <u>On the Existence and Finiteness of the Set of Special Representations Associated to a Complex Orbit</u>

Let (P, A) $(P = MN)$ be a standard p-pair of G. Fix $\sigma_0 \in \omega_0 \in {}^0\mathcal{E}(M)$. For any $\nu \in \mathfrak{a}_{\mathbb{C}}^*$ let ω_ν denote the class of the representation $\sigma_\nu = \sigma_0 \chi_\nu$ $\chi_\nu(m) = q^{\langle\sqrt{-1}\nu, H(m)\rangle}$ $(m \in M)$. Write π_{P,ω_ν} for the induced representation $\mathrm{Ind}_P^G(\delta_P^{\frac{1}{2}}\sigma_\nu)$, $\nu \in \mathfrak{a}_{\mathbb{C}}^*$.

Let $\mathcal{O}_{\mathbb{C}}$ denote the complex orbit $\omega_0 \cdot \mathfrak{a}_{\mathbb{C}}^* \subset {}^0\mathcal{E}_{\mathbb{C}}(M)$. For any $\nu \in \mathfrak{a}_{\mathbb{C}}^*$ write χ_{ω_ν} for the central exponent of ω_ν. Let $\chi = \chi_{\omega_0}$. Write $\mathcal{O}_{\mathbb{C}}(\chi)$ for the set of all $\omega_\nu \in \mathcal{O}_{\mathbb{C}}$ such that $\chi_{\omega_\nu} = \chi$.

A class $\omega_\nu \in \mathcal{E}_{\mathbb{C}}(\chi)$ $(\nu \in \mathfrak{a}_{\mathbb{C}}^*)$ is called <u>special with respect to</u> G if in the composition series for π_{P,ω_ν} there occur representations of class $\mathcal{E}_2(G)$.

The main result of this section will be:

<u>Theorem</u> 5.4.5.1. The set $\mathcal{E}_{\mathbb{C}}(\chi)$ contains only finitely many points which are special with respect to G.

Let $\Lambda(\omega_0)$ denote the set of all $\nu \in \mathfrak{a}^*$ such that $\omega_{\sqrt{-1}\nu}$ is special. By Lemma 5.4.1.5, $0 \notin \Lambda(\omega_0)$. Furthermore, if $\nu \neq 0$ and $\pi_{P,\omega_{\sqrt{-1}\nu}}$ is <u>irreducible</u>, then $\omega_{\sqrt{-1}\nu}$ is <u>not special</u>. (<u>Proof</u>: Since a discrete series representation has a unitary central exponent, we may assume ν is orthogonal to $\mathfrak{z}$. Since $\pi_{P,\omega_{\sqrt{-1}\nu}}$ is equivalent to $\pi_{\overline{P},\omega_{\sqrt{-1}\nu}}$, $\chi_{\omega\sqrt{-1}\nu}$ is an exponent with respect to both (P,A) and $(\overline{P},A)$. Clearly, either $|\chi_{\omega_{\sqrt{-1}\nu}}(a)| \to \infty$ as $a \underset{P}{\to} \infty$ or as $a \underset{\overline{P}}{\to} \infty$. Either alternative combined with, say, Lemma 4.5.3(2) implies that $\pi_{\omega_{\sqrt{-1}\nu}}$ is not even tempered.)

Let us discuss the case $\dim A/Z = 1$. In this case $[W] = 1$ or 2. If $[W] = 1$, then Theorem 5.4.4.1 implies that $\pi_{P,\omega_{\sqrt{-1}\nu}}$ is irreducible. The discussio above proves that, in this case, $\omega_{\sqrt{-1}\nu}$ is not special. We shall give precise results for the case $[W] = 2$. Recall that $\pi_{P,\omega_{\sqrt{-1}\nu}}$ is reducible if and only if $\mu(\omega_0 : \sqrt{-1}\nu)$ is not defined (Lemma 5.4.2.4), in which case the length of a composition series is precisely two (Theorem 5.4.4.1).

<u>Lemma</u> 5.4.5.2. Let $\dim A/Z = 1$, $[W] = 2$. Let $\nu \in \mathfrak{a}^{*+}$ and assume ν orthogor to $\mathfrak{z}$. If $\pi_{P,\omega_{\sqrt{-1}\nu}}$ is reducible, then the irreducible quotient representation is of class $\mathcal{E}_2(G)$.

<u>Proof.</u> Let π denote the irreducible quotient representation of $\pi_{P,\omega_{\sqrt{-1}\nu}}$. Since P and $\overline{P}$ are conjugate, it follows from Corollary 5.3.2.2 that (P,A) is the only

proper standard p-pair of G with respect to which π has class exponents. By Corollary 5.4.4.3 both components of $\pi_{P,\omega_{\sqrt{-1}\nu}}$ have class exponents with respect to (P,A). Since the corresponding Jacquet space has a composition series of length two, and since the mapping from $\pi_{P,\omega_{\sqrt{-1}\nu}}$ to its Jacquet representation is exact, each composition factor has exactly one class exponent with respect to (P,A). Since π occurs as a quotient of $\pi_{P,\omega_{\sqrt{-1}\nu}}$, Theorem 3.3.1 implies that $\omega_{\sqrt{-1}\nu} \in \mathcal{E}_\pi(P,A)$; therefore, $\{\omega_{\sqrt{-1}\nu}\} = \mathfrak{X}_\pi(P,A)$. Since $\nu \in \mathfrak{a}^{*+}$, $|\chi_{\omega_{\sqrt{-1}\nu}}(a)| = q^{-\langle\nu, H(A)\rangle}$, where ν is a positive multiple of $\alpha \in \Sigma^0(P,A)$. Theorem 4.4.4, therefore, implies that π belongs to the discrete series of G.

Theorem 5.4.5.3. Let $\dim A = 1$. If ω_0 is unramified, then $\Lambda(\omega_0) = \phi$. The set $\Lambda(\omega_0)$ is finite and does not contain zero. If $\nu \in \Lambda(\omega_0)$, then so is $-\nu$. The set $\Lambda(\omega_0)$ consists exactly of the points $\nu \in \mathfrak{a}^*$ such that $\pi_{P,\omega_{\sqrt{-1}\nu}}$ is reducible.

Proof. We have remarked that $0 \notin \Lambda(\omega_0)$ and that $\nu \in \Lambda(\omega_0)$ implies $\pi_{P,\omega_{\sqrt{-1}\nu}}$ reducible. Thus, by Lemma 5.4.5.2, $\nu_0 \in \Lambda(\omega_0)$ if and only if $\pi_{P,\omega_{\sqrt{-1}\nu_0}}$ is reducible, if and only if $\mu(\omega_0 : \nu)$ is not analytic at $\nu = \sqrt{-1}\nu_0$ (Lemma 5.4.2.4). Thus, $\Lambda(\omega_0)$ is finite (Lemma 5.4.2.5).

Let us show that, if ω_0 is unramified, then $\Lambda(\omega_0) = \phi$. Assume that $\nu_0 \in \Lambda(\omega_0)$. Let $\pi_0 \prec \pi_{P,\omega_{\sqrt{-1}\nu_0}}$ and assume that $C(\pi_0) \in \mathcal{E}_2(G)$. In this case $\mathcal{E}_{\pi_0}(P,A) \subsetneq \{\omega_{\sqrt{-1}\nu_0}, (\omega_{\sqrt{-1}\nu_0})^s\} = \mathfrak{X}_{\pi_{P,\omega_{\sqrt{-1}\nu_0}}}(P,A) = \mathfrak{X}_{\pi_{P,\omega_{\sqrt{-1}\nu_0}}}(\bar{P},A)$. Assume $\mathcal{E}_{\pi_0}(P,A) = \{\omega_{\sqrt{-1}\nu_0}\}$. Then Corollary 3.2.6 implies that $\mathfrak{X}_{\pi_0}(\bar{P},A) = \{(\omega_{\sqrt{-1}\nu_0})^*\}$. Since $\omega_{\sqrt{-1}\nu_0}$ is not unitary, $(\omega_{\sqrt{-1}\nu_0})^* = (\omega_{\sqrt{-1}\nu_0})^s$. On the other hand, $(\omega_{\sqrt{-1}\nu_0})^* = (\omega_0^*)_{-\sqrt{-1}\nu_0} = \omega_{-\sqrt{-1}\nu_0}$. Since

$(\omega_{\sqrt{-1}\nu_0})^s = (\omega_0^s)_{-\sqrt{-1}\nu_0}$, it follows that $\omega_0^s = \omega_0$. Thus, if $\omega_{\sqrt{-1}\nu_0}$ is special, ω_0 is ramified.

Since $\mu(\omega_0^s : s\nu) = \mu(\omega_0 : -\nu)$, ω_0 ramified, we see that $\nu_0 \in \Lambda(\omega_0)$ if and only if $-\nu_0 \in \Lambda(\omega_0)$. We have already noted that

$$\Lambda(\omega_0) = \{\nu \in \mathfrak{a}^* - (0) \mid \pi_{P,\omega_{\sqrt{-1}\nu}} \text{ is reducible}\}.$$

Corollary 5.4.5.4. Assume that $\dim \underset{\sim}{A}/\underset{\sim}{Z} = 1$. Then $\mathcal{O}_{\mathbb{C}}(\chi)$ contains only finitely many special points. If $\mathcal{O}$ contains no ramified point, then $\mathcal{O}_{\mathbb{C}}$ contains no special point.

Proof. There are at most finitely many ramified points $\omega_0 \in \mathcal{O}(\chi)$. Associated with each ramified point there are only finitely many special points, so $\mathcal{O}_{\mathbb{C}}(\chi)$ can contain only finitely many special points, none if the set of ramified points in $\mathcal{O}(\omega)$ is empty.

Now we drop the assumption $\dim \underset{\sim}{A}/\underset{\sim}{Z} = 1$. Let $\Sigma_r(G, A)$ denote the set of reduced A-roots. Given $\alpha \in \Sigma_r(G, A)$ let $\underset{\sim}{A}_\alpha = (\ker \alpha)^0$ be the largest subtorus of $\underset{\sim}{A}$ contained in the kernel of the root character α. Set $A_\alpha = \underset{\sim}{A}_\alpha \cap G$ and $M_\alpha = Z_G(A_\alpha)$. Then A is a standard torus of M_α. If $P \in \mathcal{P}(A)$ and $\alpha \in \Sigma^0(P, A)$, then $(P, A)_{\{\alpha\}} = (P', A')$ $(P' = M'N')$, where $A' = A_\alpha$ and $M' = M_\alpha$ and $N' \subset N$. We set $({}^*P, A) = (M_\alpha \cap P, A)$ $({}^*P = {}^*M\,{}^*N = M \cdot N \cap M_\alpha = M \cdot N_\alpha)$. There is a reflection $s_\alpha \in W(G/A)$ associated to α if and only if the group $W(M_\alpha/A)$ contains two elements. In the following we sometimes identify $\mathfrak{a}$ and $\mathfrak{a}^*$ via a W-invariant scalar product.

To any $\omega \in {}^0\mathcal{E}_{\mathbb{C}}(M)$ there corresponds a unique pair $(\omega_0, \nu_\omega) \in {}^0\mathcal{E}(M) \times \mathfrak{a}$ such that $\omega = (\omega_0)_{\sqrt{-1}\nu_\omega}$. Fix $\omega_0 \in {}^0\mathcal{E}(M)$ and let $\Lambda(\omega_0, \alpha)$ denote the set of all real numbers λ such that $\omega = (\omega_0)_{\sqrt{-1}\lambda\alpha}$ is special with respect to M_α, where $\alpha \in \Sigma_r(G, A)$. We note that Corollary 5.4.5.4 implies that $\Lambda(\omega_0, \alpha)$ is a finite set which does not contain zero; moreover, $\Lambda(\omega_0, \alpha) = -\Lambda(\omega_0, \alpha)$.

Lemma 5.4.5.5. Let P_1 and P_2 be adjacent elements of $\mathcal{P}(A)$. Let $\alpha \in \Sigma^0(P_1, A)$ and $-\alpha \in \Sigma^0(P_2, A)$. Let $\omega \in \mathcal{E}_{\mathbb{C}}(\chi)$, $\omega = (\omega_0)_{\sqrt{-1}\nu_\omega}$ $(\nu_\omega \in \mathfrak{a}^*)$. Assume there is an element $f \in \mathcal{A}(G)$ such that:

1) $\underset{\sim}{f}_P \in {}^0\mathcal{A}(M, \tau_M)$ for all $P \in \mathcal{P}(A)$;

2) $\underset{\sim}{f}_{P_1,\omega} \neq 0$, $\underset{\sim}{f}_{P_2,\omega} = 0$.

Then:

1) $s_\alpha \in W(G/A)$; 2) $s_\alpha \omega_0 = \omega_0$; and 3) $|\alpha|^{-2}\langle\nu_\omega, \alpha\rangle \in \Lambda(\omega_0, \alpha)$.

Proof. We argue by induction on $\dim A/Z = \ell$.

If $\ell = 1$, $\mathcal{P}(A) = \{P, \bar{P}\}$ and hypothesis 2) implies $[W(A)] = 2$ (Corollary 5.4.4.6). Indeed, 2) implies that ω is special with respect to G, so all three conclusions follow from Theorem 5.4.5.3.

For $\ell \geq 2$ we consider $(P', A') = (P_1, A)_{\{\alpha\}} = (P_2, A)_{\{-\alpha\}}$ $(P' = M'N')$. Then $({}^*P_i, A) = (M' \cap P_i, A)$ is a maximal p-pair of $M' = M_\alpha$. Let f be as in the hypothesis. Then $\underset{\sim}{f}_{P_i} = (\underset{\sim}{f}_{P'})_{*P_i}$ $(i = 1, 2)$, so $\underset{\sim}{f}_{P_i,\omega} = (f_{P'})_{*P_i,\omega}$. By the induction hypothesis $f_{P'}$ satisfies 1), 2), and 3) with respect to M_α, ${}^*P_1$ and ${}^*P_2$, so $s_\alpha \in W(M_\alpha/A) \subset W(G/A)$, $s_\alpha\omega_0 = \omega_0$, and $|\alpha|^{-2}\langle\nu_\omega, \alpha\rangle \in \Lambda(\omega_0, \alpha)$.

Let $\mathfrak{a}$ be the real Lie algebra of A and let $\mathfrak{a}'$ be the regular points of $\mathfrak{a}$. For any chamber $\mathfrak{C} \subset \mathfrak{a}'$ there is a unique parabolic subgroup $P_{\mathfrak{C}} \in \mathcal{P}(A)$ such that $\Sigma(P_{\mathfrak{C}}, A) = \{\alpha \in \Sigma(G, A) \mid \alpha|\mathfrak{C} > 0\}$. The correspondence between chambers $\mathfrak{C} \subset \mathfrak{a}'$ and elements $P_{\mathfrak{C}} \in \mathcal{P}(A)$ is, as we know (§0.5), a bijection. Given a simple root $\alpha \in \Sigma^0(P_{\mathfrak{C}}, A)$ there is a unique chamber $\mathfrak{C}_\alpha$ adjacent to $\mathfrak{C}$ such that $-\alpha \in \Sigma^0(P_{\mathfrak{C}_\alpha}, A)$.

Let $\sigma \in \omega \in {}^0\mathcal{E}_{\mathbb{C}}(M)$ and $P \in \mathcal{P}(A)$. Set $\pi = \pi_{P,\omega} = \mathrm{Ind}_P^G(\delta_P^{\frac{1}{2}}\sigma)$. Let (V^0, τ^0) denote $C^\infty(K \times K)$ with the usual double representation of K defined on it. Let $0 \neq f \in \mathcal{A}(\pi, \tau^0)$. Set $f_{\mathfrak{C}} = f_{P_{\mathfrak{C}}}$ and $f_{\mathfrak{C},\omega'} = f_{P_{\mathfrak{C}},\omega'}$ ($\omega' \in {}^0\mathcal{E}_{\mathbb{C}}(M)$). (We know that, for any chamber $\mathfrak{C} \subset \mathfrak{a}'$, $f_{\mathfrak{C}} \in {}^0\mathcal{A}(M, \tau_M)$ (Corollary 5.4.4.6); moreover, $f_{\mathfrak{C}} \neq 0$ and $f_{\mathfrak{C}} = \sum_{s \in W(A)} f_{\mathfrak{C},\omega^s}$.) Let

$$\mathfrak{C}(\omega) = \{\mathfrak{C} \mid f_{\mathfrak{C},\omega} \neq 0\}.$$

Then $\mathfrak{C}(\omega) \neq \phi$, since $f_{\mathfrak{C},\omega^s} \neq 0$ implies $sf_{s^{-1}\mathfrak{C},\omega} \neq 0$. As an immediate consequence of Lemma 5.4.5.5 we obtain the following:

<u>Lemma</u> 5.4.5.6. Suppose $\mathfrak{C} \in \mathfrak{C}(\omega)$ and $\alpha \in \Sigma^0(P_{\mathfrak{C}}, A)$. Then $\mathfrak{C}_\alpha \in \mathfrak{C}(\omega)$ unless $s_\alpha \in W$, $s_\alpha\omega_0 = \omega_0$, and $|\alpha|^{-2}\langle\nu_\omega, \alpha\rangle \in \Lambda(\omega_0, \alpha)$.

Let $\Phi_0(\omega)$ be the set of reduced A-roots α for which there exists a chamber $\mathfrak{C} \in \mathfrak{C}(\omega)$ such that $\alpha \in \Sigma^0(P_{\mathfrak{C}}, A)$ and $\mathfrak{C}_\alpha \notin \mathfrak{C}(\omega)$. We call $\Phi_0(\omega)$ the set of ω-singular roots. We intend to show that, if $\omega \in {}^0\mathcal{E}_{\mathbb{C}}(M)$ is special with respect to G, then $\Phi_0(\omega)$ contains ℓ linearly independent roots, where $\ell = \dim A/Z$.

<u>Theorem</u> 5.4.5.7. Let π as above. Let $0 \neq f \in \mathcal{A}(\pi)$ and assume that $\int_{{}^0G} |f(x)|^2 dx < \infty$. Then $\Phi_0(\omega)$ contains ℓ linearly independent roots.

<u>Proof</u>. Let $\mathfrak{a}(\omega) = \{H \in \mathfrak{a} \mid \alpha(H) = 0$ for all $\alpha \in \Phi_0(\omega)\}$. Certainly $\mathfrak{z} \subset \mathfrak{a}(\omega)$.

We shall show that $\mathfrak{a}(\omega) = \mathfrak{z}$. Assume $\dim \mathfrak{a}(\omega)/\mathfrak{z} \geq 1$. Choose $H_0 \in \mathfrak{a}(\omega)$ such that $H_0 \neq 0$ and $H_0 \perp \mathfrak{z}$. Let $\mathfrak{a}'' = \{H \in \mathfrak{a} \mid \prod_{\alpha \in \Phi_0(\omega)} \alpha(H) \neq 0\}$. Then $\mathfrak{a}'' \supset \mathfrak{a}'$ and each connected component of $\mathfrak{a}''$ contains a union of chambers. Fix $\mathfrak{C} \in \mathfrak{C}(\omega)$. Let B denote the connected component of $\mathfrak{a}''$ which contains $\mathfrak{C}$. Then $B + \mathfrak{a}(\omega) = B$, since if $H \in B$ and $H_0 \in \mathfrak{a}(\omega)$, then $\alpha(H_0+H) = \alpha(H)$ for any $\alpha \in \Phi_0(\omega)$. Let $\mathfrak{C}_1, \ldots, \mathfrak{C}_r$ be the distinct chambers in B. Then, clearly, $\mathfrak{C}_i \in \mathfrak{C}(\omega)$, $i = 1, \ldots, r$.

We know that $f_{\mathfrak{C}_i,\omega} \neq 0$. Let $\chi_\omega \in \mathcal{X}(A)$ be the central exponent of $f_{\mathfrak{C}_i,\omega}$. By Theorem 4.4.4, $|\chi_\omega(a)| = q^{-\langle \nu_\omega, H(a)\rangle} \to 0$, $a \xrightarrow[\bar{\mathfrak{C}}_i \cap {}^0G]{} \infty$, $i = 1, \ldots, r$.

Let $H(t) = H_i \pm tH_0$. Then $H(t) \in \bar{B}$, so $H(t) \in \bar{\mathfrak{C}}_i$ for some $i = 1, \ldots, r$. On the other hand, choosing the sign of t properly, we have $\lim_{t \to \infty} \langle \nu_\omega, H(t)\rangle \to +\infty$, and this gives a contradiction.

<u>Proof of Theorem</u> 5.4.5.1. Let $\mathcal{O}_{\mathbb{C}}$ be an orbit in ${}^0\mathcal{E}_{\mathbb{C}}(M)$. Fix $\chi \in \hat{Z}$. We shall show that $\mathcal{O}_{\mathbb{C}}(\chi)$ contains only finitely many special points.

Let $\omega = (\omega_0)_{\sqrt{-1}\nu_\omega}$ ($\nu_\omega \in \mathfrak{a}^*$) be special. Then, by Theorem 5.4.5.7, $\Phi_0(\omega)$ contains ℓ linearly independent roots $\{\alpha_1, \ldots, \alpha_\ell\}$; the reflections $\{s_{\alpha_1}, \ldots, s_{\alpha_\ell}\} \subset W(G/A)$; $s_{\alpha_j}\omega_0 = \omega_0$; and $|\alpha_i|^{-2}\langle \nu_\omega, \alpha_i\rangle \in \Lambda(\omega_0, \alpha_i)$. On the other hand, $\mathcal{O}_{\mathbb{C}}(\chi)$ contains only finitely many points ω_0 which are fixed by ℓ reflections. For each $\alpha_i \in \Phi_0(\omega)$ we have only finitely many possibilities for $|\alpha_i|^{-2}\langle \nu_\omega, \alpha_i\rangle$ in $\Lambda(\omega_0, \alpha_i)$. Since there are ℓ linearly independent roots in $\Phi_0(\omega)$, this implies that there are only finitely many possibilities for $\nu_\omega \in \mathfrak{a}^*$. This combined with the finiteness of the set of fixed points $\omega_0 \in \mathcal{O}(\chi)$ implies

Theorem 5.4.5.1.

We conclude this subsection with an interesting and easy consequence of Lemma 5.4.5.6 and Theorem 5.4.5.7. For simplicity in its formulation and proof we assume that $Z = (1)$. The reader will see immediately that this result has a more general formulation.

Let A be a standard torus, $M = Z_G(A)$, and $\omega_0 \in {}^0\mathcal{E}(M)$, as before. Assume that there are reflections with respect to ℓ linearly independent elements of $\Sigma_r(G,A)$ in $W(\omega_0)$. In this context we will say that $\omega_{\sqrt{-1}\nu}$ is a <u>real class</u> if $\nu \in \mathfrak{a}^* - (0)$.

<u>Corollary</u> 5.4.5.8. If $\omega_{\sqrt{-1}\nu}$ is special with respect to G, then $\omega_{\sqrt{-1}\nu}$ is a real class.

<u>Proof.</u> The case $\dim A = 1$ is contained in Theorem 5.4.5.3. More generally, it follows from Lemma 5.4.5.6 and Theorem 5.4.5.7 that, if $\omega_{\sqrt{-1}\nu}$ is special, then ν lies in the intersection of ℓ distinct hyperplanes. Each of these hyperplanes is defined by equations with real coefficients, so their intersection, a point, lies in $\mathfrak{a}^*$.

§5.5. <u>Wave Packets, the Plancherel Measure, and Intertwining Operators</u>.

Fix a smooth unitary double representation (V,τ) of K which possesses an involution and satisfies associativity conditions (§1.12). Let A be an A_0-standard torus of G and let $M = Z_G(A)$.

§5.5.1. <u>The Algebras of Wave Packets</u>.

The real Lie algebra $\mathfrak{a}^* = X(A) \otimes_{\mathbb{Z}} \mathbb{R}$ of A operates on $\mathcal{E}_2(M)$.

ach orbit of $\mathfrak{a}^*$ in $\mathcal{E}_2(M)$ is a compact connected real manifold. The group (A) = W(G/A) operates on the set of orbits as well as on $\mathcal{E}_2(M)$ itself. We rite $W(\mathcal{O})$ for the stabilizer in W(A) of the orbit $\mathcal{O} \subset \mathcal{E}_2(M)$, i.e., $(\mathcal{O}) = \{s \in W(A) | \omega^s \in \mathcal{O}\}$, where $\omega \in \mathcal{O}$ is arbitrary. Fix a base-point $\omega_0 \in \mathcal{O}$ ıd set $W(\omega_0) = \{s \in W(\mathcal{O}) | \omega_0^s = \omega_0\}$; $W(\omega_0)$ is called the ramification or fixing roup of ω_0. We write $\omega_\nu = (\omega_0)_\nu$ for $\nu \in \mathfrak{a}^*$ (cf. §5. 3. 1).

Let $L^* = L_M^* = \{\nu \in \mathfrak{a}^* | \langle \nu, H(M) \rangle \subset \frac{2\pi\mathbb{Z}}{\log q}\}$, $L_A^* = \{\nu \in \mathfrak{a}^* | \langle \nu, H(A) \rangle \subset$ $\frac{\pi\mathbb{Z}}{gq}\}$, and $L^*(\omega_0) = \{\nu \in \mathfrak{a}^* | \omega_\nu = \omega_0\}$. Then $L_M^* \subseteq L^*(\omega_0) \subseteq L_A^*$. We also write $= \mathfrak{F}_M = \mathfrak{a}^*/L_M^*$, $\mathfrak{F}(\omega_0) = \mathfrak{a}^*/L^*(\omega_0)$, and $\mathfrak{F}_A = \mathfrak{a}^*/L_A^*$. The action of W(A) $\mathfrak{a}^*$ stabilizes the lattices L_A^* and L_M^*; $W(\mathcal{O})$ stabilizes $L^*(\omega_0)$. Thus, W(A) ts on $\mathfrak{F}$ and $\mathfrak{F}_A$ and $W(\mathcal{O})$ on $\mathfrak{F}(\omega_0)$. There exist natural projections $^* \to \mathfrak{F} \to \mathfrak{F}(\omega_0) \to \mathfrak{F}_A$, and we use these projection mappings, sometimes without plicitly saying so, to pull back functions. Similarly, we often expect the reader understand that we really should have pulled back a point of a quotient space a representative, or pushed down by a projection mapping. Modulo the above veats, we write $\omega^s = \omega_{\nu(s)}$ with $\nu(s) \in \mathfrak{F}(\omega_0)$, $\mathfrak{F}$, or $\mathfrak{a}^*$ $(s \in W(\mathcal{O}))$.

The function $\mu(\omega_0 : \nu)$ (cf. §5. 3. 5) may be regarded as a function on (ω_0). Recall that $\mu(\omega_0^s : s\nu) = \mu(\omega_0 : \nu)$ $(s \in W(A),\ \nu \in \mathfrak{F}(\omega_0))$; moreover, it clear that $\mu(\omega_{\nu_1} : \nu) = \mu(\omega_0 : \nu + \nu_1)$ $(\nu, \nu_1 \in \mathfrak{F}(\omega_0))$. Note that there corresponds any standard torus A' such that $Z \subseteq A' \subseteq A$ a centralizer M' and a function $_{A'}(\omega_0 : \nu)$, also defined on $\mathfrak{F}(\omega_0)$. In particular, $\mu_M(\omega_0 : \nu) = 1$ and $_0 : \nu) = \mu_G(\omega_0 : \nu)$. Recall also the constant $\gamma(M'/M)$ which is uniquely sociated to (M', A') and (M, A) as above (cf. §§5. 3. 4 and 5. 4. 3). We write $= \gamma(G/M)$; $\gamma(G/G) = 1$.

It is significant that the Eisenstein integrals and c-functions which correspond to $\mathfrak{A}(\omega_0)$ are, in general, defined on $\mathfrak{F}$ but not necessarily on $\mathfrak{F}(\omega_0)$. This will complicate some of the ideas we shall discuss in this section.

Write $L(\omega_0) = \mathcal{A}(\omega_0, \tau_M)$ and $L(\omega_0, P) = \mathcal{A}(\omega_0, \tau_P)$ for any $P \in \mathcal{P}(A)$ (cf. §5.2.1). Write $L^\infty(\omega_0) = C^\infty(\mathfrak{F}) \otimes L(\omega_0)$ and $L^\infty(\omega_0, P) = C^\infty(\mathfrak{F}) \otimes L(\omega_0, P)$. The space $L^\infty(\omega_0)$ is an algebra with respect to the multiplication derived, via the tensor product, from pointwise multiplication in $C^\infty(\mathfrak{F})$ and convolution with respect to M/A on $L(\omega_0)$. The space $L^\infty(\omega_0, P)$ is obviously, for any $P \in \mathcal{P}(A)$, a subalgebra of $L^\infty(\omega_0)$. We write $\alpha \cdot \beta$ for the product of α and $\beta \in L^\infty(\omega_0)$. We call $L^\infty(\omega_0)$ the space of <u>wave forms on</u> M associated to ω_0, $L^\infty(\omega_0, P)$ the space of <u>wave forms on</u> P associated to ω_0.

We wish next to define subspaces $I^\infty(\omega_0) \subset L^\infty(\omega_0)$ and $I^\infty(\omega_0, P) \subset L^\infty(\omega_0, P)$, of $L^*(\omega_0)$-<u>invariant wave forms on</u> M <u>and</u> P. The wave packet map from $L^\infty(\omega_0)$ or $L^\infty(\omega_0, P)$ into the Schwartz space on M or G will factor, by the natural projections, through the spaces of $L^*(\omega_0)$-invariant wave forms, so perhaps these spaces should be regarded as the proper analogue of the spaces of wave forms for real groups.

Recall that, if $\psi \in L(\omega_0)$, we write $\psi_\nu = \psi\chi_\nu \in L(\omega_\nu)$ $(\nu \in \mathfrak{a}^*)$ (i.e., $\psi_\nu(m) = \psi(m)q^{\sqrt{-1}\langle\nu, H(m)\rangle}$ $(m \in M)$). Similarly, set $\alpha(\nu : m)_\nu = \alpha(\nu : m)\chi_\nu(m)$ for $\alpha \in L^\infty(\omega_0)$. Then $\alpha(\nu)_\nu \in L(\omega_\nu)$ for all $\nu \in \mathfrak{a}^*$.

Next for any $\alpha \in L^\infty(\omega_0)$ and $\nu_1 \in \mathfrak{a}^*$ we set $\alpha^{\nu_1}(\nu : m) = \alpha(\nu+\nu_1 : m)_{\nu_1}$ (i.e., $\alpha^{\nu_1}(\nu) = \alpha(\nu+\nu_1)\chi_{\nu_1}$) $(\nu \in \mathfrak{a}^*$, $m \in M)$. The mapping $\alpha \mapsto \alpha^{\nu_1}$ is obviously a bijection of $L^\infty(\omega_0)$ onto $L^\infty(\omega_{\nu_1})$. In particular, if $\nu_1 \in L^*(\omega_0)$, then $\alpha \mapsto \alpha^{\nu_1}$ maps $L^\infty(\omega_0)$ to $L^\infty(\omega_0)$, so far as we know not always trivially.

Define

$$I\alpha = [L^*(\omega_0) : L^*]^{-1} \sum_{\nu_1 \in L^*(\omega_0)/L^*} \alpha^{\nu_1}$$

r any $\alpha \in L^\infty(\omega_0)$. Then I projects $L^\infty(\omega_0)$ and $L^\infty(\omega_0, P)$ on subspaces which e denote, respectively, $I^\infty(\omega_0)$ and $I^\infty(\omega_0, P)$. These are the spaces of $L^*(\omega_0)$-variant wave forms mentioned above. It is obvious that $I^\infty(\omega_0)$ is a subalgebra $L^\infty(\omega_0)$, that $I^\infty(\omega_0, P)$ is, for any $P \in \mathcal{P}(A)$, a subalgebra of $I^\infty(\omega_0)$.

Before proceeding further, let us topologize our spaces of wave forms. et $\{\alpha^j\}_{j=1}^\infty$ be a sequence in $L^\infty(\omega_0)$. To say that $\lim_{j\to\infty}\alpha^j = \alpha \in L^\infty(\omega_0)$ will mean at the following conditions are fulfilled:

There exists an open compact subgroup K_0 of M such that, for all j sufficiently large and all $\nu \in \mathfrak{F}$ and $m \in M$, we have $\alpha^j(\nu : k_1mk_2) = \alpha^j(\nu : m)$ $k_1, k_2 \in K_0$). (In other words, $\alpha^j(\nu)$ lies in a finite-dimensional subspace of (ω_0), independent of ν.)

) For each fixed $m_0 \in M$ the sequence $\alpha^j(\nu : m_0)$ is convergent with respect to he topology induced by the Schwartz topology on $C^\infty(\mathfrak{F})$. (Recall that $\mathfrak{F}$ is compact; convergence in the Schwartz topology on $C^\infty(\mathfrak{F})$ means that, for each component of $\alpha^j(\nu : m_0)$, we have uniform convergence of the function and all partial derivatives.)

It is clear that $L^\infty(\omega_0, P)$, $I^\infty(\omega_0)$, and $I^\infty(\omega_0, P)$ are closed subalgebras of $L^\infty(\omega_0)$.

Now we proceed to define wave packets. Given $\alpha = \sum_{j=1}^r \alpha_j\psi_j \in L^\infty(\omega_0)$ $\alpha_j \in C^\infty(\mathfrak{F})$ and $\psi_j \in L(\omega_0)$), we know how to define, for any $\nu \in \mathfrak{a}^*$, the Eisenstein integral

$$E(P : \alpha(\nu) : \nu) = \sum_{j=1}^r \alpha_j(\nu)E(P : \psi_j : \nu)$$

(cf. §§5.2.2 and 5.3.5). The mapping $\alpha \mapsto \alpha(\nu) \mapsto E(P : \alpha(\nu) : \nu) \in \mathcal{A}(C_M^G(\omega_\nu, \tau)$ also factors through the projection of $L^\infty(\omega_0)$ on $L^\infty(\omega_0, P)$. As a mapping from $L^\infty(\omega_0, P)$ to $\bigcup_{\nu \in \mathfrak{F}(\omega_0)} \mathcal{A}(C_M^G(\omega_\nu), \tau)$, the Eisenstein integral is injective.

Let us normalize the Lebesgue measures $d\nu$ on $\mathfrak{F}$ and $\mathfrak{F}(\omega_0)$ such that $\int_{\mathfrak{F}(\omega_0)} d\nu = \int_{\mathfrak{F}} d\nu = 1$. Given $\alpha \in L^\infty(\omega_0)$, we define the <u>wave packet</u> $E(P : \alpha) = \int_{\mathfrak{F}} \mu(\omega_0 : \nu) E(P : \alpha(\nu) : \nu) d\nu$. Observing that $E(P : \alpha) = E(P : \alpha^{\nu_1})$ for any $\nu_1 \in L^*(\omega_0)$, we see that

$$E(P : \alpha) = E(P : I\alpha)$$
$$= \int_{\mathfrak{F}(\omega_0)} \mu(\omega_0 : \nu) E(P : I\alpha(\nu) : \nu) d\nu.$$

In the above, it is implicit that P is a parabolic subgroup of G and, hence, that $\mu(\omega_0 : \nu) = \mu_G(\omega_0 : \nu)$.

Recall that $\mathcal{C}(G, \tau) = C(G, \tau) \cap (\mathcal{C}(G) \otimes V)$.

<u>Theorem</u> 5.5.1.1. The mapping $\alpha \mapsto E(P : \alpha)$ is a continuous linear mapping of $L^\infty(\omega_0)$ into $\mathcal{C}(G, \tau)$.

<u>Proof.</u> Linearity is of course clear, so it suffices to take $\alpha = \alpha_1 \psi_1$ ($\alpha_1 \in C^\infty(\mathfrak{F})$, $\psi_1 \in L(\omega_0)$), to show first that $E(P : \alpha) \in \mathcal{C}(G, \tau)$ and second that any sequence $\{\alpha^j\} = \{\alpha_1^j \psi_1\}$ which converges to zero in $L^\infty(\omega_0)$ yields a null sequence $\{E(P : \alpha^j)\} \subset \mathcal{C}(G, \tau)$.

<u>Step</u> 1. We begin with the case $(P, A) = (G, Z)$, $\mu(\omega_0 : \nu) = 1$, $\mathfrak{a}^* = \mathfrak{z}^*$, and $\mathfrak{F} = \mathfrak{F}_G$, $E(P : \alpha) = \int_{\mathfrak{F}} \alpha_1(\nu) q^{\langle \sqrt{-1}\nu, H_G(x) \rangle} d\nu \psi_1(x)$ $(x \in G)$. For any $r > 0$ we obtain from Theorem 4.5.10 a constant $C_r > 0$ such that $\|\psi_1(x)\|_V \leq C_r \Xi(x)(1+\sigma_*(x))^{-r}$. Choosing a base $\beta_1, \ldots, \beta_r$ for $\mathfrak{z}^*$, we know that $1+\sigma(z) \asymp \max_{1 \leq i \leq r} (|\langle \beta_i, H_G(z) \rangle| + 1)$.

riting $\nu \in \mathfrak{F}$ in the form $\Sigma\nu_i\beta_i$ $(\nu_i \in \mathbb{R})$, letting D_i denote partial differentia-
on with respect to the i-th variable, and integrating by parts, we see that
$\int_{\mathfrak{F}} D_i\alpha_1(\nu)q^{<\sqrt{-1}\nu, H_G(x)>} d\nu \,|\prec (|<\beta_i, H_G(x)>|+1)|\int_{\mathfrak{F}}\alpha_1(\nu)q^{<\sqrt{-1}\nu, H_G(x)>} d\nu\,|$. Since
$\int_{\mathfrak{F}} D_i\alpha_1(\nu)q^{<\sqrt{-1}\nu, H_G(x)>} d\nu\,| \prec \int_{\mathfrak{F}} |D_i\alpha_1(\nu)|d\nu$, we conclude, via successive applica-
ons of the operators D_i, that there are constants $C'_r(x) > 0$ such that
$\int_{\mathfrak{F}}\alpha_1(\nu)q^{<\sqrt{-1}\nu, H_G(xz)>} d\nu\,| \leq C'_r(x)(1+\sigma(z))^{-r}$, for every $x \in G$, $z \in Z$, and $r > 0$.
hoosing a set of representatives S for the finite set $G/Z\,{}^0G$ such that, if
$_0 \in S$, then $\sigma_*(x) = \sigma(x_0)$, we set $C'_r = \max_{x_0 \in S} C'_r(x_0)$. Letting $C''_r = C_r C'_r$, noting
at $(1+\sigma(z))(1+\sigma(x_0)) \geq (1+\sigma(x_0 z))$, and multiplying inequalities we conclude that
$(P : \alpha) \in \mathcal{C}(G, \tau)$.

Now let $\alpha_1^j(\nu) \to 0$, $j \to \infty$, with respect to the Schwartz topology on
$= \mathfrak{F}_G$, i.e., every p.d. of α_1^j tends uniformly to zero. Then, arguing as
ove, one sees that, for every positive integer j and $r > 0$, there is a positive
nstant $C'^j_r \prec \max_i \int_{\mathfrak{F}} |D_i^r \alpha_1^j(\nu)|d\nu \to 0$, $j \to \infty$, and such that

$\int_{\mathfrak{F}}\alpha_1^j(\nu)q^{<\sqrt{-1}\nu, H_G(x_0 z)>} d\nu\,| \leq C'^j_r(1+\sigma(z))^{-r}$ $(x_0 \in S,\ z \in Z)$. It follows that
$E(P : \alpha^j : x)\|_V \leq C_r^j\, \Xi(x)(1+\sigma(x))^{-r}$ for all $x \in G$, where $0 < C_r^j \to 0$, $j \to \infty$, so
$E(P : \alpha^j)\}$ is a null sequence in $\mathcal{C}(G, \tau)$.

tep 2. In order to begin an induction on the p-rank of (P, A) let us next consider
e case $\dim A/Z = 1$ and $E(P : \alpha : x) = \int_{\mathfrak{F}} \mu(\omega_0 : \nu)\alpha_1(\nu)E(P : \psi_1 : \nu : x)d\nu$. Obvi-
usly, $E(P : \alpha) \in C(G, \tau)$; we must, first, show that $E(P : \alpha)$ satisfies the in-

equalities which define the Schwartz space. We shall omit a proof of continuity, as a slight variation of the argument that $E(P:\alpha) \in \mathcal{C}(G,\tau)$ proves continuity. Since K is compact and τ continuous, we need check the inequalities for the Schwartz space only for $x = m \in M_0^+$. Since there is a constant c such that $0 < c < \delta_{P_0}^{\frac{1}{2}}(m)\Xi(m)$ $(m \in M_0^+)$, it suffices to show that, given $r > 0$, we may find a constant $C_r > 0$ such that $\| \delta_{P_0}^{\frac{1}{2}}(m)E(P:\alpha:m)\|_V \leq C_r(1+\sigma(m))^{-r}$ for all $m \in M_0^+$.

We may write $M_0^+ = ({}^0G \cap M_0^+)SZ$, where S is a finite set. We also have the decomposition $\mathfrak{F}_M = \mathfrak{F}_G + \mathfrak{F}_{{}^0G\cap M}$. In order to be able to argue essentially as in Step 1 it is sufficient to know that

$$\left\| \int_{\mathfrak{F}_{{}^0G\cap M}} \mu(\omega_0 : \nu_1 + {}^0\nu)\alpha(\nu_1 + {}^0\nu)\delta_{P_0}^{\frac{1}{2}}(m)E(P:\psi_1:\nu_1+{}^0\nu : m)d\,{}^0\nu \right\|_V \leq C_r(1+\sigma_*(m))^{-r}$$

$(m \in M_0^+)$ for some constant $C_r > 0$ depending upon $r > 0$ but independent of $\nu_1 \in \mathfrak{F}_G$. For $m = {}^0m x_0 z \in (M_0^+ \cap {}^0G)SZ$ with ${}^0m$ outside a compact subset of ${}^0G \cap M_0^+$, there is a p-pair (P', A') $(P' = M'N'; (G,Z) \neq (P',A') \succ (P_0, A_0)$; and ${}^*P_0 = P_0 \cap M')$ such that $\delta_{{}^*P_0}^{\frac{1}{2}}(m)E_{P'}(P:\psi_1:\nu:m) = \delta_{P_0}^{\frac{1}{2}}(m)E(P:\psi_1:\nu:m)$. If (P', A') is not associated to (P, A), then $E_{P'}(P:\psi_1:\nu:m) \to 0$, $m \underset{P'}{\to} \infty$, uniformly exponentially fast, so the required majorization is more than satisfied (also use boundedness of all p.d.'s of α_1). If (P', A') is associated to (P, A), then, since $\|f_{P^y}(m^y)\|_V = \|f_P(m)\|_V$ $(f \in \mathcal{A}(G,\tau),\ y \in K)$, it suffices to replace (P', A') by $(P'', A) = (P'^y, A'^y)$ and check the majorization for $E_{P''}(P:\psi:\nu:m^y)$.

We have

$$\int_{\mathfrak{F}_{0_{G\cap M}}} \mu(\omega_0 : \nu_1+{}^0\nu)\alpha_1(\nu_1+{}^0\nu)\delta^{\frac{1}{2}}_{*_{P_0}}(m)E_{P'}(P : \psi_1 : \nu_1+{}^0\nu : m)d^0\nu \|_V$$

$$\| \int_{\mathfrak{F}_{0_{G\cap M}}} \mu(\omega_0 : \nu_1+{}^0\nu)\alpha_1(\nu_1+{}^0\nu)\delta^{\frac{1}{2}}_{*_{P_0^y}}(m^y)E_{P''}(P : \psi_1 : \nu_1+{}^0\nu : m^y)d^0\nu \|_V$$

$$\leq \| \int_{\mathfrak{F}_{0_{G\cap M}}} \mu(\omega_0 : \nu_1+{}^0\nu)\alpha_1(\nu_1+{}^0\nu)\delta_{*_{P_0^y}}(m^y) \sum_{s\in W(A)} (c_{P''|P}(s : \omega_0 : \nu_1+{}^0\nu)\psi_1)(m^y)$$

$$q^{<\sqrt{-1}s^0\nu, H(m^y)>} d^0\nu \|_V.$$

Since $\mu(\omega_0 : \nu)c_{P''|P}(s : \omega_0 : \nu)$ is holomorphic, one easily sees that Step 2 reduces to Step 1.

Step 3. Now we let (P, A) have p-rank $\ell > 1$ and assume the theorem is true for all p-pairs of smaller p-rank. We shall in this case also prove only that $E(P : \alpha) \in \mathcal{C}(G, \tau)$. The reader can easily modify the argument given here in order to check continuity.

Because it is more convenient we shall actually prove a slightly more general statement. We consider wave packets of the form $E^0(P : \alpha) = \int_{\mathfrak{F}} E^0(P : \alpha(\nu) : \nu)d\nu$ where $\alpha \in \mathcal{L}^\infty(\omega_0) = L^\infty(\omega_0)$ or $\mathcal{L}^\infty(\omega_0, P) = \mathcal{L}(\omega_0, P) \otimes C^\infty(\mathfrak{F})$. Ostensibly, the space of wave packets of the former sort comprise a subspace of the present space of wave packets, i.e., to get wave packets of the former sort, the function $\alpha \in \mathcal{L}^\infty(\omega_0, P)$ that one takes in $E^0(P : \alpha)$ must vanish on certain hyperplanes. Actually, the Plancherel's formula (Corollary 5.5.2.2) implies that the spaces are the same. To set the stage for our more general discussion we remark first that in the case of p-rank zero (Step 1) there is no difference between

$E(P:\alpha)$ and $E^0(P:\alpha)$ whereas the only change in the proof that $E^0(P:\alpha) \in \mathcal{C}(G,\tau)$ in the case of p-rank one (Step 2) boils down to observing that $c^0_{P''|P}(s:\omega_0:\nu)$ is holomorphic on $\mathfrak{F}$ (Corollary 5.2.4.5 implies this); essentially the argument is the same.

To show that $\| \delta^{\frac{1}{2}}_{P_0}(m)E^0(P:\alpha:m)\|_V \leq C_r(1+\sigma(m))^{-r}$ for some $C_r > 0$ depending on $r > 0$ for all $m \in M_0^+$ we proceed as in Step 2 and find ourselves checking that $\| \int_{\mathfrak{F}_{0_{G\cap M}}} E^0_{P'}(P:\psi_1:\nu_1+\nu:m)\delta_{{}^*P_0}(m)^{\frac{1}{2}} d^0\nu \|_V$ decays as $m \underset{P'}{\to} \infty$, $m \in M_0^+$, for any $(P',A') \succ (P_0,A_0)$. If $W(A|A') = \phi$, then, as in Step 2, the decay is uniformly exponentially fast, so our inequalities will certainly be satisfied. If $W(A|A') \neq \phi$, then without loss of generality (again as in Step 2), we may replace (P',A') by (P'',A). In this case, we have $\delta^{\frac{1}{2}}_{{}^*P_0}(m)E^0_{P''}(P:\psi_1:\nu:m)$ $= \delta^{\frac{1}{2}}_{{}^*P_0}(m)_w E^0_{P''}$ + terms which decay exponentially fast, and therefore, may be neglected. Now Theorem 5.3.5.3(5) enables us to immediately get down to our induction hypothesis. Indeed, substituting the expressions for the constant term given in the cited theorem, we obtain wave packets for lower p-rank parabolic pairs. Thus, we may apply the induction hypothesis to complete the proof of Theorem 5.5.1.1.

Given $P_1, P_2 \in \mathcal{P}(A)$, $s \in W(A)$, and $\nu \in \mathfrak{F}$, we have

$$ {}^0c_{P_2|P_1}(s:\omega_0:\nu) : L(\omega_0) \to L(\omega_0, P_1) \to L(\omega_0^s, P_2) \hookrightarrow L(\omega_0^s)$$

(§5.3.5). Define

$$c^{P_2|P_1}(s:\omega_0) : L^\infty(\omega_0) \to L^\infty(\omega_0^s)$$

by setting

$$({}^{P_2|P_1}c(s:\omega_0)\alpha)(\nu) = {}^0c_{P_2|P_1}(s:\omega_0:s^{-1}\nu)\alpha(s^{-1}\nu)$$

$$= \sum_{j=1}^{r} \alpha_j(s^{-1}\nu)\,{}^0c_{P_2|P_1}(s:\omega_0:s^{-1}\nu)\psi_j$$

$\alpha = \sum_{j=1}^{r} \alpha_j \psi_j \in L^\infty(\omega_0)$. Sometimes we also regard ${}^{P_2|P_1}c(s:\omega_0)$ as a mapping n $L^\infty(\omega_0, P_1)$ to $L^\infty(\omega_0^s, P_2)$. It is easy to see that ${}^{P_1|P_2}c(s:\omega_0)$ maps $I^\infty(\omega_0)$ $I^\infty(\omega_0^s)$, $I^\infty(\omega_0, P_1)$ to $I^\infty(\omega_0^s, P_2)$. Note that $E(P_2 : {}^{P_2|P_1}c(s:\omega_0)\alpha) = E(P_1 : \alpha)$ any $\alpha \in L^\infty(\omega_0)$.

mma 5.5.1.2. Let $P_1, P_2, P_3 \in \mathcal{P}(A)$ and $s, t \in W(A)$. Then ${}^{P_3|P_1}c(st:\omega_0) =$ ${}^{|P_2}(s:\omega_0^t)\,{}^{P_2|P_1}c(t:\omega_0)$.

of. This follows immediately from Theorem 5.3.5.3 (1).

mma 5.5.1.3. Let $P \in \mathcal{P}(A)$ and $\omega_0 \in \mathcal{E}_2(M)$ as before. Let $f \in \mathcal{C}(M, \tau_M)$ $\psi \in L(\omega_0)$. Define $\alpha(\nu)_\nu = f * \psi_\nu$. Then $\alpha \in L^\infty(\omega_0)$.

of. We have

$$(f * \psi_\nu)(m) = \int_M f(x)\psi(x^{-1}m)\chi_\nu(x^{-1}m)dx$$

$$= \chi_\nu(m)\int_M f(x)\chi_{-\nu}(x)\psi(x^{-1}m)dx.$$

any fixed $\nu \in \mathfrak{a}^*$ it is clear that $f(x)\chi_{-\nu}(x) \in \mathcal{C}(M, \tau_M)$. Therefore, for each he integral defines an element of $L(\omega_0)$. Indeed, it is obvious that the function in a finite-dimensional subspace of $L(\omega_0)$ which is independent of ν.

It is sufficient to show that, for fixed $m_0 \in M$, $\alpha(\nu : m_0)$ is a C^∞-function of $\nu \in \mathfrak{a}^*$. To prove this one can, in fact, verify that the integral of any partial derivative is the partial derivative of the integral. This verification, which we omit, depends essentially only on the observation that $(1+\sigma(x))^r f(x)$ belongs to the Schwartz space of M for any $r > 0$.

Corollary 5.5.1.4. Let $P \in \mathcal{P}(A)$, $f \in \mathcal{C}(G,\tau)$, and $\alpha \in L^\infty(\omega_0)$. Then $f^P * \alpha(\nu)_\nu = \alpha'(\nu)_\nu$, where $\alpha' \in L^\infty(\omega_0)$, and $f * E(P : \alpha) = E(P : \alpha')$.

Proof. That $\alpha' \in L^\infty(\omega_0)$ follows from Lemma 5.5.1.3. By Lemma 5.2.2.1 we have, any fixed $\nu \in \mathfrak{F}$, $f{*}E(P : \alpha(\nu) : \nu) = f{*}E(P : \alpha(\nu)_\nu : 0) = E(P : f^P * (\nu)_\nu : 0) = E(P : \alpha'(\nu)_\nu : 0)$. Since all integrals in sight are absolutely convergent, we may conclude that $f * E(P : \alpha) = E(P : \alpha')$.

Our next goal is Theorem 5.5.1.9. Those readers who are interested only in the case $\omega_0 \in {}^0\mathcal{E}(M)$ may skip the next two lemmas, which prove a technical result needed for the general case.

Recall that, for any $f \in \mathcal{A}(G,\tau)$ and semi-standard p-pair (P', A') $(P' = M'N')$ of G, we may regard $f_{P'}$ as a function defined on all of G such that $f_{P'}(km'n') = \tau(k)f_{P'}(m')$ $(k \in K,\ m' \in M',\ n' \in N')$ (cf. §5.1, especially Lemma 5.1.11). We write $\overline{N}'$ for the unipotent radical of the p-subgroup opposite to P'.

Lemma 5.5.1.5. Let $P \in \mathcal{P}(A)$ and $\alpha \in L^\infty(\omega_0)$. Let (P', A') be a maximal semi-standard p-pair of G. Then

$$\int_{\overline{N}'} \delta_{P'}^{-\frac{1}{2}}(\overline{n}')(E_{P'} - {}_wE_{P'})(P : \alpha(\nu) : \nu : \overline{n}'x')d\overline{n}'$$

onverges absolutely for all $\nu \in \mathfrak{F}$ and $x' \in M'$. The value of the integral is ero. For any fixed $x' \in M'$ the convergence is also uniform with respect to $\in \mathfrak{F}$.

roof. Without loss of generality we may assume that (P', A') is standard, that $', A') \succ (P, A) \succ (P_1, A_1)$ $(P_1 = M_1N_1)$, and that there exists $\sigma_1 \in \omega_1 \in {}^0\mathcal{E}(M_1)$ and in the dual real Lie algebra $\mathfrak{a}_1^*$ of A_1 such that $\sigma_0 \in \omega_0$ occurs as a quotient presentation of $\mathrm{Ind}_{P_1 \cap M}^{M}(\delta^{\frac{1}{2}}_{P_1 \cap M}\sigma_1, \sqrt{-1}\nu_1)$. In this case, by Theorem 4.4.4, is a linear combination of the elements of $\Sigma^0(P_1 \cap M, A_1)$ with all positive coefcients. Let $\pi_\nu = \mathrm{Ind}_P^G(\delta_P^{\frac{1}{2}}\sigma_\nu)$, $\nu \in \mathfrak{F}$. Then every element of $\mathfrak{X}_{\pi_\nu}(P', A')$ is of the form $_{,\nu} = (\chi_{\omega_1, \sqrt{-1}\nu_1 + \nu})^s | A'$, where $s \in W(G/A)$ and $s\nu_1 \in +\mathfrak{a}_1^*$. Clearly, $\eta_{s,\nu}$ is itary if and only if $s\nu_1$ is a linear combination of the elements of $\Sigma^0(P_1 \cap M', A_1)$; herwise, $|\eta_{s,\nu}(a')| = |\chi^s_{\omega_1, \sqrt{-1}s\nu_1}(a')| = q^{-\langle s\nu_1, H(a')\rangle} \to 0$, $a' \xrightarrow[P']{} \infty$ uniformly on $_0$. It follows that there exists $\epsilon > 0$ such that

$$q^{\langle \rho_{P'}, H(a')\rangle}(E_{P'} - {}_wE_{P'})(P : \alpha(\nu) : \nu : x'a') \to 0, \quad a' \xrightarrow[P']{} \infty, \text{ for all } x' \in M' \text{ and}$$

$\in \mathfrak{F}$.

Let ${}^*P_0 = P_0 \cap M' = M_0{}^*N_0$. Since $\rho_{P_0} = \rho_{P'} + \rho_{{}^*P_0}$, it follows at, for any $\bar{n}' \in \bar{N}'$, $\langle \rho_{P_0}, H_{P_0}(\bar{n}')\rangle = \langle \rho_{P'}, H_{P'}(\bar{n}')\rangle + \langle \rho_{{}^*P_0}, H_{P_0}(\bar{n}')\rangle$. rite $\bar{n}' = km'n' \in \bar{N}'$ $(k \in K,\ m' \in M',\ n' \in N')$. We may write $' = a'{}^*m'$, where $a' \in A'^+$ and ${}^*m' \in M'_T$ for some $T > 0$ independent of $\in \bar{N}'$ (cf. §4.4 for the definition of G_T). Recall that $_{P'}, H_{P'}(\bar{n}')\rangle \asymp 1 + \sigma(\bar{n}') \to +\infty$, $\|\bar{n}'\| \to +\infty$ $(\bar{n}' \in \bar{N}')$ (Corollary 4.3.2). Clearly, $_{{}^*P_0}, H_{P_0}(m')\rangle = \langle \rho_{{}^*P_0}, H_{P_0}({}^*m')\rangle$. It follows from the remarks of the preeding paragraph that there is a positive constant C such that

$$\|(E_{P'} - {}_wE_{P'})(P : \alpha(\nu) : \nu : \bar{n}'x')\|_V \leq Cq^{-\epsilon<\rho_{P'}, H_{P'}(\bar{n}')>} \|(E_{P'} - {}_wE_{P'})(P{:}\alpha(\nu){:}\nu{:}\,{}^*m'x'$$

for all $\bar{n}' \in \bar{N}'$. In order that we may apply Theorem 4.3.7 and conclude that the integral converges absolutely and uniformly, it suffices to show that there exist positive constants C' and r such that

$$\delta^{\frac{1}{2}}_{{}^*P_0}({}^*m')\|(E_{P'} - {}_wE_{P'})(P : \alpha(\nu) : \nu : {}^*m'x'\|_V \leq C'(1+\sigma(\bar{n}'))^r \text{ for all } \bar{n}' \in \bar{N}'.$$

For this note that

$$\|(E_{P'} - {}_wE_{P'})(P : \alpha(\nu) : \nu : {}^*m'x')\|_V = \|(E_{P'} - {}_wE_{P'})(P : \alpha(\nu) : \nu : k_1{}^*m'x'k_2)\|_V$$

for all $k_1, k_2 \in K_{M'}$. Using the Cartan decomposition for the group M', we may assume that ${}^*m'' = k_1{}^*m'x'k_2 \in M_0 \cap M'_T$. Moreover, we may assume that $<\alpha, H({}^*m'')> \geq 0$ for all $\alpha \in \Sigma({}^*P_0, A_0)$. It follows from Theorem 2.6.1 that there is a compact subset $\mathcal{K}$ of $M_0 \cap M'_T$ such that, if ${}^*m'' \notin \mathcal{K}$, then $(E_{P'} - {}_wE_{P'})\cdot$ $(P : \alpha(\nu) : \nu : {}^*m'') = \delta^{-\frac{1}{2}}_{{}^*Q}({}^*m'')(E_{P'} - {}_wE_{P'})_{{}^*Q}(P : \alpha(\nu) : \nu : {}^*m'')$, where $({}^*Q, B)$ is a standard p-pair of (M', A'). There is a standard p-pair (Q, B) of G such that $(P', A') \succ (Q, B)$ and such that ${}^*Q = Q \cap M'$ (cf. §0.3). Moreover, $(E_{P'})_{{}^*Q} = E_Q$, so every exponent associated to $(E_{P'} - {}_wE_{P'})_{{}^*Q}$ is an exponent corresponding to E and (Q, B). It thus follows from Lemma 4.5.3 that

$$\delta^{\frac{1}{2}}_{{}^*P_0}({}^*m'')\|(E_{P'} - {}_wE_{P'})(P : \alpha(\nu) : \nu : {}^*m'x')\|_V \leq C'(1+\sigma({}^*m'x'))^r$$

for some C' and r. (Recall that we may assume $\sigma(k_1{}^*m'x'k_2) = \sigma({}^*m'x')$.) Since $\delta_{{}^*P_0}({}^*m'') \asymp \mu(K_{M'}{}^*m'x'K_{M'}) \asymp \mu(K_{M'}{}^*m'K_{M'}) \succ \delta_{{}^*P_0}({}^*m')$, we may replace $\delta_{{}^*P_0}({}^*m'')$ by $\delta_{{}^*P_0}({}^*m')$. We want to replace $\sigma({}^*m'x')$ by $\sigma(\bar{n}')$. We argue instead that we may ignore the "polynomial factor" altogether. On the one hand, $1+\sigma({}^*m'x') \asymp 1+\sigma({}^*m')$ and $1+\sigma({}^*m') \leq (1+\sigma(m'))(1+\sigma(a'))$. On the other nd, it is clear that

$\frac{\epsilon}{2}<\rho_{P'}, H_{P'}(\overline{n}')>$ $(1+\sigma(a'))(1+\sigma(m')) \to 0$, $\|\overline{n}'\| \to +\infty$ (roughly speaking, $\|\overline{n}'\| =$ x($\|m'\|$, $\|a'\|$)). Replacing ϵ by $\frac{\epsilon}{2}$, we take $r = 0$.

Now we shall show that the value of the integral is zero. If not, then there exist elements $\alpha \in L^{\infty}(\omega_0)$ for which

$$0 \neq \int_{\mathfrak{F}} \int_{\overline{N}'} \mu(\omega_0 : \nu)\delta_{P'}^{-\frac{1}{2}}(\overline{n}')(E_{P'} - {}_wE_{P'})(P : \alpha(\nu) : \nu : \overline{n}' \cdot)d\overline{n}'\, d\nu$$

(note that the double integral exists -- $\int_{\mathfrak{F}} \int_{\overline{N}'} = \int_{\overline{N}'} \int_{\mathfrak{F}}$). We claim that the mapping which sends $\alpha \in L^{\infty}(\omega_0)$ to the above double integral is a continuous linear mapping from $L^{\infty}(\omega_0)$ into $\mathcal{C}(M', \tau_{M'})$.

The linearity being obvious, we must check that the image lies in $\mathcal{C}(M', \tau_{M'})$ and that the mapping is continuous. Note, first, that

$$\alpha \mapsto E^{\overline{P}'}(P : \alpha) = \int_{\overline{N}'} \int_{\mathfrak{F}} \mu(\omega_0 : \nu)\delta_{P'}^{-\frac{1}{2}}(\overline{n}')E_{P'}(P : \alpha(\nu) : \nu : \overline{n}' \cdot)d\nu d\overline{n}'$$

does define a continuous linear mapping from $L^{\infty}(\omega_0)$ to $\mathcal{C}(M', \tau_{M'})$. To establish our claim it suffices to check that $\alpha \mapsto \int_{\overline{N}'} \delta_{P'}^{-\frac{1}{2}}(\overline{n}')\int_{\mathfrak{F}} \mu(\omega_0 : \nu)_w E_{P'} \cdot$ $(P : \alpha(\nu) : \nu : \overline{n}' \cdot)d\nu d\overline{n}'$ also defines a continuous mapping to $\mathcal{C}(M', \tau_{M'})$. This follows almost immediately from the fact that the inside integral defines a wave packet on M' together with Corollary 4.3.8.

To prove that the integral of the lemma is zero we shall show that anything else contradicts the continuity of the above mapping from $L^{\infty}(\omega_0)$ to $\mathcal{C}(M', \tau_{M'})$.

Let $C(\pi_\nu) = C_M^G(\omega_\nu)$ and let $\chi_\nu \in \mathfrak{X}_{\pi_\nu}(P', A')$, $\chi_\nu \notin \hat{A}'$. Consider the integral $\int_{\overline{N}'} \delta_{P'}^{-\frac{1}{2}}(\overline{n}')E_{P', \chi_\nu}(P : \alpha(\nu) : \nu : \overline{n}'x')d\overline{n}'$. It is of course sufficient to show that this integral converges absolutely and is zero for any χ_ν as above. Let $\eta_{\nu,i} \in \mathfrak{X}_{\pi_\nu}(P', A'), \eta_{\nu,i} \notin \hat{A}', \eta_{\nu,i} \neq \chi_\nu$, and let d_i be the multiplicity of $\eta_{\nu,i}$, $i = 1, \ldots, r$. Then there exist functions $\alpha_1, \ldots, \alpha_r \in C_c^\infty(A')$ such that $(\rho(\alpha_i) - \eta_{\nu,i}(\alpha_i))^{d_i} E_{P',\eta_{\nu,i}} = 0$ and such that the polynomial operators

$$T_{\nu,i} = (\chi_\nu(\alpha_i) - \eta_{\nu,i}(\alpha_i))^{-1}(I + N_{\nu,i} + \ldots + N_{\nu,i}^{d_i-1}),$$

where $N_{\nu,i} = \dfrac{\chi_\nu(\alpha_i) - \rho(\alpha_i)}{\chi_\nu(\alpha_i) - \eta_{\nu,i}(\alpha_i)}$, satisfy

$$\prod_{i=1}^{r} \{(\rho(\alpha_i) - \eta_{\nu,i}(\alpha_i))^{d_i} T_{\nu,i}^{d_i}\} E_{P',\chi_\nu} = E_{P',\chi_\nu}$$

(cf. §5.3.3). Since integration from the left over $\bar{N}'$ commutes with the action of this operator on $E_{P'} - {}_wE_{P'}$, we may conclude that $\int_{\bar{N}'} \delta_{P'}^{-\frac{1}{2}}(\bar{n}') E_{P',\chi_\nu}(P : \alpha(\nu) : \nu : \bar{n}'x')d\bar{n}'$ converges absolutely. Indeed, letting d be the multiplicity of χ_ν, we have $\int_{\bar{N}'} \delta_{P'}^{-\frac{1}{2}}(\bar{n}')(\rho(a') - \chi_\nu(a'))^{d-1} E_{P',\chi_\nu}(P : \alpha(\nu) : \nu : \bar{n}'x')d\bar{n}'$ absolutely convergent. We may assume that the value of this integral is nonzero and seek a contradiction.

Fix $\nu = \nu_0$. There is no loss in generality if we assume that there is a neighborhood U of ν_0 in $\mathfrak{F}$ such that the multiplicity d of χ_ν is constant. The points where this condition holds constitute an open dense, h-dense subset of $\mathfrak{F}$. To extend the desired conclusion from this special case to the general case one can reason as follows. Take $\alpha(\nu) = \psi$ (independent of ν). Let $E_{P',J} = \sum_{i=1}^{s} E_{P',\eta_{\nu,i}}$, where $\eta_{\nu_0,i} = \chi_{\nu_0}$ for all $i = 1, \ldots, s$ and $E_{P',J}$ is analytic near $\nu = \nu_0$ (cf. §5.3.3). The integral $\int_{\bar{N}'} \delta_{P'}^{-\frac{1}{2}}(\bar{n}') E_{P',J}(P:\psi:\nu:\bar{n}'x')d\bar{n}'$ being absolutely and uniformly convergent near ν_0, defines a holomorphic function ear ν_0. Assuming the special case which we shall consider in detail below, we ee that this holomorphic function is zero on an h-dense set and, hence, identically ero near ν_0. This proves that the special case implies the general case.

Make the assumption given in the first sentence of the preceding para- raph. Set

$$_\alpha(x') = \int_U \int_{\overline{N}'} \mu(\omega_0:\nu)\delta_{P'}^{-\frac{1}{2}}(\overline{n}')(\rho(a')-\chi_\nu(a'))^{d-1} E_{P',\chi_\nu}(P:\alpha(\nu):\nu:\overline{n}'x')d\overline{n}'d\nu$$

or $\alpha \in C_c^\infty(U) \otimes L(\omega_0)$. Clearly, $\alpha \mapsto f_\alpha$ is a continuous mapping from $_c^\infty(U) \otimes L(\omega_0)$ into $\mathcal{C}(M', \tau_{M'})$. Consider $\rho(a')f_\alpha$, $a' \in A'$. On the one hand, $(a')f_\alpha = f_{\chi_\nu(a')\alpha}$, where $\chi_\nu(a')\alpha \to 0$ in $C_c(U) \otimes L(\omega_0)$ as $a' \underset{P'}{\longrightarrow} \infty$. On the other and, the semi-norm (cf. §4.4)

$$\begin{aligned}\nu_r(\rho(a')f_\alpha) &= \sup_{x' \in M'} \|f_\alpha(x'a')\| \Xi_{M'}^{-1}(x')(1+\sigma(x'))^r \\ &= \sup_{x' \in M'} \|f_\alpha(x')\| \Xi_{M'}^{-1}(x')(1+\sigma(x'a'^{-1}))^r \\ &\to \infty, \; a' \underset{P'}{\longrightarrow} \infty.\end{aligned}$$

his contradicts the assumption that $\alpha \mapsto f_\alpha$ is a nonzero mapping. We may con- lude that the integral of the lemma has value zero.

Lemma 5.5.1.6. Let $P \in \mathcal{P}(A)$ and $\alpha \in L^\infty(\omega_0)$. Let (P', A') be any semi- tandard p-pair. Then

$$E^{\overline{P}'}(P : \alpha : m') = \int_{\overline{N}'} \int_{\mathfrak{F}} \delta_{P'}^{-\frac{1}{2}}(\overline{n}')_w E_{P'}(P : \alpha(\nu) : \nu : \overline{n}'m')d\nu d\overline{n}'.$$

Proof. If (P', A') is a maximal p-pair, then this lemma follows immediately rom the preceding lemma and Lemma 5.1.11 via Fubini's Theorem. Let im $A'/Z = \ell > 1$. We shall argue by induction on ℓ.

Let $(P'', A'')(P'' = M''N'')$ be any p-pair of G such that $(G, Z) \underset{\neq}{>} (P'', A'') \underset{\neq}{>} (P', A')$. Let ${}^*P' = P' \cap M''$. Then ${}^*P' = M' \, {}^*N'$, where ${}^*N' = N' \cap M''$, and $N' = {}^*N' N''$. In the same way $\bar{N}' = {}^*\bar{N}' \bar{N}''$. Note that $\delta_{P'}(\bar{n}') = \delta_{P'}(\bar{n}'' \, {}^*\bar{n}') = \delta_{P''}(\bar{n}'')\delta_{{}^*P'}({}^*\bar{n}')$.

Recall that $E^{\bar{P}'}(P : \alpha) = (E^{\bar{P}''}(P : \alpha))^{{}^*\bar{P}'}$, the convergence of the integrals being absolute. Applying the induction hypothesis, we obtain

$$E^{\bar{P}''}(P : \alpha : m') = \int_{\bar{N}''} \int_{\mathfrak{F}} \delta_{P''}^{-\frac{1}{2}}(\bar{n}'')\mu(\omega_0 : \nu)\, {}_wE_{P''}(P : \alpha(\nu) : \nu : \bar{n}''m')d\nu d\bar{n}''$$

$$= \int_{\bar{N}''} \int_{\mathfrak{F}} \delta_{P''}^{-\frac{1}{2}}(\bar{n}'')\mu(\omega_0 : \nu)\, \tau(\kappa'')\, {}_wE_{P''}(P : \alpha(\nu) : \nu : \mu''(\bar{n}'')m')d\nu d\bar{n}''.$$

Next we note that $\int_{\mathfrak{F}} \mu(\omega_0 : \nu)\, {}_wE_{P''}(P : \alpha(\nu) : \nu : \mu''(\bar{n}'')m'\, d\nu$ is the value of a wave packet on M''. To see this recall that $\mu_G(\omega_0 : \nu) = \mu_{M''}(\omega_0 : \nu)h(\nu)$, where h is holomorphic; write $\mu(\omega_0 : \nu)E(P : \alpha(\nu) : \nu) = E^0(P : \alpha'(\nu) : \nu)$ and apply Theorem 5.3.5.3 (5).

We may now apply the induction hypothesis again.

$$(E^{\bar{P}''}(P : \alpha : m'))^{{}^*\bar{P}'} = \int_{{}^*\bar{N}'} \int_{\bar{N}''} \int_{\mathfrak{F}} \delta_{P''}^{-\frac{1}{2}}(\bar{n}'')\mu(\omega_0 : \nu)\, {}_wE_{P''}(P : \alpha(\nu) : \nu : \bar{n}'' \, {}^*\bar{n}'\, m')\, d\nu d\bar{n}'' d^*\bar{n}'$$

$$= \int_{\bar{N}''} \int_{{}^*\bar{N}'} \int_{\mathfrak{F}} \delta_{P''}^{-\frac{1}{2}}(\bar{n}'')\delta_{{}^*P'}^{-\frac{1}{2}}({}^*\bar{n}')\mu(\omega_0 : \nu)\, {}_w({}_wE_{P''})_{{}^*P'}(P : \alpha(\nu) : \nu : \bar{n}'' \, {}^*\bar{n}'\, m')d\nu d^*\bar{n}' d\bar{n}''$$

To complete the proof apply Corollary 4.5.9 in order to replace ${}_w({}_wE_{P''})_{{}^*P'}$ by ${}_wE_{P'}$.

<u>Corollary</u> 5.5.1.7. Let $\alpha \in L^\infty(\omega_0)$. Let A' be a standard torus of G and as-

ume that $W(A|A') = \phi$. Let $P' \in \mathcal{P}(A')$, then $E(P : \alpha)^{P'} = 0$.

roof. Using Lemma 5.5.1.6 (or Lemma 5.1.11, if $\omega_0 \in {}^0\mathcal{E}(M)$), we have

$$E^{P'}(P : \alpha : m') = \int_{N'} \int_{\mathfrak{F}} \mu(\omega_0 : \nu) \delta_{\bar{P}'}^{-\frac{1}{2}}(n') {}_wE_{\bar{P}'}(P : \alpha(\nu) : \nu : n'm') d\nu dn'.$$

orollary 4.5.12 (1) implies that ${}_wE_{\bar{P}'} = 0$. Therefore, $E(P : \alpha)^{P'} = 0$.

orollary 5.5.1.8. Let A and A' be as in the preceding corollary. Let $_0 \in \mathcal{E}_2(M)$ and $\omega'_0 \in \mathcal{E}_2(M')$, $M' = Z_G(A')$. Let $\alpha \in L^\infty(\omega_0)$ and $\alpha' \in L^\infty(\omega'_0)$. nen $E(P : \alpha) * E(P : \alpha') = 0$.

roof. Use Corollary 5.5.1.7 and Corollary 5.5.1.4.

heorem 5.5.1.9. Let $P_1, P_2 \in \mathcal{P}(A)$ and $\alpha \in L^\infty(\omega_0)$. Then

$$E(P_1 : \alpha)^{P_2} = \gamma E(M : \sum_{s \in W(G/A)} c^{P_2|P_1}(s : \omega_0)\alpha)$$

$$= \gamma \int_{\mathfrak{F}} \sum_{s \in W(G/A)} (c^{P_2|P_1}(s : \omega_0)\alpha)(\nu)\chi_\nu d\nu,$$

here $\chi_\nu(m) = q^{\langle\sqrt{-1}\nu, H(m)\rangle}$ $(m \in M)$.

roof. We have $P_i = MN_i$ $(i = 1,2)$ and

$$E(P_1 : \alpha : m)^{\bar{P}_2} = \int_{\bar{N}_2} E(P_1 : \alpha : \bar{n}m)\delta_{\bar{P}_2}^{-\frac{1}{2}}(m)d\bar{n}$$

$$= \int_{\bar{N}_2} \int_{\mathfrak{F}} \mu(\omega_0 : \nu)\, \delta_{\bar{P}_2}^{-\frac{1}{2}}(m)E(P_1 : \alpha(\nu) : \nu : \bar{n}m)d\nu d\bar{n}$$

$$= \int_{\bar{N}_2} \int_{\mathfrak{F}} \mu(\omega_0 : \nu)\, \delta_{\bar{P}_2}^{-\frac{1}{2}}(m)\delta_{P_2}(\bar{n}m)^{-\frac{1}{2}} E_{P_2}(P_1 : \alpha(\nu) : \nu : \bar{n}m)d\nu d\bar{n},$$

in which the last relation follows via Lemma 5.1.11. We recall that the assumptions underlying Lemma 5.1.11 imply that $E_{P_2}(P_1 : \alpha(\nu) : \nu : x) =$
$\tau(\kappa_2(x))E_{P_2}(P_1 : \alpha(\nu) : \nu : \mu_2(x))$, where $\mu_2^{-1}(x)\kappa_2^{-1}(x)x \in N_2$ and $\mu_2(x) \in M$, $\kappa_2(x) \in K$. Thus, applying Lemma 5.1.11, Lemma 5.5.1.6, and Theorem 5.3.5.4, we obtain

$$E(P_1 : \alpha : m)^{\overline{P}_2} = \int_{\overline{N}_2} \int_{\mathfrak{F}} \mu(\omega_0 : \nu)\delta_{P_2}(\overline{n})^{-\frac{1}{2}} \tau(\kappa_2(\overline{n})) \sum_{s \in W(A)} c_{P_2|P_1}(s : \omega_0 : \nu)\alpha(\nu : \mu_2(\overline{n})m)q^{\langle\sqrt{-1}s\nu, H(\mu_2(\overline{n})m)\rangle} d\nu d\overline{n}$$

$$= \sum_{s \in W(A)} \int_{\overline{N}_2} \int_{\mathfrak{F}} \mu(\omega_0 : s^{-1}\nu)\tau(\kappa_2(\overline{n}))c_{P_2|P_1}(s : \omega_0 : s^{-1}\nu)\alpha(s^{-1}\nu : \mu_2(\overline{n})m) q^{\langle\sqrt{-1}\nu, H(m)\rangle + \langle\sqrt{-1}\nu - \rho_2, H_2(\overline{n})\rangle} d\nu d\overline{n}$$

$$= \gamma \sum_{s \in W(A)} \int_{\mathfrak{F}} \mu(\omega_0 : s^{-1}\nu)(c_{P_2|\overline{P}_2}(1 : \omega_0^s : \overline{\nu}))^* c_{P_2|P_1}(s : \omega_0 : s^{-1}\nu) \alpha(s^{-1}\nu : m)\chi_\nu(m)d\nu.$$

We leave to the reader the task of justifying the last equation above. To do this, one can move the integration off $\mathfrak{F}$ into a region where changing the order of integration is permissible, then apply a uniform convergence argument to justify the equality for integration on $\mathfrak{F}$. Changing notation, we write

$$E(P_1 : \alpha : m)^{P_2} =$$

$$\gamma \sum_{s \in W(A)} \int_{\mathfrak{F}} \mu(\omega_0 : s^{-1}\nu)(c_{\overline{P}_2|P_2}(1 : \omega_0^s : \overline{\nu}))^* c_{\overline{P}_2|P_1}(s : \omega_0 : s^{-1}\nu)\alpha(s^{-1}\nu : m)\chi_\nu(m)d\nu.$$

ince $c_{\bar{P}_2|P_1}(s:\omega_0:s^{-1}\nu) = c_{\bar{P}_2|\bar{P}_2}(1:\omega_0^s:\nu)^0 c_{\bar{P}_2|P_1}(s:\omega_0:s^{-1}\nu)$

$$= c_{\bar{P}_2|\bar{P}_2}(1:\omega_0^s:\nu)^0 c_{\bar{P}_2|P_2}(1:\omega_0^s:\nu)^0 c_{P_2|P_1}(s:\omega_0:s^{-1}\nu)$$

$$= c_{\bar{P}_2|P_2}(1:\omega_0^s:\nu)^0 c_{P_2|P_1}(s:\omega_0:s^{-1}\nu),$$

e have $(c_{\bar{P}_2|P_2}(1:\omega_0^s:\bar{\nu}))^* c_{\bar{P}_2|P_1}(s:\omega_0:s^{-1}\nu) = \mu^{-1}(\omega_0^s:\nu)^0 c_{P_2|P_1}(s:\omega_0:s^{-1}\nu)$,

applying Theorem 5.3.5.2. Recalling that $\mu(\omega_0^s:\nu) = \mu(\omega_0:s^{-1}\nu)$, we obtain

$$(P_1:\alpha:m)^{P_2} = \gamma\int_{\mathfrak{F}} \sum_{s\in W(A)} {}^0c_{P_2|P_1}(s:\omega_0:s^{-1}\nu)\alpha(s^{-1}\nu:m)\chi_\nu(m)d\nu,$$

o the theorem is proved.

In the proof of the following lemma, we normalize the Lebesgue easures on both $\mathfrak{F} = \mathfrak{F}_M$ and $\mathfrak{F}_A$. We also decompose a normalized aar measure on M such that $dm = dm^* da$, where dm^* is a normalized aar measure on M/A and $da = d^0a \sum_{a\in A/^0A}$ with $\int_{^0A} d^0a = 1$ $(^0A = A\cap {}^0M)$.

For $s\in W(\mathfrak{a})$ set $\omega_{\nu(s)} = \omega_0^s$ $(\nu(s)\in\mathfrak{F})$. Note that $(\omega_\nu)^s = (\omega_0^s)_{s\nu}$ $(s\in W(A),\ \nu\in\mathfrak{F})$.

emma 5.5.1.10. Let $P_i\in\mathcal{P}(A)$ and $\alpha_i\in I^\infty(\omega_0)$ $(i=1,2)$. Then:

$(P_1:\alpha_1) * E(P_2:\alpha_2(\nu):\nu)$

$$= E(P_2:\sum_{s\in W(\mathfrak{a})}[L_A^*:L^*(\omega_0)]^{-1}\gamma(c^{P_2|P_1}(s:\omega_0)\alpha_1\cdot\alpha_2^{\nu(s)})(\nu-\nu(s):\nu-\nu(s))$$

nd $E(P_1:\alpha_1(\nu):\nu) * E(P_2:\alpha_2)$

$$= E(P_1:\sum_{s\in W(\mathfrak{a})}[L_A^*:L^*(\omega_0)]^{-1}\gamma(\alpha_1^{\nu(s)}\cdot c^{P_1|P_2}(s:\omega_0)\alpha_2)(\nu-\nu(s):\nu-\nu(s)).$$

Proof. The proofs of the two relations go similarly, so we prove only the first. Since $E(P_1 : \alpha_1) \in \mathcal{C}(G,\tau)$ and $E(P_2 : \alpha_2(\nu) : \nu)$ satisfies the weak inequality $(\nu \in \mathfrak{F})$ (§4.5), the convolution product exists. Using the Schwartz space analogue of Lemma 5.2.2.1 and recalling that $\alpha_2(\nu)_\nu = \alpha_2(\nu)\chi_\nu$, we obtain $E(P_1 : \alpha_1) * E(P_2 : \alpha_2(\nu) : \nu) = E(P_2 : E(P_1 : \alpha_1)^{P_2} \underset{M}{*} \alpha_2(\nu)_\nu : 0)$. Therefore,

$$E(P_1 : \alpha_1)^{P_2} \underset{M}{*} \alpha_2(\nu)_\nu = \int_{\mathfrak{F}} \gamma \sum_{s \in W(G/A)} (c_{P_2|P_1}(s : \omega_0)\alpha_1)(\nu')_{\nu'} \, d\nu' \underset{M}{*} \alpha_2(\nu)_\nu$$

$$= \int_{\mathfrak{F}_A} \sum_{\substack{s \in W(A) \\ \nu_1 \in L_A^*/L_M^*}} \gamma [L_A^* : L_M^*]^{-1} (c_{P_2|P_1}(s : \omega_0)\alpha_1^{\nu_1})(\nu')_{\nu'} \, d\nu' \underset{M}{*} \alpha_2(\nu)_\nu$$

$$= \gamma [L_A^* : L_M^*]^{-1} \int_{M/A} dx^* \int_{{}^0A} d^0a \sum_{a \in A/{}^0A} \int_{\mathfrak{F}_A} \sum_{\substack{s \in W(A) \\ \nu_1 \in L_A^*/L_M^*}} (c_{P_2|P_1}(s : \omega_0)\alpha_1^{\nu_1})(\nu' : xa)_{\nu'} \, \alpha_2(\nu : a^{-1}x^{-1}m)_\nu \, d\nu'$$

$$= \gamma [L_A^* : L_M^*]^{-1} \sum_{\substack{s \in W(\mathfrak{a}) \\ \nu_1 \in L^*(\omega_0)/L_M^*}} c_{P_2|P_1}(s : \omega_0)\alpha_1^{\nu_1}(\nu - \nu(s))_{\nu - \nu(s)} \underset{M/A}{*} \alpha_2(\nu$$

$$= \sum_{s \in W(\mathfrak{a})} \gamma [L_A^* : L^*(\omega_0)]^{-1} (c_{P_2|P_1}(s : \omega_0)\alpha_1 \cdot \alpha_2^{\nu(s)})(\nu - \nu(s))_{\nu - \nu(s)}.$$

Next we introduce the convention of writing $E(P : \alpha) = \sum_{j=1}^{r} E(P : \alpha^j)$ for $\alpha = \sum_{j=1}^{r} \alpha^j$ with $\alpha^j \in L^\infty(\omega_j)$ $(\omega_j \in \mathcal{E}_2(M),\ j = 1, \ldots, r)$.

Corollary 5.5.1.11. Let $\alpha_i \in I^{\infty}(\omega_0)$ $(i = 1,2)$. Then

$$E(P_1 : \alpha_1) * E(P_2 : \alpha_2) = E(P_2 : \sum_{s \in W(\Theta)} \gamma[L_A^* : L^*(\omega_0)]^{-1} c^{P_2|P_1}(s : \omega_0)\alpha_1 \cdot \alpha_2^{\nu(s)})$$

$$= E(P_1 : \sum_{s \in W(\Theta)} \gamma[L_A^* : L^*(\omega_0)]^{-1} \alpha_1^{\nu(s)} \cdot c^{P_1|P_2}(s : \omega_0)\alpha_2).$$

Proof. Checking only the first relation, we have (setting $\gamma' = \gamma[L_A^* : L^*(\omega_0)]^{-1}$)

$$E(P_1 : \alpha_1) * E(P_2 : \alpha_2)(g) = \int_G E(P_1 : \alpha_1 : x) \int_{\mathfrak{F}} \mu(\omega_0 : \nu) E(P_2 : \alpha_2(\nu) : \nu : x^{-1}g) d\nu dx$$

$$= \int_{\mathfrak{F}} \mu(\omega_0 : \nu) E(P_2 : \sum_{s \in W(\Theta)} \gamma' (c^{P_2|P_1}(s : \omega_0)\alpha_1 \cdot \alpha_2^{\nu(s)})(\nu - \nu(s)) : \nu - \nu(s) : g) d\nu$$

$$= \sum_{s \in W(\Theta)} \gamma' \int_{\mathfrak{F}} \mu(\omega_0^s : \nu) E(P_2 : c^{P_2|P_1}(s : \omega_0)\alpha_1 \cdot \alpha_2^{\nu(s)}(\nu) : \nu : g) d\nu$$

$$= E(P_2 : \sum_{s \in W(\Theta)} \gamma' c^{P_2|P_1}(s : \omega_0)\alpha_1 \cdot \alpha_2^{\nu(s)} : g).$$

Corollary 5.5.1.12. Let $P_i \in \mathcal{P}(A)$ and $\alpha_i \in L^{\infty}(\omega_i)$ $(\omega_i \in \mathcal{E}_2(M),\ i = 1,2)$. If $\Theta(\omega_1) \neq \Theta(\omega_2^s)$ for all $s \in W(A)$, then $E(P_1 : \alpha_1) * E(P_2 : \alpha_2) = 0$.

We have already observed that the wave packet mapping factors through $I^{\infty}(\omega_0)$. Recall (§5.3.1) that $M(\omega_0) = \{m \in M \mid \langle L^*(\omega_0), H(m) \rangle \subset \frac{2\pi \mathbb{Z}}{\log q}\}$.

Lemma 5.5.1.13. As a mapping with domain $I^{\infty}(\omega_0)$, the wave packet mapping $\alpha \mapsto E(M : \alpha)$ is injective.

Proof. Let $\alpha \in L^{\infty}(\omega_0)$. Then we may write $\alpha(\nu : m) = \sum_{x \in M/{}^0M} \chi_{-\nu}(x) \sum_{j=1}^{r} c_j(x)\psi_j(m)$ with $c_j(x) \in \mathbb{C}$ and $\psi_j \in \mathcal{A}(\omega_0, \tau_M)$. Since $\alpha \in I^{\infty}(\omega_0)$ if and only if $\alpha(\nu + \nu_1 : m)_{\nu_1} = \alpha(\nu : m)$ for all $\nu_1 \in L^*(\omega_0)$, if and only if

$\sum_{j=1}^{r} c_j(x)\psi_j(m)\chi_{\nu_1}(mx^{-1}) = \sum_{j=1}^{r} c_j(x)\psi_j(m)$ for every $x \in M$, $m \in M$, we see that $\alpha \in I^\infty(\omega_0)$ if and only if $\sum_{j=1}^{r} c_j(x)\psi_j(m) \neq 0$ implies $m \in xM(\omega_0)$. On the other hand, it follows easily from Lemma 5.3.1.4 that $\sum_{j=1}^{r} c_j(x)\psi_j(m) \neq 0$ for some $m \in xM(\omega)$ if and only if the same is true with $m \in x\,{}^0M$. Thus, we may assume that $\sum_{j=1}^{r} c_j(m)\psi_j(m) \neq 0$. We conclude that

$$E(M : \alpha : m) = \int_{\mathfrak{F}} \alpha(\nu : m)\chi_\nu(m)d\nu = \sum_{j=1}^{r} c_j(m)\psi_j(m) \neq 0.$$

The final results of this section relate the algebra structure on the spaces of wave forms to an algebra structure on the spaces of wave packets. Before formulating these results, we introduce some additional notation.

Given any finite set of classes $\omega_i \in \mathcal{E}_2(M)$ and elements $\alpha^i \in I^\infty(\omega_i)$ $(i = 1, \ldots, r)$, we write $\Sigma\alpha^i \sim 0$ if $E(M : \Sigma\alpha^i) = 0$. If $\omega_i = \omega_0$ for all $i = 1, \ldots, r$, then, according to Lemma 5.5.1.13, $\Sigma\alpha^i \sim 0$ if and only if $\Sigma\alpha^i = 0$. If $\alpha^1 \in I^\infty(\omega_{\nu_1})$ and $\alpha^2 \in I^\infty(\omega_{\nu_2})$ $(\nu_1, \nu_2 \in \mathfrak{F}(\omega_0))$, then $\alpha^1 - \alpha^2 \sim 0$ if and only if $(\alpha^1)^{\nu_2 - \nu_1} = \alpha^2$. We regard $\sim$ as defining an equivalence on $\bigcup_{\omega \in \mathcal{E}_2(M)} I^\infty(\omega)$.

Now let $\alpha \in I^\infty(\omega_0, P)$, $P \in \mathcal{P}(A)$. We say that α is $W(\mathfrak{O})$-<u>invariant</u> if $c^{P|P}(s : \omega_0)\alpha \sim \alpha$ for all $s \in W(\mathfrak{O})$. Write $I^\infty(\omega_0, W(\mathfrak{O}), P)$ for the space of $W(\mathfrak{O})$-invariant wave forms associated to ω_0 and P. There is a natural projection of $I^\infty(\omega_0, P)$ on $I^\infty(\omega_0, W(\mathfrak{O}), P) : \alpha \mapsto [W(\mathfrak{O})]^{-1} \sum_{s \in W(\mathfrak{O})} c^{P|P}(s : \omega_0)\alpha^{-\nu(s)}$. Corollary 5.5.1.11 implies that $I^\infty(\omega_0, W(\mathfrak{O}), P)$ is a subalgebra of $I^\infty(\omega_0, P)$.

We have seen that the wave packet mapping $\alpha \mapsto E(P : \alpha)$ factors through the projections $L^\infty(\omega_0) \to L^\infty(\omega_0, P) \to I^\infty(\omega_0, P)$. It is obvious that it may, in fact, be regarded as a mapping from $I^\infty(\omega_0, W(\mathfrak{O}), P)$ to $\mathcal{C}(\mathfrak{O}, \tau)$.

eorem 5.5.1.14. The mapping $\alpha \mapsto \gamma^{-1}[W(\mathfrak{O})]^{-1}[L_A^* : L^*(\omega_0)]E(P : \alpha)$ defines isomorphism of the algebra $I^\infty(\omega_0, W(\mathfrak{O}), P)$ onto $\mathcal{C}(\mathfrak{O}, \tau)$.

oof. Apply Corollaries 5.5.1.11 and 5.5.1.12 and Lemma 5.5.1.13.

rollary 5.5.1.15. (1) For any $\alpha \in L^\infty(\omega_0)$, if $E^P(P : \alpha) = 0$, then $E(P : \alpha) = 0$. For any $\alpha \in I^\infty(\omega_0, W(\mathfrak{O}), P)$, if $E^P(P : \alpha) = 0$, then $\alpha = 0$.

Let $V^0 = C^\infty(K \times K)$ and let τ^0 be the usual double representation of on V^0. Let $\pi_\nu = \mathrm{Ind}_P^G(\delta_P^{\frac{1}{2}}\sigma_\nu)$ $(\sigma_\nu \in \omega_\nu, P \in \mathcal{P}(A))$. Let σ_0 act in a vector ace U and write $\mathcal{H}$ for the space of all smooth U-valued functions on K such at $\beta(pk) = \sigma_0(p)\beta(k)$ $(p \in P \cap K, k \in K)$. Then, for any $\nu \in \mathfrak{a}^*$, we may regard as the representation space for π_ν.

It follows from the definition of wave packets and from Theorem 5.2.2.4 at, given $\alpha = \sum_{j=1}^r \alpha_j(\nu)\psi_{T_j} \in L^\infty(\omega_0, P)$ $(T_j \in \mathrm{End}^0(\mathcal{H})$, cf. §5.2.1 for the definition ψ_T, $T \in \mathrm{End}^0(\mathcal{H}))$, we may write

$$E(P : \alpha : k_1 : x : k_2) = \int_{\mathfrak{F}} \mu_G(\omega_0 : \nu) \sum_{j=1}^r \alpha_j(\nu) \mathrm{tr}(T_j \pi_\nu(k_1 x k_2)) d\nu$$

$$j^{-1}E(P : \alpha)(x) = \int_{\mathfrak{F}} \mu_G(\omega_0 : \nu) \sum_{j=1}^r \alpha_j(\nu) \mathrm{tr}(T_j \pi_\nu(x)) d\nu .$$

nce the space of all wave packets of the form $E(P : \alpha)$ $(\alpha \in L^\infty(\omega_0))$ is $C_c(G, \tau)$ able, the space of all functions of the form $j^{-1}E(P : \alpha)$ is $C_c^\infty(G)$ and translation able. We write $\mathcal{C}(\mathfrak{O})$ for this subspace of $\mathcal{C}(G)$. It is obviously independent of choice of $P \in \mathcal{P}(A)$. For (V, τ) any double representation of K we write $(\mathfrak{O}, \tau) = C(G, \tau) \cap (\mathcal{C}(\mathfrak{O}) \otimes V)$. Obviously, $\mathcal{C}(\mathfrak{O}, \tau)$ is the space of all "wave ckets of type τ associated to $\mathfrak{O}$".

§5.5.2. The Plancherel Formula.

Let T be a distribution on G and let $f \in C_c^\infty(G)$. We write $(T,f) = (T,f)_G = T(\bar{f})^-$, where the "-" denotes complex conjugation. If T is represented by a function F_T such that $F_T(x^{-1}) = \bar{F}_T(x)$, then $(T,f) = \int_G F_T(x^{-1})f(x)dx = (F_T * f)(1)$. If U is a complex vector space, set $(T, f \otimes u) = (T,f)u$ $(u \in U)$. Then T extends to a linear mapping of $C_c^\infty(G) \otimes U$ to U. If T is tempered, we may regard T as a mapping of $\mathcal{C}(G) \otimes U$ to U. Write θ_{ω_ν} for the character and $d(\omega_\nu)$ for the formal degree of the class $\omega_\nu \in \mathcal{E}_2(M)$ $(\nu \in \mathfrak{a}^*)$. Let Θ_{ω_ν} denote the character of $C_M^G(\omega_\nu)$ (a not necessarily irreducible class).

The following theorem is sometimes called the Plancherel's formula for wave packets. Recall that $\int_{\mathfrak{F}} d\nu = 1$ and that [ ] denotes the cardinality of a finite set.

Theorem 5.5.2.1. Let $\alpha \in L^\infty(\omega_0)$, $P \in \mathcal{P}(A)$. Then

$$E(P:\alpha:1) = [W(\mathcal{O})]^{-1}\gamma(G/M)^{-1}[L_A^* : L^*(\omega_0)]\int_{\mathfrak{F}} d(\omega_0)\mu(\omega_0:\nu)(\Theta_{\omega_\nu}, E(P:\alpha))_G d\nu.$$

Proof. Fix an open normal subgroup K_0 of K. Without loss of generality, we may assume that $\alpha \in I^\infty(\omega_0, W(\mathcal{O}), P)$ and that $E(P:\alpha) \in \mathcal{C}(G,\tau) \cap C(G/\!/K_0 : V)$.

Let $V^0 = C^\infty(K \times K)$ with τ^0 the standard double representation of K on V^0. Let $\sigma \in \omega_0$ and let $\pi = \mathrm{Ind}_P^G(\delta_P^{\frac{1}{2}}\sigma)$ act in the space $\mathcal{H}$. Let $I = I(K_0) \in \mathrm{End}(\mathcal{H})$ be the operator which projects $\mathcal{H}$ on the subspace $\mathcal{H}_{K_0}$ consisting of all K_0-fixed vectors in $\mathcal{H}$. The function $d(\omega_0)\psi_I(k_1 : m : k_2) = d(\omega_0)\mathrm{tr}(\sigma(m)\kappa_I(k_2 : k_1))$ is an idempotent in the algebra $\mathcal{A}(\sigma, \tau_P^0)$ and $j^{-1}E(P:\psi_I:\nu) * f(1) = (\Theta_{\omega_\nu}, f)_G$ for any $f \in \mathcal{C}_{K_0}(G)$, hence, any $f \in \mathcal{C}_{K_0}(G) \otimes V$, where (V,τ) is an arbitrary smooth unitary double representaon of K satisfying the usual conditions.

Define $\alpha^I(\nu) = d(\omega_0)\psi_I$ for all $\nu \in \mathfrak{F}$. Assume for the moment that $(V,\tau) = (V^0,\tau^0)$. Then $\alpha^I \in I^\infty(\omega_0, W(\mathfrak{O}), P)$ and it follows from Theorem 5.5.1.14 that $[W(\mathfrak{O})]\gamma[L_A^* : L^*(\omega_0)]^{-1}E(P:\alpha) = E(P:\alpha^I) * E(P:\alpha)$. Thus,

$[W(\mathfrak{O})]\gamma[L_A^* : L^*(\omega_0)]^{-1}E(P:\alpha:1:1:1)$

$$= \int_{\mathfrak{F}} \mu(\omega_0 : \nu) d(\omega_0)(E(P:\psi_I:\nu) * E(P:\alpha))(1:1:1)d\nu$$

$$= \int_{\mathfrak{F}} \mu(\omega_0 : \nu) d(\omega_0)(j^{-1}E(P:\psi_I:\nu) * j^{-1}E(P:\alpha))(1)d\nu.$$

This implies the theorem.

Corollary 5.5.2.2. Let U be a vector space and $f \in \mathcal{C}(\mathfrak{O}) \otimes U$. Then $f(g) = [W(\mathfrak{O})]^{-1}\gamma^{-1}[L_A^* : L^*(\omega_0)]\int_{\mathfrak{F}} \mu(\omega_0 : \nu) d(\omega_0)(\Theta_{\omega_\nu}, \rho(g)f)d\nu$. In particular, for any double representation (V,τ) and any $\alpha \in L^\infty(\omega_0)$ with respect to (V,τ),

$$E(P:\alpha:g) = [W(\mathfrak{O})]^{-1}\gamma^{-1}[L_A^* : L^*(\omega_0)]\int_{\mathfrak{F}} \mu(\omega_0 : \nu) d(\omega_0)(\Theta_{\omega_\nu}, \rho(g)E(P:\alpha))d\nu.$$

Proof. The second statement follows from the first, since $E(P:\alpha) \in \mathcal{C}(\mathfrak{O}) \otimes V$. For the first statement it is sufficient to observe that, for (V^0,τ^0),

$$j^{-1}E(P:\alpha)(g) = [W(\mathfrak{O})]^{-1}\gamma^{-1}[L_A^* : L^*(\omega_0)]\int_{\mathfrak{F}} \mu(\omega_0 : \nu) d(\omega_0)(\Theta_{\omega_\nu}, \rho(g)j^{-1}E(P:\alpha))d\nu.$$

In the case $g = 1$ this relation follows directly from Theorem 5.5.2.1. For g arbitrary it follows via the additional observation that $j\rho(g)j^{-1}E(P:\alpha)$ is a wave packet.

Corollary 5.5.2.3. Let $\alpha \in I^\infty(\omega_0)$. Then

$$\alpha(\nu : m)_\nu = d(\omega_0)[L_A^* : L^*(\omega_0)](\theta_{\omega_\nu}, \rho(m)E(M:\alpha)).$$

Proof. $E(M:\alpha:m) = \int_{\mathfrak{F}} \alpha(\nu : m) q^{<\sqrt{-1}\nu, H(m)>} d\nu$

$$= \int_{\mathfrak{F}} d(\omega_0)[L_A^* : L^*(\omega_0)](\theta_{\omega_\nu}, \rho(m)E(M:\alpha:1))d\nu \text{ (Corollary 5.5.2.2).}$$

Since $\alpha \mapsto E(M:\alpha)$ maps $I^{\infty}(\omega_0)$ injectively into $\mathcal{C}(M,\tau_M)$ (Lemma 5.5.1.13), it suffices to check that $d(\omega_0)[L_A^* : L^*(\omega_0)](\theta_{\omega_\nu}, \rho(m)E(M:\alpha))_{-\nu} \in I^{\infty}(\omega_0)$. This is a consequence of Lemma 5.5.1.3 and the obvious periodicity of θ_{ω_ν}.

§5.5.3. The Commuting Algebra Theorem.

In this subsection we use the algebra structure on the space of wave packets to characterize the commuting algebras of certain induced representations (Theorem 5.5.3.2).

Lemma 5.5.3.1. Let $P_i \in \mathcal{P}(A)$ and $\psi_i \in L(\omega_0, P_1)$, $i = 1,2$. Then

$${}^0c_{P_2|P_1}(s:\omega_0:\nu)(\psi_1 \underset{M/A}{*} \psi_2) = {}^0c_{P_2|P_1}(s:\omega_0:\nu)\psi_1 \underset{M/A}{*} {}^0c_{P_2|P_1}(s:\omega_0:\nu)\psi_2$$

for every $s \in W(A)$ and $\nu \in \mathfrak{a}^*$.

Proof. Fix $s \in W(A)$. Choose $\nu_0 \in \mathfrak{F}(\omega_0)$ such that $(\omega_{\nu_0})^t = \omega_{\nu(t)+t\nu_0} \neq \omega_{\nu_0}$ for all $t \in W(\mathfrak{O})$, $t \neq 1$. Since $\mathfrak{F}(\omega_0)$ is a Hausdorff space, we may (and do) choose a neighborhood U of ν_0 such that $U \cap (tU+\nu(t)) = \phi$ for all $t \in W(\mathfrak{O})$, $t \neq 1$.

A function $f_0 \in C^{\infty}(\mathfrak{F}(\omega_0))$ has a Fourier series development $f_0(\nu) = \sum_{M(\omega_0)/{}^0M} \chi_{-\nu}(y)c(y)$, where $c(y) \in \mathbb{C}$. We choose $f_0 \in C_c^{\infty}(U)$ such that $f_0(\nu_0) = 1$. Next, observe that the function

$$\beta_i(\nu:m) = [L^*(\omega_0):L^*]^{-1} \sum_{\substack{x \in M/M(\omega_0) \\ \nu_1 \in L^*(\omega_0)/L^*}} \chi_{-(\nu+\nu_1)}(x)\psi_{i,\nu_1}(m)$$

s an element of $I^{\infty}(\omega_0, P_1)$ such that $\beta_i(0 : m) = \psi_i(m)$ $(i = 1,2)$. Define $\alpha^i(\nu : m) = f_0(\nu)\beta_i(\nu : m)$. Then $\alpha^i \in I^{\infty}(\omega_0, P_1)$ and

$$c^{P_2|P_1}(t_1 : \omega_0)\alpha^1 \cdot c^{P_2|P_1}(t_2 : \omega_0)\alpha^2 = 0 \quad (t_i \in W(A),\ t_1 \neq t_2).$$

Using Corollary 5.5.1.11, we see that

$E(P_1 : \alpha^1) * E(P_1 : \alpha^2) = \gamma[L_A^* : L^*(\omega_0)]^{-1} E(P_1 : \alpha^1 \cdot \alpha^2)$, since

$$\sum_{t \in W(\mathcal{O})} c^{P_1|P_1}(t : \omega_0)\alpha^1 \cdot \alpha^2 = c^{P_1|P_1}(1 : \omega_0)\alpha^1 \cdot \alpha^2 = \alpha^1 \cdot \alpha^2.$$

Applying Theorem .5.1.9, we obtain

$$(E(P_1 : \alpha^1) * E(P_1 : \alpha^2))^{P_2} = \gamma^2 [L_A^* : L^*(\omega_0)]^{-1} E(M : \sum_{t \in W(A)} c^{P_2|P_1}(t : \omega_0)(\alpha^1 \cdot \alpha^2)).$$

imilarly, Theorems 4.4.3 and 5.5.1.9 and Corollary 5.5.1.11 imply that

$$\begin{aligned} (E(P_1 : \alpha^1) * E(P_1 : \alpha^2))^{P_2} &= E(P_1 : \alpha^1)^{P_2} * E(P_1 : \alpha^2)^{P_2} \\ &= \gamma^2 E(M : \sum_{t_1 \in W(A)} c^{P_2|P_1}(t_1 : \omega_0)\alpha^1) * E(M : \sum_{t_2 \in W(A)} c^{P_2|P_1}(t_2 : \omega_0)\alpha^2) \\ &= \gamma^2 [L_A^* : L^*(\omega_0)]^{-1} E(M : \sum_{t_1 \in W(A)} c^{P_2|P_1}(t_1 : \omega_0)\alpha^1 \cdot \sum_{t_2 \in W(A)} c^{P_2|P_1}(t_2 : \omega_0)\alpha^2). \end{aligned}$$

hus, by Corollary 5.5.1.12,

$$\sum_{t_1 \in W(\mathcal{O})} c^{P_2|P_1}(st_1 : \omega_0)\alpha^1 \cdot \sum_{t_2 \in W(\mathcal{O})} c^{P_2|P_1}(st_2 : \omega_0)\alpha^2 \sim \sum_{t \in W(\mathcal{O})} c^{P_2|P_1}(st : \omega_0)(\alpha^1 \cdot \alpha^2);$$

n fact, by construction,

$$\sum_{t \in W(\mathcal{O})} c^{P_2|P_1}(st : \omega_0)\alpha^1 \cdot c^{P_2|P_1}(st : \omega_0)\alpha^2 \sim \sum_{t \in W(\mathcal{O})} c^{P_2|P_1}(st : \omega_0)(\alpha^1 \cdot \alpha^2).$$

ince $c^{P_2|P_1}(st : \omega_0)\alpha^i \sim (c^{P_2|P_1}(st : \omega_0)\alpha^i)^{-s\nu(t)} \in C_c^{\infty}(stU + s\nu(t)) \otimes L(\omega_0^s, P_2)$ and $(tU + \nu(t)) \cap sU = \phi$ $(t \neq 1)$, Lemma 5.5.1.13 implies that

$$c^{P_2|P_1}(s:\omega_0)\alpha^1 \cdot c^{P_2|P_1}(s:\omega_0)\alpha^2 = c^{P_2|P_1}(s:\omega_0)(\alpha^1\cdot\alpha^2).$$

Consequently, at $\nu = s^{-1}\nu_0$, we have

$$ {}^0c_{P_2|P_1}(s:\omega_0:s^{-1}\nu_0)\psi_1 \underset{M/A}{*} {}^0c_{P_2|P_1}(s:\omega_0:s^{-1}\nu_0)\psi_2$$

$$= {}^0c_{P_2|P_1}(s:\omega_0:s^{-1}\nu_0)(\psi_1 \underset{M/A}{*} \psi_2).$$

Analytic continuation implies that this relation holds at all $\nu \in \mathfrak{F}$.

Now we proceed to the formulation and proof of the main theorem. Fix $P \in \mathcal{P}(A)$ and set ${}^0c(s)\psi = {}^0c_{P|P}(s:\omega_0:0)\psi$ $(\psi \in L(\omega_0, P),\ s \in W(\omega_0))$. Then $s \mapsto {}^0c(s)$ is both a unitary representation of $W(\omega_0)$ on $L(\omega_0, P)$ (§5. 3. 5) and also an algebra automorphism of $L(\omega_0, P)$ (Lemma 5. 5. 3. 1). Let $(V^0, \tau^0) = (C^\infty(K\times K), \tau^0)$ be as usual and recall that there is a unitary algebra anti-isomorphis $T \mapsto d(\omega_0)\psi_T$ from $\mathcal{T}(P)$ to $L(\omega_0, P)$ (Lemma 5. 2. 4. 1). For any open compact subgroup K_0 of G we obtain, by restriction, an automorphism of $\mathcal{T}_{K_0}(P) = E_{K_0}\mathcal{T}(P)$: of course also using the correspondence between $\mathcal{T}(P)$ and $L(\omega_0, P)$. Applying the Skolem-Noether theorem, we define a projective unitary representation $s \mapsto \gamma(s)$ of $W(\omega_0)$ on $\mathcal{H}_{K_0}(P)$ ($\mathcal{H}(P)$ the representation space of $\pi_{P,\omega_0} = \mathrm{Ind}_P^G(\delta_P^{\frac{1}{2}}\sigma_0)$, $\sigma_0 \in \omega_0$) by setting ${}^0c(s)\psi_T = \psi_{\gamma(s)T\gamma(s^{-1})}$, $T \in \mathcal{T}_{K_0}(P)$. Notice, in particular, that the adjoint $\gamma(s)^*$ of $\gamma(s)$ exists and, up to a scalar factor, $\lambda_s \in \mathbb{C}^\times$, $|\lambda_s| = 1$, $\gamma(s)^* = \lambda_s\gamma(s^{-1})$. Let $\Gamma_{K_0} = \mathrm{span}\{\gamma(s) \mid s \in W(\omega_0)\}$. Since each operator $\gamma(s)$ is determined up to a constant factor, Γ_{K_0} is well defined as a vector space and, in fact, Γ_{K_0} is an algebra. Indeed, Γ_{K_0} is self-adjoint, so semisimple.

eorem 5.5.3.2. **For any open compact subgroup K_0 of G which is sufficiently mall the algebra Γ_{K_0} is isomorphic to the commuting algebra of the unitary rep- esentation π_{P,ω_0}.**

Clearly, $\dim \Gamma_{K_0} \leq \dim \Gamma_{K_1}$, if $K_1 \subseteq K_0$ and $\dim \Gamma_{K_0} \leq [W(\omega_0)]$ for l K_0, so the dimension stabilizes at some point. It will be relatively easy to see at there is an injective mapping of Γ_{K_0} into the commuting algebra of π_{P,ω_0}. order to prove that the mapping is surjective it is sufficient to know that Γ_{K_0} exactly the commuting algebra of $\pi_{P,\omega_0}(\mathcal{C}_{K_0}(G))$. This we shall explain more lly later.

In order to prove Theorem 5.5.3.2 we shall study the space $L(\omega_0, 0, P) = $ $\psi \in L(\omega_0, P) \mid {}^0c(s)\psi = \psi$ for all $s \in W(\omega_0)\}$. Clearly, $L(\omega_0, 0, P)$ is both a sub- gebra of $L(\omega_0, P)$ and a subrepresentation space for $W(\omega_0)$. The mapping $\prod : \psi \mapsto [W(\omega_0)]^{-1} \sum_{s \in W(\omega_0)} {}^0c(s)\psi$ is obviously the orthogonal projection of $L(\omega_0, P)$ n $L(\omega_0, 0, P)$. The functional equations for the Eisenstein integral (Theorem 3.5.3 (1)) imply that $E(P : \psi) = E(P : \prod\psi)$ for any $\psi \in L(\omega_0, P)$. Combined with heorem 5.2.2.7, this implies that $E(P) \mid L(\omega_0, 0, P)$ is surjective to $\mathcal{A}(\pi_0, {}^0\tau)$ $\pi_0 = \pi_{P,\omega_0}$). We intend to characterize $L(\omega_0, 0, P)$ both as the orthogonal com- ement of the kernel of $E(P)$ in $L(\omega_0, P)$ and as corresponding, under the bijec- on $T \mapsto \psi_T$, to the Fourier transform $\pi_0(\mathcal{C}(G))$ of $\mathcal{C}(G)$ relative to π_0. Our eorem will follow from these characterizations.

First we define the following mapping:

$\mapsto j(\alpha) = \underset{\sim}{\alpha} \mapsto \underset{\sim}{\alpha}^P \mapsto \underset{\sim}{\alpha}_0^P = d(\omega_0)(\theta_{\omega_0}, \rho(m)\underset{\sim}{\alpha}^P)_M$. The mapping $j : \mathcal{C}(G) \to \mathcal{C}(G, {}^0\tau)$ is the restriction to $\mathcal{C}(G)$ of the bijection (§1.12) from $C^{\infty}(G)$ to $C(G, \tau^0)$; it is an isomorphism of algebras. The second arrow is the homomorphism from $\mathcal{C}(G, \tau^0)$ to $\mathcal{C}(M, \tau_P^0)$ discussed in §4.4. The third arrow obviously has image in $L(\omega_0, P)$. We shall show, in the next lemmas, that $\alpha \mapsto \alpha_0^P$ is a surjective homomorphism of algebras from $\mathcal{C}(G)$ to $L(\omega_0, 0, P)$.

Lemma 5.5.3.3. Let $\alpha \in \mathcal{C}(G)$ and $\psi \in L(\omega_0, P)$. Then $(E(P : \psi), \underset{\sim}{\alpha})_G = (\psi, \alpha_0^P)_{M/A}$.

Proof. $(E(P : \psi), \underset{\sim}{\alpha}) = \int_M (\psi(m), \underset{\sim}{\alpha}^P(m))_{V^0} dm$ (generalized Lemma 5.2.2.2; cf. §4.4-5)

$$= \int_{M/A} (\psi(m), (\chi_{\omega_0} \underset{A}{*} \underset{\sim}{\alpha}^P)(m))_{V^0} dm^*$$

$$= \int_{M/A} ((d(\omega_0)\theta_{\omega_0} \underset{M/A}{*} \psi)(m), (\chi_{\omega_0} \underset{A}{*} \underset{\sim}{\alpha}^P)(m))_{V^0} dm^*$$

$$= \int_{M/A} (\psi(m), (d(\omega_0)\theta_{\omega_0} \underset{M}{*} \underset{\sim}{\alpha}^P)(m))_{V^0} dm^*$$

$$= (\psi, \alpha_0^P)_{M/A}.$$

Lemma 5.5.3.4. The mapping $\alpha \mapsto \alpha_0^P$ is a surjective homomorphism of algebras from $\mathcal{C}(G)$ to $L(\omega_0, 0, P)$. Let $\alpha'(x) = \alpha(x^{-1})$ $(x \in G)$. Then $\alpha_0^P = d(\omega_0)\psi_{\pi_0(\alpha')}$. The mapping $E(P)|L(\omega_0, 0, P)$ is injective.

Proof. We have already observed that, for any $\alpha \in \mathcal{C}(G)$, $\alpha_0^P \in L(\omega_0, P)$. Let us first show that, if $\psi \in L(\omega_0, 0, P)$, then there exists $\alpha \in \mathcal{C}(G)$ such that $\alpha_0^P = \psi$. For this choose a neighborhood U of $\omega_0 = 0$ in $\mathfrak{F}(\omega_0)$ such that $U \cap (sU + \nu(s)) = \phi$ for all $s \in W(\Theta) - W(\omega_0)$. Fix $\beta_1 \in C_c^{\infty}(U)$ such that $\beta_1(0) = \gamma^{-1}[W(\omega_0)]^{-1}[L_A^* : L^*(\omega_0)]$.

et $\beta(\nu : m) = \beta_1(\nu)[L^*(\omega_0) : L^*_M]^{-1} \sum_{\substack{x \in M/M(\omega_0) \\ \nu_1 \in L^*(\omega_0)/L^*_M}} \chi_{-(\nu+\nu_1)}(x)\psi_{\nu_1}(m)$. Then

$\in I^\infty(\omega_0, P)$. By Theorem 5.5.1.9, $E(P:\beta)^P = E(M: \sum_{s \in W(A)} \gamma c^{P|P}(s:\omega_0)\beta)$.

herefore,

$$j^{-1}E(P:\beta:m)_0^P = \sum_{s \in W(\mathfrak{O})} \gamma d(\omega_0)(\theta(\omega_0), \rho(m)E(M: c^{P|P}(s:\omega_0)\beta^{-\nu(s)} : 1))$$

$$= [W(\omega_0)]^{-1} \sum_{s \in W(\omega_0)} {}^0c(s)\psi(m) \qquad \text{(Corollary 5.5.2.3)}$$

$$= \psi(m).$$

hus, $L(\omega_0, 0, P)$ lies in the image of $\alpha \mapsto \alpha_0^P$.

Now let $\alpha \in \mathcal{C}(G)$. Then Lemma 5.5.3.3 implies that $(E(P:\psi), \underset{\sim}{\alpha}) =$ (ψ, α_0^P) $(\psi \in L(\omega_0, P))$. Since $E(P:\psi) = E(P: \prod\psi)$, it follows that $\prod\alpha_0^P = \alpha_0^P$. hus, $\alpha_0^P \in L(\omega_0, 0, P)$ and, taking $\psi = \alpha_0^P$, we see that $E(P)|L(\omega_0, 0, P)$ is ajective.

Write $\alpha_0^P = \psi_S$. We shall show that $S = \pi_0(\alpha')d(\omega_0)$. This will nply that the mapping is a homomorphism and complete the proof of the emma. We have:

$$(\psi_T, \alpha_0^P)_{M/A} = (\psi_T, \psi_S)_{M/A} = \psi_T \underset{M/A}{*} \psi_S(1) = d(\omega_0)^{-1}\psi_{ST}(1) = d(\omega_0)^{-1}\mathrm{tr}(ST)$$

or any $T \in \mathcal{T}(P)$ (cf. §5.2.4). On the other hand,

$$(E(P:\psi_T), \underset{\sim}{\alpha})_G = (E(P:\psi_T) * \underset{\sim}{\alpha})(1)$$

$$= j(E(P:T) * \alpha)(1)$$

$$= j(E(P:\pi_0(\alpha')T))(1) \text{ (Lemma 5.2.1.4)}$$

$$= \mathrm{tr}(\pi_0(\alpha')T) \cdot \text{(Lemma 5.2.1.1)}.$$

hus, $S = d(\omega_0)\pi_0(\alpha')$, by Lemma 5.5.3.3.

The proof of Theorem 5.5.3.2 now proceeds formally via Wedderburn's classical theorem: The double commutator Γ'' of a finite dimensional semi-simple matrix algebra Γ equals Γ. We know that $L(\omega_0, 0, P)$ is precisely the orthogonal complement of $\ker E(P)$ in $L(\omega_0, P)$ and that $\psi_T \in L(\omega_0, 0, P)$ if and only if $T \in \pi_0(\mathcal{C}(G))$. Since it is clear that the commutator of $\pi_0(\mathcal{C}(G))$ is the commuting algebra of π_0, it is sufficient, in order to prove Theorem 5.5.3.2, to establish that, with $\Gamma = \Gamma_{K_0}$ as before, $\Gamma' \cong \pi_0(\mathcal{C}_{K_0}(G))$. However, since ${}^0c(s)\psi_T = \psi_T$ for all $s \in W(\omega_0)$ if and only if $\gamma(s)T = T\gamma(s)$, $T \in \mathcal{T}_{K_0}(P)$, we see that $\psi_T \in L(\omega_0, 0, P)$ if and only if $T \in \Gamma'$, if and only if $T \in \pi_0(\mathcal{C}_{K_0}(G))$. Thus, $\Gamma_{K_0} = \pi_0(\mathcal{C}_{K_0}(G))'$.

An application:

Let (P_0, A_0) $(P_0 = M_0N_0)$ be a minimal p-pair of G. Let $\pi_0 = \mathrm{Ind}_{P_0}^G(\delta_{P_0}^{\frac{1}{2}})$. Note that the function Ξ (cf. §4.2) used in defining the Schwartz space of G is an element of $\mathcal{A}(\pi_0)$. Obviously, the class "1" $\in {}^0\mathcal{E}(M_0) \subset \mathcal{E}_2(M_0)$ is ramified, so one cannot use Bruhat theory to establish the following:

Proposition 5.5.3.5. The representation π_0 is irreducible.

Proof. We argue by induction on $\ell = \dim A_0/Z$.

Assume first that $\ell = 1$, as this case will require a separate argument. We observe that $\Xi(m) \neq \delta_{P_0}^{-\frac{1}{2}}(m)$ $(m \in M_0^+)$. This implies that the exponent "1" $\in \mathcal{X}_{\pi_0}(P_0, A_0)$ is not simple. Corollary 5.4.2.3 implies that π_0 is irreducible

Now assume $\ell > 1$. For the induction we apply Theorem 5.3.5.3 (4). In order to prove that π_0 is irreducible, it suffices, according to Theorem

5.5.3.2, to show that ${}^0c_{P_0|P_0}(s : \text{"1"} : 0) = 1_{L(\text{"1"}, P_0)}$ for all $s \in W(\text{"1"}) = W(G/A_0) = W$. Since $s \mapsto {}^0c_{P_0|P_0}(s : \text{"1"} : 0)$ defines a representation of W on $L(\text{"1"}, P_0)$, it is sufficient to check the above relation for s a simple reflection. Let $\underset{\sim}{A}' = \{a \in \underset{\sim}{A}_0 | a^s = a\}^0$. Let $(P', A') \succ (P_0, A_0)$ $(P' = M'N')$. Set $({}^*P_0, A_0) = (P_0 \cap M', A_0)$, a p-pair of M'. By the induction hypothesis, since $\dim A_0/A' = \ell - 1 \geq 1$, we know that $\mathrm{Ind}_{{}^*P_0}^{M'}(\delta_{{}^*P_0}^{\frac{1}{2}}) = {}^*\pi$ is irreducible. Set ${}^*W = \{t \in W | t|A' = 1_{A'}\} = \{1, s\}$. Recall or observe that $L(\text{"1"}, P_0) \subset L(\text{"1"}, {}^*P_0)$. Since ${}^*\pi$ is irreducible, ${}^0c_{{}^*P_0|{}^*P_0}(s : \text{"1"} : 0) = 1_{L(\text{"1"}, {}^*P_0)}$. Therefore, ${}^0c_{P_0|P_0}(s : \text{"1"} : 0) = 1_{L(\text{"1"}, P_0)}$, so π_0 is irreducible.

Corollary 5.5.3.6. The exponent $\text{"1"} \in \mathfrak{X}_{\pi_0}(P_0, A_0) = \mathfrak{X}_{\Xi}(P_0, A_0)$ has multiplicity exactly $[W(G/A_0)]$.

Proof. This follows easily from the fact that, since π_0 is irreducible, the representation $\bar{\pi}_0$ on $\mathcal{H}/\mathcal{H}(\bar{P}_0)$ has only one quotient representation of class $\text{"}\delta_{P_0}^{-\frac{1}{2}}\text{"}$ (Theorem 1.7.10). Apply Theorem 5.4.1.1 and Corollary 3.2.5 (3).

§5.5.4. An Almost Explicit Plancherel's Formula for G of Semisimple Rank One.

Our goal in this subsection is a proof that, for groups of semisimple rank one, the algebra $\mathcal{C}(G)$ is the orthogonal direct sum of the algebras $\mathcal{C}(\mathcal{O})$ of wave packets. In other words, only tempered representations occur in the Plancherel's formula for such G. The results will be more precise than in the case of real groups in that we shall see that $C_c^\infty(G)$ itself splits into orthogonal ideals, one per complex orbit $\mathcal{O}_{\mathbb{C}}$ in the supercuspidal representations of a Levi factor (of course, modulo certain equivalence relations). It is hoped (and expected) that these results provide a model for the general case.

Let π be an irreducible unitary representation of G in a Hilbert space If $\mathcal{H}$ contains a dense π-stable subspace which is, under π, a unitary admissible representation in the sense of §§1.3 and 1.5, then we sometimes abuse language to say that π is admissible. Bernstein ([1a]) has proved that every irreducible unitary representation of G is admissible in this sense.

The following lemma provides the basis for a proof, given in this subsection, that for groups of semisimple rank one, all irreducible unitary representations are admissible.

Lemma 5.5.4.1. Suppose that $C_c^\infty(G)$ is the direct sum of two orthogonal ideals, one being ${}^0C_c^\infty(G)$. Then every irreducible unitary representation of G is admissible.

Proof. Let K_0 be any open compact subgroup of G and let $E_0 = E_{K_0}$, the identity element of $C_c(G/\!/K_0)$. The hypothesis implies that $E_0 = {}^0E_0 + E_1$ with ${}^0E_0 \in {}^0C_c(G/\!/K_0)$ and $E_1 \perp {}^0C_c(G/\!/K_0)$. Let ${}^0Z = \bigcap_{\chi \in X(Z)} \ker|\chi|$ be the compact part of the split component Z of G and note that $E_0 \underset{{}^0Z}{*} {}^0\chi = 0$ except for finitely many elements ${}^0\chi \in {}^0\widehat{Z}$. Recall that ${}^0\mathcal{E}_{\mathbb{C}}(G)$ consists of disjoint orbits. For any orbit $\mathcal{O}_{\mathbb{C}} = \mathcal{O}_{\mathbb{C}}(\omega)$ we may choose $\pi \in \omega$ and assume that π acts in a vector space $\mathcal{H}$. Then $\pi_\nu \in \omega_\nu$ is defined to act in $\mathcal{H}$ such that $\pi_\nu(g) = \pi(g)\chi_\nu(g)$, where $\chi_\nu(g) = q^{\langle\sqrt{-1}\nu, H_G(g)\rangle}$ $(g \in G,\ \nu \in \mathfrak{z}_{\mathbb{C}}^*)$. Clearly, to each orbit $\mathcal{O}_{\mathbb{C}}$ there corresponds a single element ${}^0\chi \in {}^0\widehat{Z}$ such that $\pi_\nu(z) = {}^0\chi(z)$ for every $\nu \in \mathfrak{z}_{\mathbb{C}}^*$

nd $z \in {}^0Z$.

Now let $\omega \in {}^0\mathcal{E}_{\mathbb{C}}(G)$ have central exponent χ. Then, for any $\psi \in \mathcal{A}(\omega)$, $\in \mathcal{Z}^*_{\mathbb{C}}$, we have $E_0 \underset{G}{*} \psi_\nu = ({}^0E_0 \underset{G}{*} \psi)_\nu = ({}^0E_{0,\chi} \underset{G/Z}{*} \psi)_\nu$, where ${}^0E_{0,\chi} = \chi \underset{Z}{*} E_0$. ince ${}^0E_{0,\chi} \in {}^0\mathcal{A}(G,\chi)$ (cf. [7c], p. 20), ${}^0E_{0,\chi}$ is orthogonal to all but a finite-imensional subspace of ${}^0\mathcal{A}(G,\chi)$. It follows that, for a given ${}^0\chi \in {}^0\hat{Z}$, the imension of ${}^0\mathcal{A}(G,\chi) \cap C(G/\!/K_0)$ is finite and independent of $\chi \in \mathfrak{X}(Z)$ such that $|{}^0Z = {}^0\chi$. From this we conclude that

$$\sup_{\chi \in \mathfrak{X}(Z)} \dim {}^0C_c(G/\!/K_0, \chi) = \sup_{{}^0\chi \in {}^0\hat{Z}} \ \sup_{\chi | {}^0Z = {}^0\chi} \dim {}^0C_c(G/\!/K_0, \chi)$$

s finite. But this is Conjecture I of [7c], p. 16, to which we now refer for a proof hat this conjecture implies the lemma.

For the remainder of the subsection, in order to eliminate uninterest-ng technical details, we assume that the split component Z of G is trivial. We ssume that $\dim A = 1$, fix a class $\omega_0 \in {}^0\mathcal{E}(M)$, and write $\mathcal{O}_{\mathbb{C}}[\mathcal{O}]$ for the orbit $\mathfrak{I}_{\mathbb{C}}(\omega_0)[\mathcal{O}(\omega_0)]$ (cf. §5.3.1).

heorem 5.5.4.2. For any $f \in C_c^\infty(G)$ there exists a function $f_{\omega_0} \in C_c^\infty(G)$ such at $f - f_{\omega_0}$ is orthogonal to $\mathcal{C}(\mathcal{O})$. More precisely, the function f_{ω_0} may be osen to be a finite sum $f_{\omega_0} = f_{\mathcal{O}} + \sum_{i=1}^{r} f_i$, where $f_{\mathcal{O}} \in \mathcal{C}(\mathcal{O})$ and $f_i \in \mathcal{A}(\pi_{\nu_i, d})$, e points $\omega_{\nu_1}, \ldots, \omega_{\nu_r}$ being special points of $\mathcal{O}_{\mathbb{C}}$ and $\pi_{\nu_i, d}$ the discrete series mponent of π_{ν_i}.

roof. Let $f \in C_c(G/\!/K_0)$, where K_0 is an open normal subgroup of K. Let $\in \mathcal{P}(A)$ and, as usual, set $\pi_\nu = \mathrm{Ind}_P^G(\delta_P^{\frac{1}{2}}\sigma_\nu)$, $\sigma_\nu \in \omega_\nu \in {}^0\mathcal{E}_{\mathbb{C}}(M)$ $(\nu \in \mathfrak{F}_{\mathbb{C}})$. If $)_{K_0}$ does not contain 1_{K_0}, then $f_{\omega_0} = 0$ suffices.

Assume $(\pi_\nu)_{K_0}$ does contain 1_{K_0}. Let σ_ν act in a vector space U and set $\mathcal{H} = \mathcal{H}(P) = \{\beta \in C^\infty(K:U) | \beta(pk) = \sigma(p)\beta(k)\ (p \in P \cap K,\ k \in K)\}$. Then, as we know, $\mathcal{H}$ may be regarded as the common representation space for π_ν for all $\nu \in \mathfrak{F}_{\mathbb{C}}$. Let $\mathcal{H}_{K_0}$ be the space of K_0-fixed vectors in $\mathcal{H}$ and write I for the projection operator on $\mathcal{H}_{K_0}$ in $\mathcal{T}(P)$. Then $I = \pi_\nu(E_{K_0})$, where $E_0 = E_{K_0}$ is the identity element in $C_c(G /\!/ K_0)$. Let ψ_I be the corresponding function in $L(\omega_0, P) = \mathcal{A}(\omega_0, \tau_P^0)$ (cf. §5.2.1). Define $\alpha \in I^\infty(\omega_0, P)$ by setting

$\alpha(\nu) = [L_A^* : L^*(\omega_0)] d(\omega_0) \gamma(G/M)^{-1} [W(\mathcal{O})]^{-1} \psi_I$.

Form the wave packet $\underset{\sim}{E}_{\mathcal{O}} = E(P:\alpha) \in \mathcal{C}(\mathcal{O}, \tau^0)$. Write $E_{\mathcal{O}} = j^{-1} \underset{\sim}{E}_{\mathcal{O}}$. It follows immediately from Theorem 5.5.1.14 that $E_{\mathcal{O}}$ is an idempotent in $\dot{\mathcal{C}}_{K_0}(G)$, i.e., that $\underset{\sim}{E}_{\mathcal{O}}^2 = \underset{\sim}{E}_{\mathcal{O}}$.

Now let us show that, by taking into consideration the special points in $\mathcal{O}_{\mathbb{C}}$, we can produce an idempotent in $C_c(G /\!/ K_0)$. Without loss of generality, assume that $W(\mathcal{O}) = W(\omega_0)$. Theorem 5.4.5.1 implies that there are at most finitely many special points in $\mathcal{O}_{\mathbb{C}}(\omega_0)$. We enumerate the (possibly empty) set of special points $\omega_{\nu_1}, \ldots, \omega_{\nu_r}$ such that $\nu_i = \nu_{i,\mathbb{R}} + \sqrt{-1}\nu_{i,I}$ with $\nu_{i,I} \in \mathfrak{a}^{*+}$. If ω_{ν_i} is a special point, then so is $\omega_{\bar{\nu}_i}$, where $\bar{\nu}_i = \nu_{i,\mathbb{R}} - \sqrt{-1}\nu_{i,I}$. If there are special points in $\mathcal{O}_{\mathbb{C}}(\omega_0)$, then $W(\omega_0)$ contains two elements; furthermore, $W(\omega_{\nu_{i,\mathbb{R}}}) = W(\omega_0)$ for each $i = 1, \ldots, r$. For any i the induced representations π_{ν_i} and $\pi_{\bar{\nu}_i}$ have equal composition series of length two; a discrete series representation $\pi_{\nu_{i,d}}$ occurs as a quotient representation of π_{ν_i} and as a subrepresentation of $\pi_{\bar{\nu}_i}$ (Theorem 4.4.4 and Lemma 5.4.4.2). If $\psi_I \in L(\omega_0, P)$ is the function introduced above, then, since $\bar{\nu}_i = s\nu_i$ $(s \in W(\omega_0),\ s \neq 1)$,

$E(P:\psi_I:\nu_i) = E(P:\psi_I:\overline{\nu}_i) = j\,\mathrm{tr}(\pi_{\nu_i}(E_0)\pi_{\nu_i}\pi_{\nu_i}(E_0))$. It follows that $E_P(P:\psi_I:\nu_i) = E_P(P:\psi_I:\overline{\nu}_i)$ and $E_{P,1}(P:\psi_I:\nu_i) = E_{P,s}(P:\psi_I:\overline{\nu}_i)$. From the fact that $\mathcal{E}_{\pi_{\nu_{i,d}}}(P,A) = \{\chi_{\omega_0,\nu_i}\}$, it follows that $\delta_P^{-\frac{1}{2}}(ma)E_{P,1}(P:\psi_I:\nu_i:ma) = \delta_P^{-\frac{1}{2}}(ma)E_{P,s}(P:\psi_I:\overline{\nu}_i:ma) = j\,\mathrm{tr}(\pi_{\nu_{i,d}}(E_0)\pi_{\nu_{i,d}}(ma)\pi_{\nu_{i,d}}(E_0))$, provided $m \in M$ is fixed and $a \in A^+(P,t)$ with $t = t(m) >> 1$.

Since $E(P:\psi_I:\nu) = j\,\mathrm{tr}(\pi_\nu(E_0)\pi_\nu\pi_\nu(E_0))$, the function $\nu \mapsto \mu(\omega_0:\nu)E(P:\alpha(\nu):\nu:m)$ is, for any fixed $m \in M$, periodic with period lattice $L^*(\omega_0)$. There exists $t > 1$ such that if m is constrained to lie in a compact subset of M and $a \in A^+(P,t)$, then $E(P:\alpha(\nu):ma) = \delta_P^{-\frac{1}{2}}(ma)E_P(P:\alpha(\nu):ma)$ for all $\nu \in \mathfrak{F}_{\mathbb{C}}(\omega_0)$ (Lemma 5.1.5). Writing $\gamma = \gamma(G/M)$ and $W = W(G/A)$, we have

$$\int_{\mathfrak{F}(\omega_0)} E(P:\alpha(\nu):ma)\mu(\omega_0:\nu)d\nu$$

$$= [L_A^*:L^*(\omega_0)]d(\omega_0)\gamma^{-1}[W]^{-1}\int_{\mathfrak{F}(\omega_0)}\delta_P^{-\frac{1}{2}}(ma)E_P(P:\psi_I:\nu:ma)\mu(\omega_0:\nu)d\nu$$

$$= [L_A^*:L^*(\omega_0)]d(\omega_0)\gamma^{-1}[W]^{-1}\int_{\mathfrak{F}(\omega_0)}\sum_{s\in W}\delta_P^{-\frac{1}{2}}(ma)c_{P|P}(s:\omega_0:\nu)\psi_I(ma)\chi_{s\nu}(ma)\,\mu(\omega_0:\nu)d\nu.$$

Since $c_{P|P}(s:\omega_0:\nu)\psi_I(m)\chi_{s\nu}(m)$ is, for any $m \in M$ and $s \in W$, a periodic function of $\nu \in \mathfrak{a}_{\mathbb{C}}^*$ with the period lattice $L^*(\omega_0)$ and since this function of ν is holomorphic on $\mathfrak{a}_{\mathbb{C}}^*$ except for isolated simple poles which are located in $\mathfrak{a}^*$ and are cancelled in the product $\mu(\omega_0:\nu)c_{P|P}(s:\omega_0:\nu)\psi_I(m)\chi_{s\nu}(m)$ by the zeros of $\mu(\omega_0:\nu)$ which occur at the same values of ν (Theorems 5.3.5.2 and 5.4.2.1), we may integrate before we sum over $s \in W$. Since $E(P:\psi_I:\nu) = E(P:\psi_I:s\nu)$ and, therefore, $E_P(P:\psi_I:\nu) = E_P(P:\psi_I:s\nu)$ for all $\nu \in \mathfrak{a}_{\mathbb{C}}^*$, we see that $c_{P|P}(s:\omega_0:\nu)\psi_I\chi_{s\nu} = c_{P|P}(1:\omega_0:s\nu)\psi_I\chi_{s\nu}$ for all $\nu \in \mathfrak{a}_{\mathbb{C}}^*$ where either side is defined. Since, furthermore, $\nu \mapsto s\nu$ defines a measure preserving isomorphism of the compact group $\mathfrak{F}(\omega_0)$, we obtain

$$[W]^{-1}\int_{\mathfrak{F}(\omega_0)}\delta_P^{-\frac{1}{2}}(ma)E_P(P:\psi_I:\nu:ma)\mu(\omega_0:\nu)d\nu$$

$$=\int_{\mathfrak{F}(\omega_0)}\delta_P^{-\frac{1}{2}}(ma)c_{P|P}(1:\omega_0:\nu)\psi_I(ma)\chi_\nu(ma)\mu(\omega_0:\nu)d\nu.$$

We wish to evaluate

$$[L_A^*:L^*(\omega_0)]d(\omega_0)\gamma^{-1}\int_{\mathfrak{F}(\omega_0)}c_{P|P}(1:\omega_0:\nu)\psi_I(ma)\chi_\nu(ma)\mu(\omega_0:\nu)d\nu$$

as a contour integral. To do this we introduce a coordinate on $\mathfrak{a}_{\mathbb{C}}^*$ and then change variables. Write β^* for the unique generator of $L^*(\omega_0)$ in $\mathfrak{a}^{*+}$. Set $\nu(t) = t\beta^*$ $(t \in \mathbb{C})$ and thus define an isomorphism of $\mathbb{C}$ onto $\mathfrak{a}_{\mathbb{C}}^*$. Write $t = t(\nu)$ for the inverse mapping. If f is a continuous function on $\mathfrak{a}_{\mathbb{C}}^*$ which is periodic with period lattice $L^*(\omega_0)$, then $\int_{\mathfrak{F}(\omega_0)}f(\nu)d\nu = \int_0^1 f(\nu(t))dt$. Now set $z = z(t) = e^{2\pi\sqrt{-1}t}$ $(t \in \mathbb{C})$. The composed mapping $\nu \mapsto z(t(\nu))$ defines a holomorphic group homomorphism of $\mathfrak{a}_{\mathbb{C}}^*$ onto $\mathbb{C}^*$. The kernel of this homomorphism is clearly $L^*(\omega_0)$. Thus, we may regard $\nu \mapsto z(\nu)$ as an isomorphism of $\mathfrak{F}_{\mathbb{C}}(\omega_0)$ onto $\mathbb{C}^*$; this isomorphism sends $\mathfrak{F}(\omega_0)$ onto the unit circle. It is clear that with f as above and with $\nu = z(\nu)$ denoting the inverse of $\nu \mapsto z(\nu)$

$$\int_{\mathfrak{F}(\omega_0)}f(\nu)d\nu = \int_0^1 f(\nu(t))dt = \frac{1}{2\pi\sqrt{-1}}\int_{|z|=1}f(\nu(z))\frac{dz}{z}$$

with the usual positive orientation for $|z| = 1$.

Now consider the function $D(\nu : ma) = \mu(\omega_0:\nu)c_{P|P}(1:\omega_0:\nu)\psi_I(ma)\chi_\nu($ This function has period lattice $L^*(\omega_0)$ for any fixed ma. Therefore, $D(\nu(z) : ma)$ is holomorphic at least on the punctured plane $z \neq 0$. For any $a \in A$ the function $\nu \mapsto \chi_\nu(a)$ has a period lattice which contains $L^*(\omega_0)$. Thus, $\chi_\nu(a) = q^{\sqrt{-1}\langle\nu, H(a)\rangle} =$

$e^{\sqrt{-1}t\langle\alpha^*, H(a)\rangle}$, where $\langle\alpha^*, H(a)\rangle = k2\pi(\log q)^{-1}$ with $k \in \mathbb{Z}$. In other words, $\nu(z)(a) = z^k$ for some $k \in \mathbb{Z}$; if $a \in A^+(P, t)$, with $t > 1$, then $k > 0$. To show that $D(\nu(z) : ma)$ is holomorphic and vanishing at $z = 0$ for $a \in A^+(P, t)$ with t sufficiently large, it is enough to show that $D(\nu(z) : ma)$ is meromorphic at $z = 0$.

For this, recall that we have a convergent integral formula for $c_{P|P}(1 : \omega_0 : \bar{\nu})^*$, though not for $c_{P|P}(1 : \omega_0 : \nu)$, for all $\nu \in \mathfrak{F}_{\mathbb{C}}(P)$, i.e., for $\nu = \nu_{\mathbb{R}} + \sqrt{-1}\nu_I$ with $\nu_I \in (\mathfrak{a}^*)^+$ (Theorem 5.3.5.4). This implies that $c_{P|P}(1 : \omega_0 : \bar{\nu}(z))^*$ is holomorphic on some finite covering of the unit disc in the z-plane. Since $\mu(\omega_0 : \nu)c_{P|P}(1 : \omega_0 : \nu) = (c_{P|P}(1 : \omega_0 : \nu)^*)^{-1}$ (Theorem 5.3.5.2) and since $D(\nu(z) : ma)$ is holomorphic on the punctured z-plane ($z \neq 0$), it follows that $D(\nu(z) : ma)$ is, for any fixed ma, meromorphic at $z = 0$. Choosing $a \in A^+(P, t)$ with t sufficiently large, we may assume both that $D(\nu(z) : ma)$ is holomorphic at $z = 0$ and that $D(\nu(z) : ma) = 0$, $z = 0$.

We conclude that $\frac{1}{2\pi\sqrt{-1}}\int_{|z|=1} D(\nu(z) : ma)\frac{dz}{z}$ equals the sum of the residues of the integral with no residue at $z = 0$, provided $a \in A^+(P, t)$, $t \gg 1$. Since $\mu(\omega_0 : \nu)c_{P|P}(1 : \omega_0 : \nu)$ is holomorphic for all $\nu \in \mathfrak{a}^*_{\mathbb{C}}$ except at the poles of $\mu(\omega_0 : \nu)$ and since these poles occur exactly at the special points,

$$\int_{\mathfrak{F}(\omega_0)} \delta_P^{-\frac{1}{2}}(ma)c_{P|P}(1 : \omega_0 : \nu)\psi_I(ma)\chi_\nu(ma)\mu(\omega_0 : \nu)d\nu$$

$$= \sum_{i=1}^{r} \delta_P^{-\frac{1}{2}}(ma)c_{P|P}(1 : \omega_0 : \nu_i)\psi_I(ma)\chi_{\nu_i}(ma)\frac{1}{z(\nu_i)}\text{residue}_{z=z(\nu_i)}(\mu(\omega_0 : \nu(z))).$$

Combining this with the fact that $E_{P'}(P : \psi_I : \nu) = 0$ for all proper p-pairs (P', A') such that $A' \not\sim A$ (Corollary 5.3.2.2), we deduce that

$$E_{\mathfrak{a}} - [L_A^* : L^*(\omega_0)]d(\omega_0)\gamma^{-1}\sum_{i=1}^{r} \delta_P^{-\frac{1}{2}}c_{P|P}(1 : \omega_0 : \nu_0)\psi_I\chi_{\nu_i}\frac{1}{z(\nu_i)}\text{residue}_{z=z(\nu_i)}(\mu(\omega_0 : \nu(z)))$$

is a compactly supported function on M_0^+.

Define

$$\mathrm{Res}_{\nu=\nu_i}\mu(\omega_0:\nu) = \left(1-\frac{z(\nu_i)}{z}\right)\mu(\omega_0:\nu(z))\Big|_{z=z(\nu_i)}$$

and set

$$f_{\nu_i} = (-1)[L_A^* : L^*(\omega_0)]d(\omega_0)\gamma^{-1}\mathrm{Res}_{\nu=\nu_i}\mu(\omega_0:\nu)\mathrm{tr}(\pi_{\nu_{i,d}}(E_0)\pi_{\nu_{i,d}}\pi_{\nu_{i,d}}(E_0)),$$

provided $\mu(\omega_0 : \nu(z))$ has a simplepole at $z = z(\nu_i)$; otherwise set $f_{\nu_i} = 0$. (We shall show in Corollary 5.5.4.3 that, in fact, the pole is always simple.)

Then $E_{\omega_0} = E_{\mathcal{O}} + \sum_{i=1}^{r} f_{\nu_i} \in C_c(G/\!\!/K_0)$.

Let us write $E_{\omega_0} = E_{\mathcal{O}} + \sum_{i=1}^{r} c_i E_{\nu_{i,d}}$, where $E_{\nu_{i,d}}$ is the identity element in $\mathcal{C}_{K_0}(G) \cap \mathcal{A}(\pi_{\nu_{i,d}})$; clearly, f_{ν_i} is a multiple of $E_{\nu_{i,d}}$. We want to show that $c_i = 1$ $(i = 1, \ldots, r)$. Since $E_{\nu_{i,d}} * E_{\mathcal{O}} = 0$ (Corollary 5.5.1.12) and $E_{K_0} * E_{\mathcal{O}} = E_{\mathcal{O}} * E_{\mathcal{O}} = E_{\mathcal{O}}$, it follows that $(E_{K_0} - E_{\omega_0}) * E_{\mathcal{O}} = 0$. Since $E_{K_0} - E_{\omega_0} \in C_c(G/\!\!/K_0)$, analytic continuation of the convolution product $0 = (E_{K_0} - E_{\omega_0}) * E(P : \psi_I : \nu)$ implies that $(E_{K_0} - E_{\omega_0}) * c_i E_{\nu_{i,d}} = 0$. On the other hand, $E_{K_0} * c_i E_{\nu_{i,d}} = c_i E_{\nu_{i,d}}$ and $c_i E_{\nu_{i,d}} * c_i E_{\nu_{i,d}} = c_i^2 E_{\nu_{i,d}}$; this implies that $c_i = 1$. Thus, $f_{\nu_i} = E_{\nu_{i,d}}$.

To complete the proof of Theorem 5.5.4.2 set $f_{\omega_0} = f*E_{\omega_0}$. Then $f_{\omega_0} = f*E_{\mathcal{O}} + \sum_{i=1}^{r} f*E_{\nu_{i,d}} = f_{\mathcal{O}} + \sum_{i=1}^{r} f_i$, as required. Obviously, $f - f_{\omega_0} \perp \mathcal{C}(\mathcal{O})$.

<u>Corollary</u> 5.5.4.3. Every pole of the function $\mu(\omega_0 : \nu(z))$ is simple. Let $\omega_{\nu_0} \in \mathcal{O}_{\mathbb{C}}(\omega_0)$ be a special point with $\pi_{\nu_{0,d}}$ the discrete series component of π_{ν_0}. The formal degree of $\pi_{\nu_{0,d}}$ is given by the formula

$$d(\pi_{\nu_{0,d}}) = (-1)[L_A^* : L^*(\omega_0)]d(\omega_0)\,\gamma^{-1}\mathrm{Res}_{\nu=\nu_0}\,\mu(\omega_0 : \nu),$$

here $\mathrm{Res}_{\nu=\nu_0}\,\mu(\omega_0 : \nu)$ is as defined in the proof of Theorem 5.5.4.2.

roof. Let $\nu_0 = \nu_{0,\mathbb{R}} + \sqrt{-1}\nu_{0,I}$ ($\nu_{0,\mathbb{R}} \in \mathfrak{a}^*$, $\nu_{0,I} \in \mathfrak{a}^{*+}$) be chosen such that (ν_0) is a pole of $\mu(\omega_0 : \nu(z))$. Since every pole of $\mu(\omega_0 : \nu(z))$ corresponds to a pecial point of $\mathcal{O}_{\mathbb{C}}(\omega_0)$ (Lemmas 5.4.2.4 and 5.4.5.2), there is no loss of generality in ssuming that ω_{ν_0} is a special point of $\mathcal{O}_{\mathbb{C}}(\omega_0)$. If $\mu(\omega_0 : \nu(z))$ does not have a simple pole $z = z(\nu_0)$, then we can construct, as in Theorem 5.5.4.2, an idempotent $E_{\omega_0} \in C_c(G/\!/K_0)$ for any K_0 such that E_{ω_0} is orthogonal to $\mathcal{A}(\pi_{\nu_{0,d}})$. This mplies that both $E_{K_0} - E_{\omega_0}$ and E_{K_0} are orthogonal to $\mathcal{A}(\pi_{\nu_{0,d}})$ and obviously eads to a contradiction.

That the above formula gives the formal degree of $\pi_{\nu_0,d}$ may be een as follows. Observe that the function $j^{-1}f_{\nu_0}$ is a multiple of the sum of the iagonal matrix coefficients with respect to an orthonormal base for $(\mathcal{H}_{\nu_0,d})_{K_0}$. he formal degree of $\pi_{\nu_0,d}$ is exactly the factor which converts such a sum f diagonal matrix coefficients into an idempotent in $\mathcal{A}(\pi_{\nu_0,d})$. As we have een, the factor above does just that.

Set $C_c^\infty(\mathcal{O}_{\mathbb{C}}) = C_c^\infty(G, \mathcal{O}_{\mathbb{C}}) = C_c^\infty(G) \cap (\mathcal{C}(\mathcal{O}) \oplus \sum_{i=1}^{r} \mathcal{A}(\pi_{\nu_i,d}))$. Similarly, et $C_c^\infty(\mathcal{O}_{\mathbb{C}}^\perp) = C_c^\infty(G, \mathcal{O}_{\mathbb{C}}^\perp) = \{f \in C_c^\infty(G) \mid f$ is orthogonal to $C_c^\infty(\mathcal{O}_{\mathbb{C}})\}$. The existence f the idempotent E_{ω_0} of Theorem 5.5.4.2 implies that $C_c^\infty(G) = C_c^\infty(\mathcal{O}_{\mathbb{C}}) \oplus C_c^\infty(\mathcal{O}_{\mathbb{C}}^\perp)$, n orthogonal direct sum.

From here on, <u>assume that</u> $\dim A_0 = 1$. In this case, any complex rbit $\mathcal{O}_{\mathbb{C}} \subset {}^0\mathcal{E}_{\mathbb{C}}(G)$ [orbit $\mathcal{O} \subset \mathcal{E}_2(G)$] consists of a single point: $\mathcal{O}_{\mathbb{C}} = \{\omega\}$, $\in {}^0\mathcal{E}(G)$ [$\mathcal{O} = \{\omega\}$, $\omega \in \mathcal{E}_2(G)$]. Two complex orbits $\mathcal{O}_{\mathbb{C}}$ and $\mathcal{O}'_{\mathbb{C}}$ in $\mathcal{E}_{\mathbb{C}}(M_0)$ are called equivalent if there is an element of $W(G/A_0)$ which permutes them. Similarly, orbits $\mathcal{O}$ and $\mathcal{O}'$ in $\mathcal{E}_2(M_0)$ are also defined to be equivalent when $W(G/A_0)$ permutes them.

Theorem 5.5.4.4. Assume $\dim \underset{\sim}{A}_0 = 1$. Then $\mathcal{C}(G) = \oplus\, \mathcal{C}(\mathcal{O})$, where $\mathcal{O}$ ranges over orbits in $\mathcal{E}_2(G)$ and equivalence classes of orbits in $\mathcal{E}_2(M_0)$. Similarly, $C_c^\infty(G) = \oplus\, C_c^\infty(\mathcal{O}_{\mathbb{C}})$, where $\mathcal{O}_{\mathbb{C}}$ ranges over orbits in ${}^0\mathcal{E}_{\mathbb{C}}(G)$ and equivalence classes of orbits in ${}^0\mathcal{E}_{\mathbb{C}}(M_0)$.

If $f \in \mathcal{C}(G)$, then $f = \Sigma f_{\mathcal{O}}$, $(f_{\mathcal{O}} \in \mathcal{C}(\mathcal{O}))$, a finite sum. Similarly, if $f \in C_c^\infty(G)$, then $f = \Sigma f_{\mathcal{O}_{\mathbb{C}}}$ $(f_{\mathcal{O}_{\mathbb{C}}} \in C_c^\infty(\mathcal{O}_{\mathbb{C}}))$, also a finite sum.

Proof. Fix an open compact normal subgroup K_0 of K. Let E_{K_0} be the identity element of $C_c(G/\!/K_0)$. To prove all of the theorem, it is obviously sufficient to show that $E_{K_0} = \sum_{i=1}^{r} E_{\omega_i}$, a finite sum of orthogonal idempotents, where $E_{\omega_i} \in C_c^\infty(\mathcal{O}_{\mathbb{C}}(\omega_i))$ with $\omega_i \in {}^0\mathcal{E}(G)$ or ${}^0\mathcal{E}(M_0)$. By [7c], p. 24, Theorem 6, it is enough to show that $E_{K_0} - \sum_{i=1}^{r} E_{\omega_i} \in {}^0C_c(G/\!/K_0)$, where $\omega_1, \ldots, \omega_r \in {}^0\mathcal{E}(M_0)$. To see this, note first that, since ${}^0M_0$ is compact, there are only finitely many orbits $\mathcal{O}_{\mathbb{C}}(\omega_1), \ldots, \mathcal{O}_{\mathbb{C}}(\omega_r)$ such that $\sigma|K_0 \cap M_0$ contains $1_{K_0 \cap M_0}$. This is clearly a necessary condition in order that E_{K_0} not be orthogonal to $C_c^\infty(\mathcal{O}_{\mathbb{C}}(\omega_i))$--i.e., $\mathrm{Ind}_{P_0}^G \sigma|K = \mathrm{Ind}_{P_0 \cap K}^K(\sigma|K \cap M_0)$, etc., apply Frobenius' reciprocity theorem.

Set ${}^0E = E_{K_0} - \Sigma E_{\omega_i}$. If ${}^0E \notin {}^0C_c^\infty(G)$, then $0 \neq {}^0\underset{\sim}{E}^{P_0} \in C_c(M_0/\!/K_0 \cap M_0, {}^0\tau_{P_0})$. However, the identity element of this algebra is a sum of wave packets $\Sigma E(M_0 : \alpha^i)$, where $\alpha^i \in I^\infty(\omega_i, W(\mathcal{O}(\omega_i)), P_0)$. We know that ${}^0\underset{\sim}{E}^{P_0} * E(M_0 : \alpha^i) = ({}^0\underset{\sim}{E} * E(P_0 : \alpha^i))^{P_0} = 0$, which contradicts ${}^0\underset{\sim}{E}^{P_0} \neq 0$ and proves the theorem.

orollary 5.5.4.5. Let $\dim \underset{\sim}{A}_0 = 1$. Then every irreducible unitary representation G is admissible.

roof. Apply Theorem 5.5.4.4 and Lemma 5.5.4.1.

orollary 5.5.4.6. Let $\dim \underset{\sim}{A}_0 = 1$. Then for any compact open subgroup K_0 of there are only finitely many discrete series representations π of G such that $\pi) \cap C(G /\!/ K_0) \neq (0)$. The space ${}^0\mathcal{C}_{K_0}(G)$ is finite-dimensional, i.e., every ısp form in the Schwartz space is Hecke finite.

References

[1] a. I. N. Bernstein, All reductive p-adic groups are tame, Functional Anal., 8 (1974), 3-5.

b. I. N. Bernstein and A. V. Zelevinskii, Induced representations of the group GL(n) over a p-adic field, Functional Anal., 10 (1976), 74-75.

[2] a. A. Borel, Linear Algebraic Groups, W. A. Benjamin (New York-Amsterdam), 1969.

b. A. Borel, et al., Seminar on Algebraic Groups and Related Finite Groups Springer Lecture Notes in Math., 131, Springer (Berlin-Heidelberg-New York), 1970.

c. A. Borel and J. Tits, Groupes réductifs, Publ. Math. I.H.E.S., 27 (1965), 55-150; Compléments, Publ. Math. I.H.E.S., 41 (1972), 253-276.

[3] N. Bourbaki, Éléments de mathématique, Fasc. XXXIV. Groupes et algèbres de Lie. Chap. IV: Groupes de Coxeter et système de Tits. Chap. V: Groupes engendrés par des réflexions. Chap. VI: Systèmes de racines, Actualités Sci. Indust., no. 1337, Hermann, Paris, 1968.

[4] a. F. Bruhat, Sur les représentations induites des groupes de Lie, Bull. Soc. Math. France, 84 (1956), 97-205.

b. F. Bruhat, Distributions sur un groupe localement compact et applicatio à l'étude des représentations des groupes p-adiques, Bull. Soc. Math. France, 89 (1961), 43-75.

c. F. Bruhat and J. Tits, Groupes réductifs sur un corps local I, Publ. Math. I.H.E.S., 41 (1972), 5-252.

[5] a. W. Casselman, The Steinberg character as a true character, Proc. Sympos. Pure Math., vol. 26, Providence, R.I., 1974, 413-417.

b. W. Casselman, Introduction to the theory of admissible representations of p-adic reductive groups, to appear.

[6] R. Godement and H. Jacquet, Zeta Functions of Simple Algebras, Springer Lecture Notes in Math., 260 (1972).

[7] a. Harish-Chandra, Automorphic Forms on Semisimple Lie Groups, Springer Lecture Notes in Math., 62 (1968).

b. Harish-Chandra, Harmonic analysis on semisimple Lie groups, Bull. Amer. Math. Soc., 76 (1970), 529-551.

c. Harish-Chandra, Harmonic Analysis on Reductive p-adic Groups, Springer Lecture Notes in Math., 162 (1970).

d. Harish-Chandra, On the theory of the Eisenstein integral, Proc. Int. Conf. Harmonic Analysis, University of Maryland, Springer Lecture Notes in Math., 266 (1972), 123-149.

e. Harish-Chandra, Harmonic analysis on reductive p-adic groups, Proc. Sympos. Pure Math., vol. 26, Amer. Math. Soc., Providence, R.I., 1974, 167-192.

f. Harish-Chandra, The Plancherel formula for reductive p-adic groups, preprint. Also see Corrections.

[8] a. R. Howe, Kirillov theory for compact p-adic groups, preprint.

b. R. Howe, Some qualitative results on the representation theory of GL(n) over a p-adic field, preprint.

c. R. Howe, The Fourier transform and germs of characters (case of GL_n over a p-adic field), Math. Ann., 208 (1974), 305-322.

[9] H. Jacquet, Représentations des groupes linéaires p-adiques, C.I.M.E., Summer School on the Theory of Group Representations and Harmonic Analysis, Montecatini (1970).

[10] I. G. Macdonald, Spherical Functions on Groups of p-adic Type, Publications of the Ramanujan Institute for Advanced Study, No. 2, Ramanujan Institute for Advanced Study, Madras, India, 1972.

[11] D. Montgomery and L. Zippin, Topological Transformation Groups, Interscience (New York), 1955.

[12] L. S. Pontrjagin, Topological Groups (2d ed.), Gordon and Breach (New York), 1966.

[13] a. G. Van Dijk, Computation of certain induced characters of p-adic groups, Math. Ann., 199 (1972), 229-240.

b. G. Van Dijk, Quasi-admissible representations of p-adic groups, Proceedings of the Summer School on Group Representations, Bolyai János Math. Soc. (Budapest, 1971), Halsted Press (New York), 1975.

[14] a. A. Weil, L'Intégration dans les Groupes Topologiques et ses Application Hermann (Paris), 1940.

b. A. Weil, Basic Number Theory (2d ed.), Springer (Berlin-Heidelberg-New York), 1973.

SELECTED TERMINOLOGY

§0.1: Zariski closed , Zariski open , Zariski dense
Ω-closed Ω-open Ω-dense

linear algebraic group
Ω-group

defined over

morphism of l.a.g.'s
Ω-morphism

connected l.a.g.

rational characters

radical
impotent radical

reductive group
semisimple group

torus
dimension of a torus
Ω-torus
Ω-split torus
anisotropic

§0.2: Cartan subgroup

reduced, semisimple, or split rank

split component

§0.3: Borel subgroup
parabolic subgroup
p-subgroup

Levi subgroup
Levi factor
Levi decomposition

parabolic pair
p-pair

§0.3 (cont'd)

parabolic rank
p-rank

special torus
$\underset{\sim}{A}_0$-standard torus

minimal p-pair
maximal p-pair

standard p-pair
semistandard p-pair

associated
$\underset{\sim}{A}$-associated

opposite parabolic subgroups
opposite p-pairs

§0.4: p-adic field

split torus

Levi decomposition

§0.5: Weyl group of A

real Lie algebra of A

dual real Lie algebra of A

weight

root character
root
A-root
reduced root
chamber
simple root

relative Weyl group of G

§0.6: A_0-good maximal compact subgroup

Cartan decomposition

normalized Haar measure

Iwasawa decomposition

§1. 4: smooth representation
quasi-character
character
smooth double representation
smooth linear functional
smooth contragredient representation
smooth antilinear functional
smooth adjoint representation

§1. 5: admissible representation
quasi-admissible representation
Hecke finite function

§1. 6: intertwining form
intertwining operator
invariant bilinear form
invariant Hermitian form

§1. 7: big (left or right) representation
little (left or right) induced representation
smooth induced representation

§1. 10: Hecke algebra
automorphic form
Hecke finite function

§1. 11: trace of operator of finite rank
discrete series

§1. 12: associativity conditions for a smooth double representation

§1. 13: character of an admissible representation
invariant distribution
distribution of positive type

§2. 2: J-supercuspidal representation
supercuspidal representation

§2. 4: dual exponent
multiplicity of a dual exponent
π-minimal p-pair
π-critical exponent

§2. 5: ramification group
unramified representation

SELECTED NOTATIONS

§0.1: Ω , $\bar{\Omega}$

$\underset{\sim}{G}$, $\underset{\sim}{G}(R)$, $\underset{\sim}{G}^0$

$\mathfrak{D}_{\underset{\sim}{G}}$, $\underset{\sim}{X}(\underset{\sim}{T})$

§0.3: $\underset{\sim}{A}_0$

$(\underset{\sim}{P}_0, \underset{\sim}{A}_0)$, $(\underset{\sim}{P}, \underset{\sim}{A})$, $(\bar{\underset{\sim}{P}}, \underset{\sim}{A})$

$(\underset{\sim}{P}, \underset{\sim}{A}) \succ (\underset{\sim}{P}', \underset{\sim}{A}')$

$\mathcal{P}(\underset{\sim}{A})$

§0.4: q , $|\ \ |$

(P, A)

$X(G)$

$H_G(x)$

$\langle \chi, H_G(x) \rangle$

^0G

§0.5: $W(A)$, $W(G/A)$, $W(A_2 | A_1)$

χ^s

$(P, A)_F$

Σ^0 , $\Sigma(P, A)$

$\mathfrak{a}^+$, $^+\mathfrak{a}$

M^+ , ^+M

A^+ , ^+A

§0.6: $H_P(kmn), \kappa(x), \mu(x), \eta(x)$

§1.1: $C^\infty(X : V)$, $C_c^\infty(X : V)$

$C_c^\infty(X)$, $C_c(X)$

$\mathfrak{D}(X : V)$

$\int_X f d\mu$

§1.2: $d_\ell x$

$\delta_G(x)$

§1.2.1: ρ (as weight)

§1.2.2: $\gamma(G/P)$, $\gamma(G/M)$

§1.3: $\sigma \subset \pi$, $\sigma \prec \pi$, $\pi_1 \sim \pi_2$, ^tT

$T^\wedge$, $\pi^\wedge(x)$, ρ, λ

§1.4: χ_∞ , $\mathfrak{X}(G)$, $\hat{G}$, $\tilde{V}$, $\tilde{\pi}$,

V^* , π^* , $C^\infty(G, Z : U)$,

$C^\infty(G, \chi : U)$, $C^\infty(G : U)_\chi$

§1.5: $\mathcal{E}_{\mathbb{C}}(G)$, $\mathcal{E}(G)$

§1.6: $\mathcal{B}(\pi_1, \pi_2)$, $I(\pi_1, \pi_2)$, $\mathcal{T}(\pi_1, \pi_2)$

$J(\pi_1, \pi_2)$, $\|v\|$, $\|v\|_H$, $S^\perp$

§1.7: σ^G , $\mathrm{Ind}_H^G \sigma$

§1.10: $\mathcal{A}(G)$, $\mathcal{A}(G : U)$, $\mathcal{A}(\pi)$

§1.11: $\mathrm{End}^0(V)$, $\mathcal{E}_2(G)$

§1.12: $C(G, \tau)$, $\mathcal{A}(G, \tau)$, $jf = \underset{\sim}{f}$

§1.13: Θ_π

§2.2: $V(N)$, $V(P)$, $^0C_c^\infty(G, Z : U)$,

$^0\mathcal{A}(G)$, $^0\mathcal{E}_{\mathbb{C}}(G)$, $^0\mathcal{E}(G)$

§2.4: $\mathcal{Y}_\pi(P, A)$, $\mathcal{E}'(G)$

§2.6: $A^+(t)$, $\gamma_P(a)$, f_P ,

$f(xa) \to L, a \underset{P}{\to} \infty$

§2.7: α^P

§3.1: $\mathfrak{X}_\pi(P, A)$, $f_{P,\chi}$, $\mathfrak{X}_f(P, A)$

§3. 2: $\mathfrak{X}_f(P, A)$, $\mathfrak{X}_\pi(P, A)$

§3. 4: $f_P \sim 0$

§4. 1: $\asymp$, $\| \ \|$, $\sigma(x)$, $\sigma_*(x)$, $\prec$

§4. 2: $\Xi(x)$

§4. 4: $\mathfrak{C}(G)$, $\mathfrak{C}_*(G)$, ${}_w\mathcal{A}(G)$, ${}_wf_P$, ${}_w\mathfrak{X}_f(P, A)$, ${}_w\mathfrak{X}_f(P, A)$, ${}_w\mathfrak{X}_\pi(P, A)$, ${}_w\mathfrak{X}_\pi(P, A)$, ${}_w^0\mathcal{A}(G)$

§4. 7: $D_G(x) = D(x)$

Library of Congress Cataloging in Publication Data

Silberger, Allan J
Introduction to harmonic analysis on reductive p-adic groups.

(Mathematical notes ; 23)
Bibliography: p.
1. Groups, Theory of. 2. Harmonic analysis.
I. Harish-Chandra. II. Title. III. Series.
QA171.S616 1979 512'.22 79-19020
ISBN 0-691-08246-4

www.ingramcontent.com/pod-product-compliance
Ingram Content Group UK Ltd.
Pitfield, Milton Keynes, MK11 3LW, UK
UKHW021006080125
453275UK00005B/228

9 780691 61136